The Anthropocene: Politik—Economics—Society—Science

Volume 17

Series editor

Hans Günter Brauch, Mosbach, Germany

More information about this series at http://www.springer.com/series/15232
http://www.afes-press-books.de/html/APESS.htm
http://afes-press-books.de/html/APESS_17.htm

Lourdes Arizpe Schlosser

Culture, International Transactions and the Anthropocene

Springer

Organización de las Naciones Unidas para la Educación, la Ciencia y la Cultura

Cátedra UNESCO de Investigación sobre Patrimonio Cultural Intangible y Diversidad Cultural
Universidad Nacional Autónoma de México

CRIM

Lourdes Arizpe Schlosser
Centro Regional de Investigaciones
Multidisciplinarias
Universidad Nacional Autonoma de Mexico
Cuernavaca, Mexico

More on this book is at: http://afes-press-books.de/html/APESS_17.htm

ISSN 2367-4024 ISSN 2367-4032 (electronic)
The Anthropocene: Politik—Economics—Society—Science
ISBN 978-3-642-41601-9 ISBN 978-3-642-41602-6 (eBook)
https://doi.org/10.1007/978-3-642-41602-6

Library of Congress Control Number: 2018942002

Copy-editing: PD Dr. Hans Günter Brauch, AFES-PRESS e.V., Mosbach, Germany
Style and language editing: Dr. Hester Higton, Durham, UK
Review editing: Mike Headon, Colwyn Bay, UK

The cover photo illustrates: *Anthropology and the Transformation of Cultures*: Lourdes Arizpe and her research team observing neo-indigenous, New Age and Buddhist rituals in the Spring celebrations at the Pyramid of the Sun in Teotihuacan, Mexico, 2013. The photo was taken by C. Amescua.
The photo on the internal title page iii portrays Assistant Director-General for Culture Lourdes Arizpe addressing a meeting on culture at UNESCO, 1997. The photo was taken by Sayah Msadek.

Printed on acid-free paper

This Springer imprint is published by the registered company Springer-Verlag GmbH, DE part of Springer Nature
The registered company address is: Heidelberger Platz 3, 14197 Berlin, Germany

To all those who believe that living is the path towards a sustainable future

For MVG

Acknowledgements

So many researchers, professional colleagues, politicians, functionaries and activists speak, deliberate and carry out transactions in the pages of this book that it is impossible to acknowledge their contributions. They are all active participants in working towards human advancement in the world of the living. I am particularly indebted to the National University of Mexico, and its previous Rector, José Narro Robles, and to my colleagues at the Centre for Multidisciplinary Research, especially to its former Director, the late Raul Bejar, and to its present Director, Margarita Velazquez Gutierrez, for their very generous support. My mentor and friend, Rodolfo Stavenhagen, gave me most important support when I began my career as an anthropologist in Mexico.

For early induction into the world of international research and policy, I wish to recognize my intellectual debt to the brilliant leaders of the Society for International Development, among them Amartya Sen, Sir Richard Jolly, Mahbub ul Haq, Sakiko Fukuda-Parr, Francis Stewart, Paul Streeten and so many others, and particularly to Louis Emmerij, who invited me to India and blew my mind with the unimagined possibilities of cultural creation, as well as the complexities of international negotiations.

My introduction to the world of international science policy I owe to many colleagues, especially to anthropologists and scientists of the International Union of Anthropological and Ethnological Sciences, and the International Science Organization. A special thanks to my colleagues of the constellation of social science associations of the International Social Science Council now World Social Science Organization, whose research work has partnered scientific communities all over the world in the last 60 years. I would like to mention my friends and colleagues of the Standing Committee on Global Human Environmental Change, Sir Robert Worcester, the late Harold Nicholson, Roberta Miller, Martin Price, Lezek Kosinski, Pilar Magannon and so many others.

I owe my high-level participation in decision-making in international cultural development policy to Federico Mayor, former Director General of UNESCO, and to Javier Pérez de Cuéllar, who initiated me into the subtle mysteries of diplomacy and persuasion. My special thanks to the members of the World Commission on Culture and Development, among them Keith Griffin and Elizabeth Jelin, and to the staff of its Secretariat, to Jérôme Bindé, Raj Isar, Jean-Ives Le Saux, Guiomar Alonso, Malick M'Baye and many others.

In the first two weeks after my appointment as Assistant Director General of UNESCO for Culture, I met more than 200 ambassadors, delegates of member states and UNESCO staff members, who offered advice, demanded changes or gave me instructions, a widening sphere of thinking and action that in fact constitutes a privilege for any human being to be involved in. Many of them became friends, close professional collaborators, political allies or critics and, through our immersion in the world of ideas, politics, diplomacy and programmatic failures and successes, they imprinted in my mind the possibility and the necessity of the Human Project. When he became Director General of UNESCO, I shared this vision closely with Kōichirō Matsuura, as well as with other distinguished UNESCO staff: Francoise Rivière, Katerina Steniou, Ann-Belinda Preis, Noriko Aikawa, Isabelle Anatole-Gabrielle, Lyndell Prott, Doudou Diene and countless others.

I would not have discovered how to set out in creating new research pathways clearly if I had not come into contact with indigenous and migrant women in Mexico, nor would I have discovered my history and place in the world were it not for my feminist friends all around the world. First of all I thank Helen Safa, whose support in helping me get started in research was invaluable, as was that of many other researchers, among them Kate Young, Lourdes Beneria and Carmen Diana Deere. The list of them would be endless so I will only mention those of our Third-World partnership in Development for Women in a New Era (DAWN)—Devaki Jain, Gita Sen, Newma Aguiar, Magali Pineda and many others.

If I were to mention the institutions that supported my work, the list would also exceed these brief acknowledgements. In addition to the National University of Mexico and the Comisión Nacional de Ciencia y Tecnología of Mexico, I will only mention here the Ford Foundation, the Rockefeller Foundation, the John G. Guggenheim Foundation, la Maison des Sciences de l'Homme, the International Labour Organization, the Wenner-Gren Foundation, the Swedish International Development Agency and institutions such as the Library of Alexandria (with thanks to its dynamic director, Ismail Sergeldin).

Special mention must be made of the London School of Economics and Political Science (LSE) in London, where I received a Ph.D. in Ethnology in 1975 under the tutorship of Professor Julian-Pitt-Rivers. Many of my classmates there, among them Olivia Harris, became me lifelong friends and colleagues. My thanks to Dr. Debra Williams, for my visit at the Department of Anthropology at LSE in the summer term of 2017 which allowed me to continue to use the extraordinary support of the Librarians at the LSE Library. I was also able to spend a term in 2004 at New York University in relation with the King Juan Carlos fellowship. My thanks for this and for their friendship to Professors Kate Stimpson, Renato Rosaldo and Mary Louise Pratt. My gratitude also to the University of Manchester for the Honorary Fellowship that

I held in 2014, thanks to the welcoming hand of John Gledhill and other colleagues in the anthropology department.

I would also like to mention the deep satisfaction that spurred me on with my tasks as I worked with successive generations of researchers in Mexico, the USA, the UK, France, India, Japan, Senegal, Brazil and Costa Rica. In recent years, it was the enthusiasm and conviction of many young anthropologists as they plunged into fieldwork with me that enabled me to experiment with concepts and methodologies —on development, on global environmental change, on migration, on intangible cultural heritage—while closely exchanging insights with indigenous and local people. In particular, I would like to mention Cristina Amescua, as rising star, as well as Edith Perez, Carolina Buenrostro and Juan Carlos Dominguez, among others I would also like to acknowledge the research and revision work carried out by two young scholars, anthropologist Felicity Errington and writer Mariana Roa Oliva.

My special thanks also go to the Springer editor of the book, Hans Günter Brauch, whose advice, insightful suggestions and patience followed this book from its inception. Crystallizing all those exchanges and experiences into a book required sustained support in keeping to rigorous argument and consistent explanation, made possible by the work of editors Hester Higton and Mike Headon.

This book required intellectual input, institutional support and time to write, all resources that are at present continually being undermined by policies geared only towards monetary profit. If my writing contributes to keep the memory of such endeavours, it is proof that international intellectual collaboration will always be possible, and a necessity, for the Human Project.

The Complexities of Culture: culture unites humankind yet may be used to affirm particular identities, to discriminate against others or as a tool to fight back against exclusion. Aztec Dancers reinventing their rituals, at the Chalma Sanctuary of Mesoamerican origin in Mexico. *Source* Photo by Lourdes Arizpe

Contents

Abbreviations

AI	Artificial Intelligence
AWG	Anthropocene Working Group
CEDAW	Convention to Eliminate All Forms of Discrimination Against Women
CIRCLE	Cultural Information and Research Liaison in Europe
COP21	21st Conference of the Parties to the UN Framework Convention on Climate Change
ECLAC	Economic Commission of Latin America and the Caribbean
GDP	Gross Domestic Product
GNP	Gross National Product
GPS	Global Positioning System
HDGEC	Human Dimensions of Global Environmental Change
HDR	Human Development Report
ICAES	International Congress of Anthropological and Ethnological Sciences
ICC	International Criminal Court
ICPHS	International Council for Philosophy and Human Sciences
ICSU	International Council of Scientific Unions
IDRC	International Decade for the Rapprochement of Cultures
IFIAS	Institute for International Assistance and Solidarity
IGBP	International Geosphere-Biosphere Programme
IGO	Intergovernmental organization
ILO	International Labour Organization
IMF	International Monetary Fund
IPCC	Intergovernmental Panel on Climate Change
ISSC	International Social Science Council
IUAES	International Union of Anthropological and Ethnological Sciences
LAWS	Lethal Autonomous Weapons Systems
MAI	Multilateral Agreement on Investment
MDG	Millennium Development Goal(s)
MPAA	Motion Picture Association of America
NATO	North Atlantic Treaty Organization

NGO	Non-governmental Organization
OAS	Organization of American States
OAU	Organisation of African Unity
OECD	Organisation for Economic Co-operation and Development
R2P	Responsibility to Protect
SDG	Sustainable Development Goal(s)
SID	Society for International Development
UNDP	United Nations Development Programme
UNESCO	United Nations Educational, Scientific and Cultural Organization
UNU	United Nations University
WCCD	World Commission on Culture and Development

Chapter 1
Introduction

1.1 Culture: The 'Dreamcatcher' of Human Experience

The extraordinary appeal of the word 'culture' is that it is the 'dreamcatcher' of the peoples of the world and, at the same time, the "golden account" of their civilizational achievements. The Navajo native peoples of the United States build a circle of twigs with a finely woven mesh inside it with coloured threads of small feathers that, hanging over one's slumber, catches dreams and foretells the future. In just such a manner, culture seems to capture so many dreams, so many questions, so many possibilities around the world. The rapid rise in the use of culture in liberation movements in the fifties and sixties and the "cultural turn" in art and critical theories from the 1980s onwards led to culture becoming a policy instrument for international development, national political management, human rights, gender equality, and ethnic and religious assertions and the transformation of identities in the second decade of the millenium. At the same time, the "golden account" in the national GDP of countries is now recognized as the economic contribution of the arts and cultural activities to overall growth—in parallel with the "green account" of environmental activities. Most importantly, many of these changes are part of the ongoing transformation of cosmopolitical models of human existence in the Earth system.

Culture is often said to be unique to human beings. Recent scientific discoveries, however, show that genetically close primates, as well as other species, use some elements of cognition, language and skills, so the question is: How have humans become a geological force in the world while thriving as culturally diverse peoples? I propose that there are two traits that allowed humans to evolve from *Homo habilis* to *Homo sapiens sapiens*: the desire to know, and the will to exchange. All peoples of the world have engaged in co-producing ideas and knowledge and exchanging them together with goods for millenia. Importantly, this permeable process, however unequal and wrought with forced conquests, in the end, has benefitted peoples all over the world. The combined effect of producing knowledge and exchanging

L. Arizpe Schlosser, *Culture, International Transactions and the Anthropocene*,
The Anthropocene: Politik—Economics—Society—Science 17,
https://doi.org/10.1007/978-3-642-41602-6_1

leads to the human characteristic that is highlighted in studies of cultural evolution: to cumulative learning and, as feminist research has shown, to great benefits in social reproduction. These gave humans a fundamental advantage in accumulating meanings and in sharing, copying and adopting these from other groups. Whom do we learn from? Other people. Who has been adventurous enough to change what had been learned? Individuals born in many diverse cultures. That is why the definition of culture chosen for this book, among the myriad others available, is that culture is made up of meanings activated through social relations that allow persons to transform their lives and their environments.

By refocusing the concept of humanity in this way, the United Nations' new term of a "transformative agenda" to achieve the Sustainable Development Goals and the United Nations 2030 agenda is set in a broader context, as can be seen at the end of this chapter. The unprecedented challenge is that people are thinking, making things and transforming their lives and the world within a single global space managed by information and communication technologies and by a planned capitalist market. As a result, cultures have indeed been changing because of the impact of scientific discoveries and their technological applications to the ways in which human beings develop notions of self, relate to each other individually, in groups and in countries, and think about the world and the cosmopolis.

Today the unprecedented challenge is that people are co-producing knowledge and making exchanges within a single global space of communication and trade. The ethical and political framework of this space is still based on the values of the Enlightment, capitalism, the nation state, democracy and human rights, which are now being recast in the new conditions of the Fourth Industrial Revolution and an electronically webbed world. As empirical research in this book will show, culture has been brought into the complex transactions having to do with cultural, ethnic and religious inclusion, fair globalization, cultural industries and reconceptualizing the role of the West and other regions in world development.

Culture, in this sense, keeps adding very favourable connotations in that it represents reflexivity, empathy and ethics. And if there is anything humanity needs today, it is that. As several participants in the meetings discussed in this book mentioned repeatedly in different meetings, "Culture speaks from the heart." Yet the recent events of brutal terrorism, the destruction of cultural heritage, the veiling of intolerance behind the benign polysemy of culture and the double helix of ethnic and religious prejudice slowly giving rise to retribalization and racism are also turning culture into a nightmare.

The ideas and processes that have generated such contrasting outcomes need to be clearly understood, an outcome this book hopes to contribute to as an intellectual history of international transactions in the field of culture from the 1990s to 2016.[1] The empirical data collected are based on the participation (including decision-making) of the author, in United Nations meetings, government cultural

---

[1]The intellectual history of cultural institutions before 1990, summarized in Chap. 2, was published in Arizpe 2004.

policy conferences, world social science organizations, global economic forums and a few national and local meetings where questions about culture were discussed. The methodological approach I have explored in this book is the "ethnography of transactions", a heuristic that, instead of relying on the end-products of international debates, i.e. institutional reports of sessions, resolutions and official documents, allows us to see who said what, where and how. The materials retrieved in this way keep the textures of debates, the full diversity of human interactions. This is "living culture", a view that we expressed in the 2000 UNESCO World Culture Report (UNESCO 2000: 45) and which leads to one of my contentions in this book. That is, that instead of focusing on 'life' as a state, we should focus on human "creative living", as will be made clear throughout the book.

1.2 Finding Patterns in International Transactions

I have aimed to do three things through laying out this material. The first aim has been to create an archive of the ways in which new theories, vocabularies and semantics of culture grew as a force in development and in international relations as part of a process of global change, for the first time incorporating all nations and cultures in an open, international dialogue. This is important for setting precedents and marking ascendant steps in a world in which people feel increasingly insecure and believe that retrenchment into past identities will bring back security. As a result, however, instead of allowing for readjustment in a new political and social setting, they have created intractable conflicts.

In proposing more incisive heuristic tools to analyse cultures at a global level, I have proposed a theoretical distinction between culture as an ontology of human becoming and culture as an instrument for political or policy applications. Since it is impossible to separate these two aspects of concept of culture completely, I have suggested making a distinction between *ontoculture*, as a philosophical discussion of the origins, nature and becoming of humans, and *laoculture* as the development of policy options to organize nations, ethnic communities, religious groups and territorial inhabitants, through governance and politics. Such a distinction would allow for the more rapid development of thinking at both the macro- and the micro-scale, in a way similar to that of macroeconomics and microeconomics, thus avoiding the conflation of these levels of analysis in the current confusing discourses on cultural rights and cultural policies. Ontoculture gives humans a capacity to create representations, ideas and actions that other life forms in the planet do not have. As Mark Pagel has insisted, humans are "wired for culture" (Pagel 2012). Thus culture is, by definition, an instrument of freedom from the constraints of geo-biological systems, which seem to be overcome to an astonishing degree as we become cyborgs, no longer only cultivators and creators but developers as well. This is why I believe that the freedom to create is by far the deepest core value of humans.

Reconfiguring heritage: the new avatar of the Mexican Day of the Dead—"La Catrina" in the James Bond film 'Spectre', and Disney's 'Coco'. This is a young woman reimagining herself at the annual fiesta of the "La Huehuenchada" ("The Elders") in Tetela del Volcan, Morelos, Mexico, 2011. *Source* Photo by Lourdes Arizpe

The second aim of the book is to try to identify the features of a worldwide transactional process that may help in setting up and sustaining world debates on sustainability which incorporate social and cultural aspects. How are these debates organized? Who generates such topics of debate? How do institutions channel such demands? Although the approach taken in this book is not anthropology of policy or public anthropology or global anthropology as frameworks of analysis, many aspects taken up in these related fields are touched upon. Instead, the approach taken in this book leans more towards an anthropology of the new cosmopolitanism (Werbner 2008). Being well aware of the theoretical and methodological obstacles to carrying out empirical analyses of global phenomena, I have used as a methodology an "ethnography of transactions".

My third aim has been to explain the emergence of the concept of culture as the most frequently used and quoted term in a surprising variety of areas of political and social life, as well as in international political and normative assignments. As made explicit further on in this Introduction, data demonstrate that the rise of the concept of culture, overtaking many previous terms, is evident in the number of political, intellectual and academic texts referring to the concept in the last quarter of the twentieth century and especially at the beginning of the millennium.

In the Conclusions, the major trends discerned in the transactions in international debates in culture from 1990 to 2016 will be summarized, although readers and social science researchers may also find other patterns of connections. I will also suggest, on the basis of my experience both as a fieldworker and as an international participant, how to clarify the new, recently politicized versions of old, polysemic concepts such as culture and civilization so that they won't turn dreams into nightmares. Additionally, I will point out the pathways that have to be explored to make sustainability less a static, conservationist/conservative maintenance task and more a transformative, proactive undertaking.

With these aims in mind, Chap. 1 presents an introduction to the main concepts and the major issues related to culture and sustainability discussed in the meetings and analyses discussed in this book. Chapter 2 sets the scene of culture and development programmes, international political movements and early studies that took up the concept of culture in migration, economic growth and social development. Chapters 3–6 present the report of the World Commission on Culture and Development and the meetings held around the world that provided the data and ideas for the report, as well as the reactions to it in international policy and social science communities. Chapter 7 describes the debates and transactions held in other institutions such as the Global Economic Forum in Davos, the Pontifical Academy in the Vatican, and the United Nations Development Programme, among others. Chapter 8 takes a more detailed look at the recent academic research and political debates on culture that have framed and led many of the discussions held internationally.

A new pathway is explored in Chap. 9 by delving more deeply into the current new understanding that "culture is becoming nature" and "nature is becoming culture". A most interesting window of opportunity has now opened in proposing that the new name for our present era be the 'Anthropocene', since humans have become a powerful force directly affecting the geo-ecology of Earth. The main issue in this field for this book is to arrive at an understanding that the main obstacles to achieving sustainability come from the ways in which humans interact with other humans in relating to the Earth system.

As an anthropologist I had become a life-long ethnographer before I had the honour of serving as Vice-President and President of several social science organizations and as Assistant-Director General of UNESCO in the Culture Sector, thus becoming a decision-maker in managing culture programmes around the world. After returning to Mexico to carry out fieldwork to safeguard intangible cultural heritage, I realized that the vast amount of evidence I had gathered through a kind of multi-sited ethnography could be usefully systematized to understand how culture was discussed and transacted for international political and policy purposes in the nineties and at the beginning of the millenium. Accordingly, notes of the oral debates in more than forty international meetings, together with major innovative ideas on culture as well as on the proposed perspective of the Anthropocene, are included in this book. Because of my own professional involvement in many of these meetings, behind the ethnographic narrative there is also an autobiographical account of how I perceived the ongoing events. This, I believe, is part of an anthropologist's role as a co-producer of interpretations of observed practices.

Beginning in the fifties, themes related to culture generated an astounding wealth of declarations and transactions on economic development, globalization, the clash or the dialogue of civilizations, cultural diversity, indigenous knowledge and intellectual property, and physical and intangible cultural heritage. The fact that many of these debates were political in nature is shown in the proliferations of '–isms': multiculturalism, cultural pluralism and what should also be called ethnicism and religionism. The proliferation of such debates, which have surfaced in many different international and national settings, indicates that people all over the world are deeply concerned with what they perceive as the basic components for a future global society: cultural survival, cultural hegemony and cultural liberty.

1.3 Why Culture?

The vast process of building a world perspective on the most important characteristic of human societies was entrusted by member states of the United Nations to the World Commission on Culture and Development[2] and presented to the UNESCO General Conference in 1995. Its report *Our Creative Diversity* led to the launch of a worldwide debate on culture and development. Subsequently, transactions on cultural policies were taken up in other United Nations agencies, the World Bank, the Global Economic Forum at Davos, the Vatican, the International Social Science Council, and other international social science organizations, while many countries set up Ministries of Culture to develop policies in many areas. At the same time, the international politics of culture which began in the fifties with the interest of developing countries in achieving their independence and creating a non-aligned political movement became entangled with a plurality of intellectual, civil society and artistic movements, fuelled also by postmodernism, that took the "cultural turn", the "cultural exemption", "cultural capitalism", and multiculturalism to many diverse areas of development. Very recently, this vast search for new cosmpolitical meaning has now landed in the virtual world with the concept of "algorithmic culture". Two other important policy areas also rely on culture as a key

[2] I was a member of the World Commission on Culture and Development before I became Assistant-Director General for Culture at UNESCO, and was subsequently asked by Mr. Javier Perez de Cuéllar, Chair of the Commission, to be in charge of its Secretariat for the writing of its Report.

concept in their interpretations: the urgent need to defend authochtonous and indigenous cultures as neo-liberal markets have incorporated vast swathes of their resource-rich territories, and the second wave of feminism which surged to become a worldwide movement of introspection into how cultures construct notions of femininity and masculinity that end up being oppressive to women. In the last few years this movement has also been extended to encompass the full range of gender identities.

Many of the threads of these discussions related to culture help explain the meteoric rise in the use of the term 'culture' during the twentieth and twenty-first centuries. In the n-grams offered by Google, which plot the uses of particular words and phrases throughout the corpus of texts written in the world's main languages that have been scanned by Google (dating from 1500 to 2008), the frequency of the word 'culture' has overtaken other important concepts used in English in reference to human society in the following years: 'mankind', 1892; 'humanity', 1897; 'tribes', 1900; 'conscience', 1901; 'democracy', 1920; 'consciousness', 1925; 'civilization'/'civilisation', 1926; 'soul', 1939; 'Indians', 1945; 'progress', 1959; 'religion', 1966; 'nation', 1971; 'spirit', 1971; and 'revolution', 1972. 'Society' and 'country' have both exhibited significant decreases in usage in comparison to 'culture' in recent years.

N-grams are, of course, a blunt analytical tool. The dataset is reliant upon optical character recognition, which can produce inaccurate renditions of text. The information is also skewed towards scientific literature, and there are issues with incorrectly dated texts. Finally, the analysis ceases in 2008, so more recent correlations cannot be charted. However, comparison can be made with the frequency of 'culture' in book titles in significant academic library catalogues. The Copac catalogue, which covers nearly one hundred major UK and Irish libraries, records 173 titles from 1800 that included the word 'culture'. Numbers were still small in 1880 (288) and 1948 (862) but had more than doubled by 1970 (2,434). Thereafter growth was almost exponentially rapid: 3,516 in 1982, 8,342 in 1995 and 16,382 in 2015. A similar pattern can be charted by searches in the Library of Congress catalogue and on worldcat.org. Again, this is a blunt tool, but it gives some sense of the phenomenal rise in the use of the term 'culture' in English as an international and scientific language mirrored in many other languages.

It is very interesting to note that, in coining the word 'meme' as the core aspect of 'culturemes', Professor Richard Dawkins probably never expected it to be taken up so enthusiastically, to such an extent that it has gone viral in social media especially among young people. Professor Dawkins defined it as a parallel form which might replicate as selfish genes do, jumping from body to body, or, in the case of culture, from statement to statement or practice to practice. In fact, if meme is taken as an abbreviation of cultureme, then the use of the word meme has not only been as meteoric as 'culture', as just mentioned, it has broken the time barrier at warp speed, projecting the twentieth-century concept of culture into the twenty-first-century microelectronic digital revolution in modes of thinking. Of course, its connotations are changing as rapidly, but because it is young people who use this catchy term, 'meme' may yet become a core concept in digital culture.

So, why has culture absorbed or overtaken all the concepts assessed in the n-gram search? And why has it become particularly prevalent in the last twenty-five years? These are the questions whose answers I try to elucidate in this book by presenting the empirical evidence I gathered during this period as a new vocabulary and semantics: consensual ideas and political platforms were constructed which are still in place but are insufficient now for the challenges ahead. I also try to focus on a better appraisal of the utility of the concept of culture for managing such diverse forms of interventions. Unfortunately, because it is such an attractive, positive idea, a reified concept of culture has become a staple of political discourses that lead back only to the same old debates and to fossilized culture policies in some countries. Many anthropologists are asking questions such as that posed by Timothy Clark, who, through the lens of ecocriticism, asks whether we have reached the limits of now anachronistic cultural conventions which we are no longer capable of reinventing (Clark 2015: 176).

At the same time, it is important is to guard against what I call the "flat culture syndrome", to paraphrase Thomas Friedmann's "flat world" metaphor that was used to signal that the world was moving towards equal participation and representation, when, in fact, it has been moving very quickly towards unprecedented levels of inequality. In terms of cultures, Christoph Brumann made a very pertinent case for taking into account the textures of cultures that could be useful for social analyses (Brumann 1999). At present, however, through the influence of multiculturalism, flat narratives about cultures abound. Yet cultures are not 'equal', because 'cultures' do not exist. They are a heuristic device used to organize a fundamentally fluid—or even liquefying, according to Zygmunt Bauman—reality in which cultural practitioners choose and decide which path they and their immediate social entourage will take. The key elements in this selection are the knowledge and cooperation that groups of cultural practitioners contribute to their environment and to the world. This must become the most important value and measure of the worth of human collective endeavours.

Indeed, this book will propose that instead of endlessly debating how values of cultural variants may possibly fit together in a global cosmopolis, a new template must be created in which the positive contribution that each culturally or religiously defined group is the measure by which their influence is legitimized at the international level. In other words, the important question to ask is not "How is each culture going to advance their interests on the world stage?" Instead, new values and normative systems must be legitimized on the basis of the question "How are the groups formed according to your values thinking and acting to ensure the sustainability of human survival in the Earth system?"

1.4 "Becoming Enlightened About Relations"

The first step that springs to mind in answer to that question is that whoever is trying to find cooperative paths to the future won't focus exclusively on singular groups—be they cultures, ethnic groups, national identities, religions—or whatever —but on the relationships between them.

We must become "enlightened about relations", Marilyn Strathern proposed at the 2014 meeting of the Association of Social Anthropologists in Edinburgh, UK. She mentioned the Dutch traveller Maria Sibilla Merian, who brought back specimens of insects from Surinam in the early eighteenth century (Strathern 2014). Professor Strathern highlighted the fact that, in contrast to other naturalists of her epoch, in all her illustrations of the natural world, Merian identified the relationships between insects, butterflies, plants, flowers, etc. as "nature's connections". In doing so, Merian remarked that what was missing in collections was process, context and transformation. Merian's drawings depict relations and interdependence between animals (alligators, caterpillars, plants): what Raymond Firth would have called an interlocking system, according to Strathern. Pictures emphasize the materiality of forms—say, butterflies—, drawing attention to these, where similarity would only be a simplified form of understanding. Thus, for Merian, different organisms could become one another. Strathern argued that the intellectual legacy of using similarity and difference is a negative one because it has led to the current European excesses about difference, and she concluded that we must move away from images to scenarios (Strathern 2014).

I contend that, in today's international debates on culture, the problem with the old classificatory schemes is that they make it impossible to understand "cultures' connections". Such connections are first made explicit in debates and then negotiated through live transactions. Producing lists of cultural elements, or 'culturemes' in ethnographic grids, or drawing cartographies of units of cultural practitioners has been vital in safeguarding people's creations in time and place but has left us with little understanding of the dynamics of cultural processes. The disconnectedness of cultures in the model of multiculturalism and the "politics of difference" puts a spike in the wheels of thinking about the co-evolution of cultures, and is responsible for conflicts that, instead of allowing necessary readjustments, create intractable clashes leading to ethnic cleansing, racism and xenophobia. Marc Augé, as early as 2002, spoke about "a new hundred years war, an internal war, civil and eminently political, whose challenge will be to know if the planetary utopia was realisable or if it will be overrun … by the alternating injunctions of religious madness and market barbarism" (Augé 2017: 102).

Needless to say, this is happening just at the time when cooperation in transforming environments is vital for recombining and reinventing cultural cosmopolitical conceptions so as to work together collectively towards a sustainable future. Understanding the dynamics of exchange and confrontation between individuals and people who choose a diversity of identities to deal with unstable conditions in their daily lives has become a burning issue.

The concerns over culture and cultural diversity may have become combustible because public and political debates on these themes have taken up confused or equivocal features of these concepts. Two in particular must be mentioned: firstly, how cultures have been turned into 'things', that is, in anthropological terms, how they have been reified. This leads to an irrational, visceral fidelity to a set of norms exalted as unique by cultural gatekeepers. Secondly, how sectarian and partisan interests hide behind the libertarian and liberal heredities of these words, driving them to become insignia of conservative movements.

1.5 Defining the 'Undecidable'

In his "Brief Cultural History of 'Culture'", Marshall Sahlins, noting the sometimes bewildering variety of the meanings of this concept, mentioned that "in the past two or three decades, however, the word has entirely escaped academic control and been taken up by peoples all over the world in an extraordinary movement of social self-consciousness—an awareness of their own way of life as a value and, above all, a political right. This sudden development of 'culturalism' is one of the most important phenomena of world history in the late 20th century" (Sahlins 1994: 1). Importantly, Sahlins had also clarified that "in addition to 'culture' as refinement and 'culture' as social life-form, since the late 19th century there has been 'culture' as the distinctive capacity of the human species" (ibid.: 7).

The polysemy of culture makes it a difficult concept to define. The 'undecidability' of this concept, as Jacques Derrida would call it (that is, the multiplicity of possible meanings), builds people's identities and evokes sentiments of ancestry, political affiliation and emotional attachment. But this can work in both positive and negative ways. Culture can be invoked for purposes of belonging—of solidarity—as well as for the destruction of others. Practically all groups that mobilize their own forces may discriminate against other groups, particularly if they use cultural, ethnic or religious commitment as justification for excluding or enslaving others. Once one group claims special rights and privileges, their interaction with other groups immediately changes into relationships of defence or attack. Muslims imposing sharia law in northern Nigeria infringe the citizen rights of Christian groups; the Serbian army's behaviour during the war in Yugoslavia opened the way for horrifying repetitions of rape and genocide which have now found more extreme and extended forms in the caliphate of Isis in Iraq and Syria. Jihadist terrorists kill in the name of the religions they have only recently taken up. Everywhere, ethnicism—that is, the use of ethnicity for political purposes—has led to the same brutal excesses that some nationalisms had formerly produced.

Nevertheless, the polysemy of the term 'culture' is, in fact, very useful in providing the broadest intellectual and political space for debating and negotiating the future relationships between peoples, nations and cultural groups. Definitions of culture, which were counted as a little over two hundred by Alfred Kroeber in 1948, have now risen to several hundred more (see Eagleton 2000; Fischer 2009). The problem, in my view, is not deciding on which definition is better, but trying to define how uses of 'culture' in the contemporary world are leading both to positive ends and to intractable and violent conflicts. A rapid appraisal shows extraordinary collective cultural achievements, as in the rebuilding of the Bridge in Bosnia in 2004, the reconstruction of the Bibliotheca Alexandrina in Egypt and the bringing to justice at the International Court of Justice in The Hague of one of the Islamic leaders responsible for the destructions of ancient manuscripts and museums in Timbuktu. At the same time, we have witnessed some of the most degrading and wilful obliteration of human achievements, as in the destruction of antiquities in Syria and in Iraq—the attacks against the museums as well as the site of Palmyra—by the Islamic State. On 13 November 2015, Paris was the scene of an atrocious terrorist attack at the Bataclan music venue; yet, just three weeks later, the same city was host to the most successful international conference on climate change (COP21) ever to have been held. This is a tribute to humanity's resilient capacity to reiterate a perceived necessity even in the face of ignorant or brutal opposition by minority groups.

Another reason for the emergence of culture as a key international issue is that, by deliberately leaving aside political philosophies in organizing the nation state or the international sphere, culture has been thrust to the fore as the term to define the political ideology of a person or a group in the new cosmopolitical space. Its use as an easy surrogate for complex political positionings explains why many debates on culture have become circular or fallacious. Furthermore, a different definition of 'culture' has emerged in recent discussions of human nature, gene-culture evolution and the Anthropocene. This topic will be taken up in Chaps. 8 and 9. Matt Ridley, in trying to summarize the evolution of culture, points to the almost perfect parallel between the evolution of DNA sequences and the evolution of written and spoken language: "Languages mutate, diversify, and evolve by descent with modification" (Ridley 2015: 79).

In the same way, cultures mutate, diversify and evolve by descent with modification. Yet this description is only based on an initial concept of cultures as single units. As applied to human sociosystems, similar to the way ecosystems are now the central concept in the natural sciences, the core perspective must take into account the way relationships enhance or deter knowledge creation, exchanges and reciprocities, and cooperation to attain group objectives, as well as transformation of the environment.

The most important thing in this respect is to highlight that cultures are not species: all the practitioners of different cultures belong to the same species *Homo*. Even as this latter concept continues to evolve, as I will posit in this book, the biosocial traits of all peoples of the world are contained in identical DNA. This is the foundation of the concept of humanity. With these conceptual tools and heuristics we can now take stock of the way in which international transactions on culture have been developed in the last two and a half decades.

1.6 Culture and International Relations

The rise of culture as an international concern since the 1950s began in the aftermath of the Second World War, for several reasons. Culture first appeared in policy in 1948, with the proclamation of the Universal Declaration of Human Rights, whose Article 27 ensures that "everyone has the right to freely participate in the cultural life of their community, to enjoy the arts and to share in scientific advancement and its benefits". Subsequently, culture became an issue in development in the 1950s, when the United Nations launched economic development programmes in decolonizing countries, with different results from those of the Marshall Plan in Europe. In Latin America, the "culture of poverty", a term coined by the anthropologist Oscar Lewis in a study of Mexican rural–urban migrants (Lewis 1959), was used to debate whether culture could explain the positive or the negative aspects of uneven development. In Africa, especially, and also in southern Asia, culture became a political issue as leaders of the liberation struggle included cultural issues in their demands for endogenous development. In left-wing movements, however, culture was viewed with distrust, since cultural difference had also been used as the fundamental argument to justify apartheid.

The involvement of international organizations in raising awareness of the importance of culture rose dramatically in the 1970s and 1980s. Through the 1972 International Convention Concerning the Protection of the World Cultural and Natural Heritage, UNESCO took the lead in programmes to protect cultural heritage. The 1982 UNESCO Intergovernmental Conference on Cultural Policy, held in Mexico City, then fostered the creation of ministries of culture in many countries. Later in the 1980s, the Group of 77, along with a number of European countries, highlighted the importance of culture as a political issue in international development and an instrument for governance through cultural policies that led to the programme of the United Nations World Decade for Culture Development from 1987 to 1997.

Also in the eighties, the intellectual "cultural turn" was launched by Fredric Jameson, celebrating 'postmodernism' as a sign of late capitalism, the death of the subject, the transformation of reality into images and the end of grand narratives (Jameson 1998). A very different kind of "cultural turn" began to take shape in the early 1990s as the conflict between Serbia, Croatia and Bosnia unveiled a new kind of civil war: "ethnic cleansing" as the emerging crest of religious fundamentalist movements espousing political goals. Another powerful cultural movement, which had been present for several decades, was also fuelled by the impending loss of the livelihoods and ways of life of indigenous and aboriginal groups in a majority of developing countries.

The UN Decade for Cultural Development mobilized many cultural groups and movements that carried out an astonishing array of traditional cultural celebrations, a revaluing of folklore and crafts, and a rediscovery of imperilled or lost productive or artistic activities. Yet these events left very little in terms of accumulated knowledge. As political pressures grew to protect heritage, recognize discriminated cultural groups and use cultural levers for economic development, the need to rethink the role of culture in its connections to contemporary political and economic challenges became clear. As a result, the UN General Assembly decided to create a knowledge-making programme through the World Commission on Culture and Development. Its work is introduced in Chap. 2 and forms the core of the ethnographic discussions of Chaps. 3–5.

Today, in the second decade of the new millennium, as the media constantly reminds us, identity politics are coming to the fore in debates on the future of governance and culture, as diverse groups invoke culture in power struggles to pursue their economic, political or social advantages. A case in point is Samuel Huntington's *Who are We? The Challenges to America's National Identity*, in which he contends that Hispanics generally, and Mexicans in particular, pose a threat to the unity of the United States in terms of language and culture (Huntington 2005). A new, more belligerent version of this position is that of Donald Trump, insisting that Mexicans should build a wall to stop themselves from migrating to the United States, a simplified version of an extremely complex situation.[3]

The weakening of nation states in globalization has brought about these heightened forms of cultural politics, fuelled by jihadism, terrorism and the large-scale movements of migrants out of impoverished or war-torn regions. This has led to serious challenges to multiculturalism in scientific studies (Barry 2001; Vertovec 2010), as well as from European leaders. Both the former UK prime minister David Cameron and Chancellor Angela Merkel of Germany have rejected this policy, the latter having explicitly declared that "multiculturalism is dead". Steven Vertovec points out that, since the beginning of the twenty-first century, this idea has been seriously challenged from a number of directions:

> From the political Right many critics now see multiculturalism as a foremost contributor to social breakdown, ethnic tension and the growth of extremism and terrorism. From the Left, where numerous commentators were long dubious of a seeming complicity with Empire and willing blindness to class-based inequalities, even previous supporters of multiculturalism came to question the model as contributing to a demise of the welfare state and the failure of public services (Vertovec 2010: 84).

He sees emergent migration trends as leading to conditions of "super-diversity" in patterns of diasporic identification and transnationalism (ibid.: 84). In spite of that, he believes that policies for social cohesion and national identity can coexist, fashioning a greater sense of cosmopolitanism, respect for others and social justice for migrants and their descendants (ibid.: 94).

[3]A very early—and opposite—view of this situation can be found in Arizpe (1981).

1.7 Culture and Globalization

Anthropologists are justifiably cautious of the sweeping generalizations about globalization that incorporate some of the outdated notions with which they had to contend in the early twentieth century. They are even more distrustful of the sudden revival of the term 'civilization', which was discarded in anthropology because of the imprecision, exclusion and hierarchization that it introduced in analysing the contemporary world. As the data show in Chap. 7, after Samuel Huntington reintroduced the term 'civilization' it took on a political life of its own, generating an avalanche of critiques against the "clash of civilizations" and a countermovement, in which I participated, of a "dialogue among civilizations". However, the analysis of the empirical data provided in this book demonstrates that, even if a term is used equivocally to obscure a partisan position, this is no reason to reject that term. This, it seems to me, applies to the use of the term 'civilization' since the nineties, as is evident in what participants have said in many meetings.

The Reflexivity of Anthropology: reflected in the mirror at the centre of the offering of flowers of the Conchero Dancers, the anthropologist (Lourdes Arizpe) reflects on the marvels of rituals, Tlayacapan, Morelos, Mexico. 2013. *Source* Photo by Lourdes Arizpe

Globalization, in contrast, rapidly became a frame of reference for policy and research from the mid-1990s. As Richard Jolly succinctly expressed it:

> The new forces of globalization are being played out on a battlefield that is far from new. Trade and conquest are its antecedents, empire and colonialism its earlier forms, inequalities in power and relationships part of the ongoing saga. The need for a level playing field is widely accepted today, but at such a general level that the phrase means quite different things to different parties. The rich and powerful argue that it will be achieved by extending the rule of law and competition, the poor and weaker argue for conscious international action and regulation to offset inherited inequalities and to provide special support and advantages to poorer and weaker countries (Jolly 1998: 1).

As will be discussed in Chap. 8, many authors in the social sciences have carried out research on culture and globalization since the 1990s, among them Robertson (1992), Inda/Rosaldo (2002), and Pieterse (2004). The most comprehensive work in this area is the "Cultures and Globalization Series", edited by Anheier/Isar (2007, 2008, 2010, 2011, 2012). The series has covered the main issues in this field, from the culture economy, through heritage, memory and identity, to conflicts and tensions. The series title is significant in that, instead of 'culture', it refers to 'cultures', consolidating the view of a plurality of cultures as the main subject in research. This has helped the development of particular regional viewpoints in the worldwide debate on the future of cultures.

Reference to the diversity of cultures gives greater texture to the flat representation of cultures, as does the more recent use of the term 'intercultural'. In both cases, however, the tendency is to go back to the reified definition of culture by assuming an ontological separateness of entities defined in a linear exchange. Anthropological terms in early studies, such as synchretism, 'mestizaje' and 'metissage', have been collapsed into that of 'hybridity' but, significantly, they all posit the same ontological separateness of culture, all describe the mixture of cultures, with few actually are intent on explaining process, context and transformation.

Many authors have warned against referring to cultures as "billiard balls" (Wolf 1982: 6), 'mosaics' (Arizpe 2014) or any other metaphors that represent them as bounded, discrete units which can be managed as objects. As a result, Thomas Eriksen optimistically believes that "the battle to escape from the straitjacket of a reified concept of culture, along with its concomitant reified ideas of identity, seems to have been won for now" (Eriksen 2003: 14). I will argue that we cannot win this battle in the abstract but we can distinguish with greater precision at what scale and for what purpose the concept of culture is being used.

Culture is now being used to depoliticize as a bland, good-wishes term, as one which comforts and alleviates tensions; however, through multiculturalism it has become a strong force for exacerbating ethnic, religious and cultural clashes (Clark 2015: 13). Additionally, it is being understood as an almost biologically inherent way of life; this is a type of "cultural fundamentalism", denounced by many anthropologists, one of the first being Veronica Stolcke (1995). I agree that these are problematic usages of the term and would add that the bland discourse on culture is now often used to gloss over fierce confrontations whose violent aftermaths are appearing ever more frequently on our screens. All the more reason to investigate how this discourse is created and how it is spreading some of its ramifications and outlooks towards the future.

Evidently, culture is being instrumentalized for different purposes. Policy decisions in governance, diplomacy and cosmopolitics require a classificatory list of more or less discernible "units of culture" that can be the target of economic policies, governance decisions and political negotiations. I learned this the hard way in UNESCO when I held the position of Assistant-Director for Culture. In one instance, I erased the reference in my speeches to what was presented as the fact that 6,000 languages are spoken in the world, because, as linguists and anthropologists well know, separating languages from dialects, variants and idiolects is simply not possible. To my consternation, I soon found out that I could not get ambassadors and delegates interested in adopting the programme of endangered languages that we were proposing unless a figure was quoted to raise awareness about the importance of this issue. Indeed, at all scales of international, national or local politics, delegates require streamlined discourses, preferably based on statistics and with information bullet points to make debates and negotiations flow faster, with simplified clarity, and more to the point. Thus, normativity, policy prescriptions and political discourse all require, as far as possible, a quantification of phenomena under scrutiny but not to the extent that these are fossilized into artefacts of stone.

A new framework for culture, then, requires an inquiry into questions of scale. There is a new surge in ideas about culture in terms of the process of the evolution of humankind in the context of geological, geographical, historical, philosophical and anthropological time. This is the level at which the discussion of the Anthropocene is taking place at present. In Chap. 9, where I turn to these debates, I propose that the term *cosmoculture* be used to refer to discussions about culture at this level of philosophical dialogue.

1.8 Anthropology and Reflexivity

The empirical data presented in this book is based on my participation and decision-making in meetings, conferences and forums that I began to record in 1988 at the 12th Congress of the International Union of Anthropological and Ethnological Sciences (IUAES) held in Zagreb, Yugoslavia. Soon after, I recorded the key ideas and arguments for the programme Human Dimensions of Global Environmental Change (HDGEC). After becoming Assistant Director-General for Culture at UNESCO in 1994, I was Chair of the Scientific Advisory Committee for the two World Culture Reports and subsequently an advisor to some of UNESCO's cultural programmes and preparatory meetings for the creation of two international conventions in 2003 and 2005. It was also possible to attend many civil society and private-sector meetings that touched on culture: as a member of the Academic Body of the Global Economic Forum at Davos, Switzerland; as President of the International Union of Anthropological and Ethnological Sciences; as President of the International Social Science Council (ISSC); as a member of the Advisory Committee of the United Nations Human Development Report; as a member of the Board of Trustees of the Library of Alexandria; and as Chair of the Board of the United Nations Institute for Social Development on concerns related to culture and social development, ethnicity and globalization and associated topics.

In all of these capacities, I participated in a vast amount of debates that I found myself recording in outlines and notes that clarified, contrasted and identified common vocabularies and patterns of ideas and proposals for our common future. These notes, in fact, had become a precious resource when chairing international social science meetings, UNESCO and United Nations meetings, or other conferences, in particular when I had to provide arguments for elaborating documents and resolutions, especially those that required consensus. It is worth noting that grants from many institutions (Guggenheim, John D. Rockefeller and Ford foundations and the Bibliotheca Alexandrina, among others) allowed me to travel to Latin American, African and Asian countries. Invitations from the United Nations, the International Labour Organization (ILO) and UNESCO and other governmental organizations made it possible for me to carry out research and participate in applied development projects. This combining of fieldwork research and being involved in applied development programmes formulation was made possible thanks to the academic liberty given to me by the National University of Mexico. At present, the cleavage which is developing between academic research and public debates on policy is delaying the dynamics of innovation and lowering the quality of both such activities.

Anthropology must develop new methods to follow the course of such debates, which are now a permanent feature of international political life as the global becomes the level of abstraction of major debates. Ulrich Beck, in his late writings, acknowledged the utility of ethnography but argued, as Michael Fischer has also insisted, that cosmopolitanism would require "new grammars, syntax, and semantic elements in new forms of metapower games and recombinant play with semantic elements such as universalism, relativism, ethnicism, nationalism, cosmopolitanism, multiculturalism, etc." (Fischer 2009: 258).

Anthropology is already in the midst of a transformation, in search of new methods to discuss world processes. Having inspired many of the recent trends in ethnographic research, George Marcus referred to the need to "reform, reinvent, or redesign the deeply engrained aesthetic and culture of method for certain kinds of research horizons in which anthropology is now involved" (Marcus 2011: 16). He specifically mentions the anthropology of globalization, the anthropology of the contemporary, and he calls for a public anthropology (ibid.). In the same book, James Ferguson stated that "this new sense of ethnographic purchase on 'the global' contributes to a high level of intellectual excitement within the discipline" (Ferguson 2011: 2).

This book is not an anthropology of policy as developed by Shore/Wright (1999); nor an anthropology of international institutions such as that carried out by Marc Abélès on the European Parliament and the International Trade Organization; nor an anthropology of global organizations (as Garsten/Nyqvist 2013). Garsten and Nyqvist bring up an interesting issue in that they mention working as insiders and outsiders of such organizations. While being an insider might seem the ideal position for an ethnographer, I do want to alert researchers to the fact that, if you wish to be effective institutionally and diplomatically in an international agency, you have to accept the constraints of being an insider, just like other staff. This may even take the form of having to sign confidentiality agreements. The balance that must be found is how to be inside these processes as an embodied participant and, at the same time, remain an anthropologist in one's mind.

This is not an easy task. In a publication for the UN Intellectual History Project in which I participated, I said that working for the United Nations was like working for a government in which all the political parties were in power at the same time: with the responsibility came constraints. In previous publications, I described my location in UNESCO as anthropological "decision-making participation", an involvement that made me understand two mechanisms that are frequently absent from analysis of process: those of diplomacy and arbitration. My position as Assistant Director-General for Culture carried the responsibility of avoiding harm to any of the participants in transactions that would have exacerbated tensions, discriminated against countries or individuals or thrown a spanner into the continued dialogue that is essential for consensus and resolution. All this in a multistorey glass building where tacit understandings, insider knowledge, innuendos and subtleties affected even the simplest diplomatic action. In the end, however, and this I will reiterate time and again throughout this book, the prodigious reality is that people coming from extremely dissimilar cultures, entangled historical interpretations and diverse individual stories, are able to work together if they put their minds to it.

I am aware of the pitfalls of drawing on my own memory, as Collins and Gallinat have discussed in their study of "ethnographic self-awareness" (Collins/Gallinat 2013: 16). As my notes were all written out, however, I hope that, in transcribing these debates, I am providing an opportunity for other anthropologists and scholars to reinterpret the data and to create different perspectives for the future.

1.9 From Global Environmental Change to the Anthropocene

The most decisive turn in debates on culture, however, has been the deep new involvement of social scientists with natural scientists working on the Earth system, which strengthens my conviction that a new cosmopolitics cannot be conceptualized without rethinking the terms ‘culture’ and ‘nature’. It brings to my mind that it was only in February 2000 that the International Geosphere–Biosphere Programme (IGBP) and the International Council of Scientific Unions (ICSU) hosted a meeting in Cuernavaca, Mexico, the town where I work in the National University of Mexico’s Centre for Multidisciplinary Studies. As I had been a member of the ISSC Standing Committee for the HDGEC programme, I had worked closely with natural scientists in the IGBP and ICSU in trying to find common theoretical and methodological grounds to set up research on the social aspects of environmental change (see Arizpe et al. 2016). At the Cuernavaca meeting, over coffee, a scientist friend remarked “You should be happy. Professor Crutzen just mentioned the human impacts on the environment. He said we are living in the Anthropocene.” It is a great achievement therefore that, only fifteen years later, Earth system science has been consolidated, leading to the creation of the Earth System Science Partnership and a formal proposal to have the Anthropocene recognized as a new geological age. The debates relating to this are summarized in Chap. 9.[4] However, while natural scientists have been hotly debating the name, duration and time of beginning of the Anthropocene over the last decade, social scientists have only very recently entered the debate arena.

Anthropogenic change: deeper and deeper into the rainforests people flee from poverty, in the wealthiest yet most unequal era in world history. Indigenous women settlers in the Lacandon rainforest in Marqués de Comillas, Chiapas, Mexico, 1993. *Source* Photo by Lourdes Arizpe

[4]For a complete history and review by numerous authors on these topics, see Brauch et al. (2016).

For all its scientific trappings, Jedediah Purdy believes that the Anthropocene is mainly a cultural idea, and its potential is political and ethical. I couldn't agree more. He calls for taking responsibility for what we make, as well as for what we destroy. "It is the starting place for a new politics of nature, a politics more encompassing and imaginative than what we have come to know as environmentalism" (Purdy 2015: 17). Similarly, Dipesh Chakrabarty, who has been very influential in opening the debate on "provincializing Europe", argues that the Anthropocene entails a "new universal history of humans" (Chakrabarty 2009: 221).

In his book on ecocriticism, Timothy Clark states bluntly that "the Anthropocene names a newly recognized context that entails a chastening recognition of the limits of cultural representation as a force of change in human affairs, as compared to the numerous economic, meteorological, geographical and microbiological factors and population dynamics, as well as scale effects, such as the law of large numbers, that arise from trying to think on a planetary scale" (Clark 2015: 21). He goes as far as to ask whether the Anthropocene is "a threshold at which art and literature touch limits to the human psyche and imagination themselves" as he discusses Barbara Kingsolver's attempt to look at the problem of scale in her 2012 novel *Flight Behaviour* (ibid.: 176). He questions whether "the human imagination is really so depressingly enclosed, able to be captivated only by immediate images of itself", and whether the psychology of narrative—of what makes a credible or compelling story—is itself a problem for representations of the Anthropocene (ibid.: 178). These questions construct a Gordian knot which ties culture to the capacity of the human mind and which opens up the startling issue of how to expand that capacity extremely rapidly.

The Anthropocene as a threshold concept opens a "politics of cognition" (to use Tom Cohen's term), where two different readings are confronted. One reading uses and adapts the inherited conception of the human—the social, cultural, and so on; according to the other, the environmental crisis questions the seeming self-evidence or coherence of such basic conceptions (Cohen, cited in Clark 2015: 20). Debating the Anthropocene from a social science perspective opens up a host of interlinkages that have to be researched. This is all the more pressing because, since 2000, increasing securitization of the world's resources has been driven by climate change, water scarcity, degradation and stress, as well as soil erosion and desertification, turning these pressures into major security dangers and concerns in the new era of Earth history (see Brauch 2009: 1311).

1.10 Culture Makes the Difference Between 'Life' and 'Living'

Within this context, I posit a distinction between life and living. Life abounds on the planet: in plants, in animals, in human beings. But "being alive", as Ingold has recently argued, is a passive state which, when activated, leads to "becoming human", a continuous process of becoming (Ingold 2011). My contention is that culture makes the difference between 'life' and 'living'. Defending life is, again, a passive act. The assumption made in this book, then, is that human beings are created through genetic and biological processes, but they become acting conscious subjects through culture and the probabilities of action that it offers. By co-producing knowledge collectively, people are able to reinterpret the world developing new ideas and practices that provide meaning for the present and the future. The only way to save our species is by securing 'living', the capacity of humans to be conscious of their own capacities for action and, hopefully, to insist on conviviability, against all divisive trends, so as to ensure sustainability.

I also contend that there is one more step to be taken in the reconceptualizing of the place of humans in the geohistory of the world. I propose adding a new term, to those of *Homo habilis* (Man as a generic term for all genders), Man the maker of tools and things, and *Homo sapiens sapiens*, Man the thinker. This is *Man the Transformer*, the Man that imagines a possibility, makes ideas, tools and ways of cooperating with others to achieve this possibility. Becoming human then means applying cognition, embodied knowledge and technologies created—and hopefully, forever controlled by humans—to achieve this possibility by transforming their environments. A possible Latin term would *Homo transfigurator* (*transfero*, to transform). This takes into account sociopolitical and economic organizations embedded in natural ecosystems but also highlights a process that has only recently appeared in the human narrative. This is *the transformation of the self*. It is contained in gene expression yet, through epigenetics, it is a transformation that is still hinged on human relationships. Will Man the Transformer be able to develop a new geo-human world in time? This is the question that is most pressing. Framing all of these challenges is the threat of the sixth greatest extinction event in the history of the planet—which would include us.

Many thresholds have to be negotiated and managed. It is no coincidence that the word 'culture' has become highly prominent just as the geo-ecological and cosmopolitical place of humans in the Earth system is shifting. The need to research and rethink this future in a framework of multiverses is daunting. At the same time, there is an urgent need for creativity and it is precisely cultural liberty that engenders creative humans. In preceding pages I have highlighted how people from extremely diverse backgrounds and intellectual traditions are able to work together, over and above diversity. The whole of the United Nations and all its international normativities, laws and resolutions, and its ceaseless search for consensus and peace, are proof that this can be done.

References

Anheier, H. & Isar, Y. R. (Eds.) (2007). *Cultures and globalization: conflicts and tensions*. Los Angeles: Sage Publications.

Anheier, H. & Isar, Y. R. (Eds.) (2008). *Cultures and globalization: the cultural economy*. Los Angeles: Sage Publications.

Anheier, H. & Isar, Y. R. (Eds.) (2010). *Cultures and globalization: cultural expression, creativity and innovation*. Los Angeles: Sage Publications.
Anheier, H. & Isar, Y. R. (Eds.) (2011). *Cultures and globalization: heritage, memory and identity*. London: Sage Publications.
Anheier, H. & Isar, Y. R. (Eds.) (2012). *Cultures and globalization: cities, cultural policy and governance*. Los Angeles: Sage Publications.
Arizpe, L. (1972). Parentesco y economía en una sociedad nahua. MA dissertation, Instituto Nacional Indigenista, Mexico City.
Arizpe, L. (1975). The migration of Mazahua Indians to Mexico City. PhD dissertation. London School of Economics and Political Science.
Arizpe, L. (1981). The rural exodus in Mexico and Mexican migration to the United States. *International Migration Review*, 15, 4, 626–650. Reprinted in: P. Brown & H. Shue (Eds.) (1983), *The border that joins: Mexican migrants and U.S. responsibility* (pp. 162–183). Totowa, NJ: Roman and Littlefield.
Arizpe, L. (1986). *Campesinado y migración*. Mexico City: Secretaría de Educación Pública.
Arizpe, L. (1989). On the social and cultural sustainability of world development. In: L. Emmerij (Ed.), *One world or several?* (pp. 207–219). Paris: OECD Development Centre.
Arizpe, L., Rao, V. & Walton, M. (Eds.) (2004). *Culture and Public Action*. Stanford, CA: Stanford University Press.
Arizpe, L. (2015). The genealogy of intangible cultural heritage. In L. Arizpe, *Culture, diversity and heritage: major studies* (pp. 101–117). Heidelberg: Springer.
Arizpe, L. & Alonso, G. (2002). *Cultura, comercio y globalización*. Buenos Aires: Consejo Latinoamericano de Ciencias Sociales. First published in English as *Culture, trade and globalization*. Human Development Report Working Papers. New York: UNDP, 2000.
Arizpe, L., Price, M. F. & Worcester, R. (2016). The first decade of initiatives for research on the human dimensions of global (environmental) change (1986–1995). In: H. G. Brauch, U. O. Spring, J. Grin & J. Scheffran (Eds.), *Handbook on sustainability transition and sustainable peace* (pp. 349–358). Heidelberg: Springer.
Augé, M. (2017). *L'Avenir des Térriens: fin de la préhistoire de l'humanité comme société planetaire*. Paris: Editions Albin Michel.
Barry, B. (2001). *Culture and equality: an egalitarian critique of multiculturalism*. London: Polity Press.
Bessire, L. & Bond, D. (Eds.) (2014). *American Ethnologist*, special issue on "Ontology".
Brauch, H. G. (2009). Summary and results: facing global environmental change and sectorialization of security. In: H. G. Brauch, Ú. O. Spring. J. Grin, et al. (Eds.), *Facing global environmental change* (pp. 1311–1331). Heidelberg: Springer.
Brauch, H.G., Oswald, U., Grin, J. & Scheffran J. (Eds.) (2016). *Handbook on sustainability transition and sustainable peace*. Heidelberg: Springer.
Brumann, C. (1999). Writing for culture: why a successful concept should not be discarded. *Current Anthropology*, supplement on culture, 40(S1), S1–S27.
Chakrabarty, D. (2009). The climate of history: four theses. *Critical Inquiry*, 35, 197–222.
Clark, T. (2015). *Ecocriticism on the edge: the Anthropocene—questions of definition*. London: Bloomsbury Academic.
Clifford, J. & Marcus, G. E. (Eds.) (1986). *Writing culture: the poetics and politics of ethnography*. Berkeley: University of California Press.
Collins, P. & Gallinat, A. (2013). The ethnographic self as resource: an introduction. In: P. Collins & A. Gallinat (Eds.), *The ethnographic self as resource: writing memory and experience into ethnography* (pp. 1–24). New York and Oxford: Berghahn Books.
Eagleton, T. (2000). *The idea of culture*. Oxford: Blackwell.
Eriksen, T. H. (Ed.) (2003). *Globalisation: studies in anthropology*. London: Pluto Press.
Ferguson, J. (2011). Novelty and method: reflections on global fieldwork. In: S. Coleman & P. von Hellermann (Eds.), *Multi-sited ethnography: problems and possibilities in the translocation of research methods* (pp. 194–207). New York: Routledge.

Fischer, M. (2009). *Anthropological futures*. Durham, NC: Duke University Press.
Garsten, C. & Nyqvist, A. (Eds.) (2013). Organisational anthropology: doing ethnography in and among complex organisations. London: Pluto Press.
Haugerud, A. (2015). From the editor: ethnography unbound. *American Ethnologist*, 42(1), 1–3.
Holmes, D. R. & Marcus G. E. (2005). Cultures of expertise and the management of globalization: toward the re-functioning of ethnography. In: A. Ong & S. J. Collier (Eds.), *Global assemblages: technology, politics and ethics as anthropological problems* (pp. 232–252). Oxford: Blackwell Publishing.
Huntington, S. (2005). *Who are we? The challenges to American national identity*. New York: Simon & Schuster.
Inda, J. X. & Rosaldo, R. (Eds.) (2002). *The anthropology of globalization: a reader*. Oxford: Blackwell.
Ingold, T. (2011). *Being alive: essays on movement, knowledge and description*. London: Routledge.
Ingold, T. (2014). "Against human nature". Paper delivered at the 2014 conference of the Association of Social Anthropologists, Edinburgh, 21 June. http://www.nomadit.co.uk/asa/asa2014/panels.php5?PanelID=2925, accessed 1 December 2016.
Jameson, F. (1998). *The cultural turn: selected writings on the postmodern, 1983–1998*. London: Verso.
Jolly, R. (1998). Globalization and global inequalities: long-term trends and policy options. Paper presented at the Human Development Report Discussion, Bellagio, Italy, May 1998.
Kroeber, A. (1948). *Anthropology: race, language, culture, psychology, pre-history*. New York: Harcourt, Brace.
Lee, R. (2003). *Life and times of cultural studies: the politics and transformation of the structures of knowledge*. Durham, NC: Duke University Press.
Lewis, O. (1959). *Five families: Mexican case studies in the culture of poverty*. New York: New American Library.
Marcus, G. (2011). Multi-sited ethnography: five or six things I know about it now. In: S. Coleman & P. von Hellermann (Eds.), *Multi-sited ethnography: problems and possibilities in the translocation of research methods* (pp. 16–32). New York: Routledge.
Marcus, G., & Rabinow, P. (2008). *Designs for an anthropology of the contemporary*. 2 vols. Durham, NC: Duke University Press.
Ortner, S. (1984). Theory in anthropology since the sixties. *Comparative Studies in Society and History*, 26(1), 126–166.
Pagel, M. (2012). *Wired for culture: origins of the human social mind*. London: W. W. Norton.
Pieterse, J. N. (2004). *Globalization and culture: global mélange*. Lanham, MD: Rowman and Littlefield Publishers, Inc.
Purdy, J. (2015). *After nature: a politics for the Anthropocene*. Cambridge, MA: Harvard University Press.
Ridley, M. (2015). *The evolution of everything: how new ideas emerge*. London: HarperCollins.
Robertson, R. (1992). *Globalization social theory and global culture*. London: Sage.
Sahlins, M. (1994). A brief cultural history of "culture". Unpublished paper presented to the World Commission on Culture and Development.
Shore, C. & Wright, S. (1999). *Anthropology of policy: critical perspectives on governance and power*. London: Routledge.
Stolcke, V. (1995). Talking culture: new boundaries, new rhetorics of exclusion in Europe. *Current Anthropology*, 36(1), 1–24.
Strathern, M. (2014). Becoming enlightened about relations. ASA Firth Lecture 2014. http://www.theasa.org/downloads/publications/firth/firth14.pdf, accessed 3 October 2016.
UNESCO (2000). World culture report 2000: cultural diversity, conflict and pluralism. Paris: UNESCO.

Vertovec, S. (2010). Towards post-multiculturalism? Changing communities, conditions and contexts of diversity. *International Social Science Journal*, 61(199), 83–95.

Werbner, P. 2008. *Anthropology and the New Cosmopolitanism: Rooted, Feminist and Vernacular Perspectives*. Oxford: Berg.

Wolf, E. (1982). *Europe and the people without history*. Berkeley: University of California Press.

"The Challenge of Cultural Diversity" by Lourdes Arizpe Schlosser, published by the National Indianist Institute in Mexico, 1977

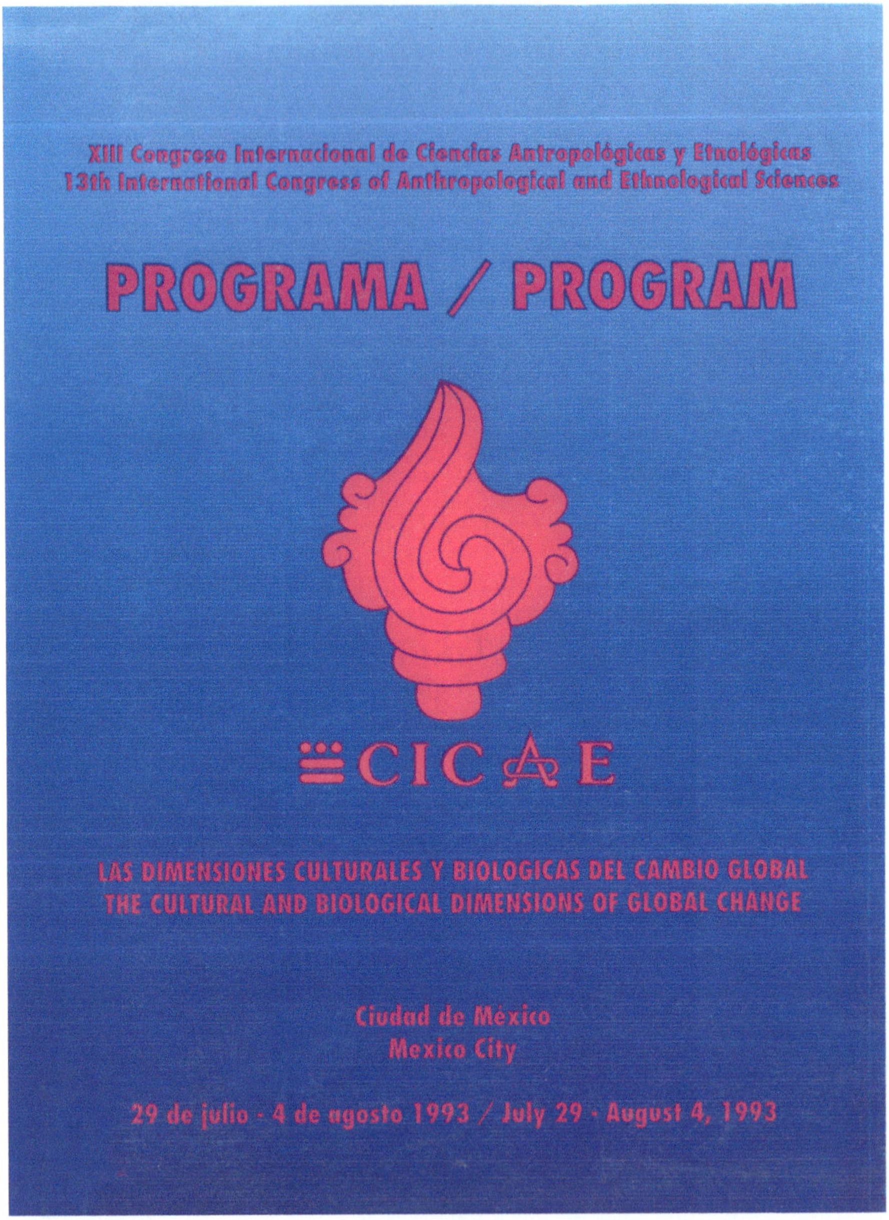

Programme of the 13th International Congress of the Union of Anthropological and Ethnological Studies (IUAES), 1993

Chapter 2
The Politicization of Culture 1947–1995

Amid the rubble of the most destructive war and genocide the world had witnessed, came the hope that it could not happen again. While the loss of life and the physical devastation of the Second World War were appalling, the real shock was the unprecedented rise of a racist and xenophobic ideology capable, on the one hand, of convincing people that they should uphold fascism, and, on the other, of perpetrating the deliberate assassination of millions of people because of hatred for their race, nationality, ideology or sexual orientation. A further shock was the obliteration of millions of people with two bombs in Hiroshima and Nagasaki, which demonstrated an unprecedented change in humanity's capacity for (self-) destruction.

2.1 The Second UNESCO General Conference in Mexico City, 1947

As part of the creation of the United Nations, one of the events held very soon after the war "to prevent wars in the minds of men" was the Second Session of the UNESCO General Conference in Mexico City in November–December 1947. In its inaugural session, Manuel Gual Vidal, Mexico's Secretary of Education, began his speech by stating that, through education, "multiple and heterogeneous particularisms which engender wars, become unified and coordinated for purposes of peace, enhancing independence, integrity and the fecund diversity of cultures" (UNESCO 1947b: 19). He stressed that the highest task of all is to transform every individual, through education, into a citizen of the world, and ended by saying, "To educate is to elevate the individual to the state of culture in the social life in which he lives" (ibid.: 21).

Jacques Maritain, one of the drafters of the UN Universal Declaration of Human Rights, in his own speech praised Mexico as one of the examples of the accord established between two different races when they are animated by the same human

L. Arizpe Schlosser, *Culture, International Transactions and the Anthropocene*,
The Anthropocene: Politik—Economics—Society—Science 17,
https://doi.org/10.1007/978-3-642-41602-6_2

ideal. Indeed, the Mexican regimes that followed the first social revolution of the century opened the door to pioneering work in many cultural areas. Such cooperation was exemplified at the session held in the National Ethnological Museum on 11 November, when exchanges were proposed in archaeological research and between museums. Many prominent Mexicans played a significant role in the Second Session. Jaime Torres Bodet, then Secretary of Foreign Relations and soon to become Director-General of UNESCO, chaired the important sessions on international relations. The Secretary of the Mexican delegation, the philosopher Samuel Ramos, proposed a division on "Philosophy and Civilizations". The distinguished composer Carlos Chavez presided at the Commission on Arts and Literature, and the astronomer Luis Enrique Erro guided the UNESCO delegations on a tour to the Astronomical Observatory in Tonantzintla, Puebla. At the same time, the pioneer archaeologist Beatriz Braniff, and other women who were members of the Alliance of Cultural Feminine and Civil Action Associations, steeped in post-revolutionary activities, were asked to accompany the wives of delegates on their visits to Mexico City sights.

In the proposed "Programme of UNESCO in 1948" presented at this Second Session, the main objective of UNESCO was stated as being to raise the standards of education, science and culture, principally through promoting education for international understanding and the free flow of ideas (UNESCO 1947c). In the field of culture, reference was made only to publication and translation of books, especially "world classics", "programmes of education for women in countries or areas where they do not now have equal rights" and "contributing to the Bill of Rights now being drafted at Lake Success" (ibid.: 24). "The preservation of folk arts and the traditions of craftsmanship" were included as a specific activity of the programme (ibid.: 10). Significantly, the Commission on Social Science adopted a resolution asking the Director-General to carry out a survey on (1) the distinctive characteristics of different cultures; (2) the concepts that different nations had of themselves and others; (3) modern techniques used to modify spiritual attitudes; and (4) the influences that either favour international understanding or, in contrast, make it difficult to promote such understanding (UNESCO 1947a).

It may therefore be said that the first use of the term 'culture' as a specific concern in world affairs was set in motion in 1948, with the proclamation of the Universal Declaration of Human Rights. The first clause of its Article 27 reads: "1. Everyone has the right freely to participate in the cultural life of the community, to enjoy the arts and to share in scientific advancement and its benefits." At the time, the transaction on culture involved specifying a few programmatic activities related to cultural production and performance, as in the proposed UNESCO Programme for 1948. Meanwhile, however, another major debate was being held in the political sphere, as governments, political thinkers and probably social scientists proposed replacing the concepts of race and national supremacy—which had led to millions of deaths—with that of culture.

In the following decades, culture became a prominent political issue through several international movements. The Non-Aligned Movement created the idea of a "Third World", as opposed to both the First World of capitalism and the Second

World of socialism, in the search for more equitable political relationships between regions. Apartheid in South Africa generated a fierce reaction to racism and led to the counter-movement of Négritude in Africa, the Caribbean, and Latin America. The 1955 Bandung Conference of African and Asian states opened the discussion about "Asian values" as an alternative value system for development.

Culture entered the transactions on economic development when anthropologist Oscar Lewis, in his 1959 pioneering study on rural migrants in Mexico City, coined the new phrase "a subculture of poverty" to denote the unstable, financially weak conditions that marginalized workers and migrants had to adapt to as economic policies subsidized industrialization (Lewis 1959). A decade later, Charles Valentine turned this concept on its head by arguing that "a culture of poverty" was, in fact, deterring economic development in underdeveloped countries (Valentine 1967).

By the 1960s, governments and non-governmental organizations had taken these debates to international multilateral organizations, with UNESCO taking the lead in programmes focused on cultural heritage, cultural policy, and the arts.

2.2 The Idea of Cultural Policy

The concept of cultural policy was first proposed in a 1969 UNESCO preliminary study, in which criteria were formally recommended to define the concept and to link culture to the fulfilment of personality and to economic and social development (UNESCO 1969a). The different political positionings on the subject became evident in the recommendations approved by the government delegations. The first recommendation stated that cultural policy should be integrated in general planning. However, while the second recommendation read "it is the duty of the State to replace private initiative which is not always capable of carrying out this [cultural policy] task successfully", the third immediately added that "one should beware of centralization and of a predominant role for the State in direct management of cultural institutions" (ibid.: 2). The document ended by reiterating that one of the main guidelines should be that literacy programmes and "cultural development" form "an indivisible whole".

The UNESCO Cultural Policy Programme generated a series of country publications which explained the main strategies and programmes of national cultural policies. From the outset, the United States made its policy statement very clear: "By refusing the central government the right to set policy, the states and private sectors are forced to adopt concepts suitable to their aims, resulting in a pluralistic approach. Diversity in cultural policy is one of the touchstones of the United States position" (UNESCO 1969b). France, in contrast, began with a flourish by expressing how disquieting it was for a liberal and pluralistic country to extend planning to culture, but justified it by saying that this in no way implied that the state sought to impose its own particular conception of culture on the French people, since "cultural activities belong to the domain of individual freedom and subjectivity" (UNESCO 1970). Importantly, as this later became a widespread cultural policy, Canada specifically promoted the idea of "multicultural policy", in order to facilitate the incorporation of indigenous peoples into all institutions of Canadian society, while respecting their cultural traditions (UNESCO 1974).

The case of Yugoslavia also attracted the attention of developing countries for its policies of "self-management and culture", which were seen as an alternative to the orthodox Soviet approach. Its self-proclaimed aim was "to replace the statist culture by the socialized culture. This is not an isolated process, restricted to culture alone, but a part of the overall self-management transformation of society" (UNESCO 1980: 29). The document defined "Yugoslav cultural pluralism"—one of the first times that this term appears in official documents—as a policy for "various cultures to interact, permeate and affect each other and conduct 'a permanent dialogue' while at the same time endeavouring to preserve and further national cultural specificities" (ibid.: 49). In the light of the brutality with which "ethnic cleansing" was deliberately and systematically applied in that country only twenty-five years later, it is evident that such a government discourse cannot erase fundamental religious and historical processes that lead to civil wars.

2.3 Culture as a Rational, Critical and Ethical Capacity

The UNESCO Medium-Term Plan of 1977–82 established a "Programme on Culture" which reflected the emerging issues related to culture and development both in international policy debates and in scientific research. The objectives of the programme were very different from those of previous programmes, which had emphasized "cultural readjustment" for development, cultural heritage conservation and support mainly for artists and artisans. The aims of the 1977–82 programme were:

1. promotion of appreciation and respect for the cultural identity of individuals, groups, nations and regions;
2. promotion of cultural identity as a means to achieve independence and solidarity;
3. promotion of cultural identity within the framework of a global development strategy;
4. promotion of respect for the cultural identity of individuals and groups, particularly those who are marginalised in developed and developing countries (UNESCO 1977).

The challenge of relating 'culture' to 'development' through government policy was taken up at the World Conference on Cultural Policies held in Mexico City in 1982. 'Mondiacult', as the Conference came to be known, established an international working concept based on a broader, anthropological definition of culture as "the whole complex of distinctive spiritual, material, intellectual and emotional features that characterize a society or social group. It includes not only the arts and letters, but also modes of life, the fundamental rights of the human being, value systems, traditions and beliefs" (UNESCO 1982: 2).

Less remarked but equally important in the Mexico City Declaration was the critical perspective reflecting the thinking of the Frankfurt School (which included Theodor Adorno and Max Horkheimer). This viewed art and culture as a critical instrument for evaluating the quality of life in a society. The Declaration states: "It is culture that gives man the ability to reflect upon himself. It is culture that makes us specifically human, rational beings, endowed with a critical judgement and a sense of moral commitment. It is through culture that we discern values and make choices" (Mondiacult 1982: 2). According to Chimène Keitner, in her 1999 informal survey of UNESCO's approach to cultural diversity, this definition of culture as a universal faculty rather than a rigid set of practices allows for greater flexibility and transcendence. Keitner adds that

> It builds the ideas of renewal, reevaluation and critical choice into the definition of culture itself … This constant attempt to balance affirmation with openness, a central pillar of the approach to culture crystallized in this document, represents an important counterweight to the excessive politicization of cultural identity at the expense of a search for common values (Keitner 1999: 19).

How relevant this observation has become in today's ultrapoliticization of identity!

In his opening speech at Mondiacult, the Mexican Minister of Education, Fernando Solana, cautioned that "In general, the triumph of production has not signified a triumph for mankind … . Contemporary societies reveal inexplicable contradictions. We are more familiar with the laws of social evolution yet are unable to put a stop to the conflict of ideologies and to overcome violence" (Mondiacult 1982: 178). Again, such words—spoken more than thirty years ago—seem even more relevant today.

2.4 Culture Enters the United Nations Development Agenda

After Mondiacult, the social and cultural aspects of development were given greater emphasis in development planning in all UN institutions. UNDP projects were created that focused on indigenous peoples, on women and development, and on community development. The World Bank also followed suit—too slowly, according to Michael Cernea, the Bank's Senior Adviser for Sociology and Social Policy until 1997. In pioneering cultural studies in the World Bank, Cernea explained that the focus had shifted because of the inconsistencies or failures of many development programmes (Cernea 1995). He pointed out that a major obstacle was the lack of a comprehensive theory of "induced development", making it difficult to convince economists—who were resistant to incorporating cultural values into their development models—of the importance of social and cultural variables.

In a 1983 working paper for the UN Centre for Regional Development, Georgette Wang, Wimal Dissayanake and Barbara Newton proposed an approach which considers development as a dynamic process in which culture, as an open system, serves as a mediator: "In other words, changes do not occur in a vacuum; they must be mediated by the culture throughout the whole process" (Wang et al. 1983: 14). They cite E. J. Leed as saying that "it is easier to find examples of a society which is integrating a new medium of communication into a culture than it is to find a new medium which is transforming those systems" (ibid.: 14–15). One example they give is the use of newspapers for finding appropriate caste partners in India.

In 1987, with the support of key European countries, the "Group of 77" (which represents a majority of developing nations) proposed a resolution declaring 1988–97 a "Decade for Culture and Development". The UN accordingly launched a World Decade for Cultural Development in 1988. The programme for this decade successfully promoted more than a thousand activities around the world, ranging from local fairs celebrating traditional gastronomy to world media music concerts, all of which ended, like any fiesta, with an aftermath of ephemeral joy, many souvenirs of cultural activities that no-one knew how to keep together and, as one government delegate once told me, a feeling of having no direction towards the future. When I arrived at UNESCO in 1994 as Assistant Director-General, this was one programme that I was asked—by practically all government ambassadors—to restructure completely.

While most of the activities of the Decade concentrated on valorizing traditional and local cultures and promoting cultural exchanges, a few of the resulting publications made interesting forays into its main topic. In a 1994 UNESCO-sponsored publication ominously titled *Culture: Hostage of Development?*, Gilbert Rist, its editor, summarized the views of the different authors by stating that it was clear from examining concrete social practices that 'culture' and 'development' could not be taken hostage in a dichotomized proposition that would make their interaction impossible. Instead, there was a two-way interaction between them so that, if 'culture' is transformed, so is 'development', which may thereby become an entirely different practice (Rist 1994a: 11). The opposition between the two concepts in fact reveals itself to be a simplification, "since the assumptions of development are themselves the expression of a culture" (ibid.: 12). This warning against oversimplification is repeated throughout the book: for example, Rist, citing Emmanuel Ndione, notes that the Western assumption that social change is everywhere determined by the pursuit of maximum profits or the accumulation of goods is not even uniformly shared in Western societies (Rist 1994b: 65).

Rist emphasizes the role of the *présupposé* (assumption) in development, namely, that it is made up of a cluster of concrete measures to improve the material well-being of populations. On the basis of this empirical assumption, everything turns around the notions of investment, pre-investment, efficiency, returns, economic growth, exploitation of resources and so on. Such applications are real if one takes as a criterion of 'reality' the fact that they can be measured with the help of conventional economic instruments. Yet "they depend, in the last analysis, on a symbolic system which—with the excuse that it escapes such measures—is considered unreal" (ibid.: 61). The central question, then, is not whether new techniques are introduced or not but how they are absorbed by cultures and societies which did not create them.

Growing environmental concerns made sustainability the central strategy for development at the end of the 1980s, and in my 1989 article "On Cultural and Social Sustainability" I argued that these two processes were a necessary component of general sustainability (Arizpe 1989). I described four analytically different movements espoused by cultural practitioners evident at that time: cultural groups that consciously and willingly tried to stay out of modernizing development; groups that were retrenching because they were being marginalized from development; cultural groups that were using culture to fight for political advantages; and groups that were using culture to protect their own national markets. It seems to me that all four trends have intensified in the last twenty-five years, while the theory and the application of this concept in public affairs have not moved an inch.

2.5 The Search for a New Vocabulary of Culture

In the 1990s four major currents began to have an impact on culture and development thinking: the return of the term 'civilizations'; the rewriting of culture as a "site of contention" in cultural studies; the rise of national policies of multiculturalism; and the postcolonial perspective on culture.

2.5.1 The "Clash of Civilizations"

The arguments that had advocated "endogenous development" and multicultural policies were suddenly subsumed in the nineties within a large-scale "clash of civilizations" couched in cultural terms but strictly political in its oversimplification and conceptual confusions. This concept was launched by Samuel Huntington's 1993 article, which immediately engendered a fierce international debate that has fostered retrenchment into ill-defined cultural and religious categorizations (Huntington 1993). Huntington began by stating that "it is my hypothesis that the fundamental source of conflict in this new world will not be primarily ideological or primarily economic. The great divisions among humankind and the dominating source of conflict will be "cultural" and "conflict" between civilizations will be the latest phase in the evolution of conflict in the modern world" (ibid.: 22). He defined civilization as "the highest cultural grouping of people and the broadest level of cultural identity people have short of that which distinguishes humans from other species" (ibid.: 24), but he blurred this definition by allowing that there were areas of overlap and occasional subdivisions within civilizations. Nevertheless, he was clear that there were fundamental differences between civilizations that had developed over centuries and that the increased globalization of the late twentieth century was making people more aware of those differences, sparking tensions between East and West, North and South.

Because of the ambiguity of his definition, Huntington constantly proposed different classifications of world civilizations in his writings. In the original article, he made the case for seven or eight main civilizations, including "Western, Confucian, Japanese, Islamic, Hindu, Slavic-Orthodox, Latin American and possibly African" (ibid.: 25). As can be seen from this list, a fundamental difficulty arose in his tendency to equate civilizations with particular religions, a difficulty which led to an avalanche of critiques of his subsequent publications. Thus "Western civilization" is established as being coterminous with Western Christianity, ignoring both the influences of Judaism and the cultural foundations provided by Egypt, Greece, Rome and Byzantium. Similarly, Huntington's reference to Confucianism can only be meant in relation to China. Yet China is a secular socialist state, whose cultural origins are not only rooted in Confucianism but are also drawn from Taoism, other religions and many regional cultures. Even if an equivalence could be found between civilization and religion, the problem remains that religions are not solid-state entities, but are made up of elements gleaned over many centuries from other local and regional cultures and other religions, with enormous variations in doctrine, liturgy, rituals and ethical standards in different continental regions.

Huntington assumed that each 'civilization' developed its core template independently of others, a supposition that is clearly untenable when viewed through the lens of human history. All boundaries around clusters of cultural, ethnic, religious and national entities signify precise territorial and political processes at a given point in time. All such bounded entities are made up of core and marginal elements held in intellectually, politically and socially hybridized communities.

Huntington's conclusion was that there could be no universal civilization but only a world of different civilizations "each of which will have to learn to coexist with the others" (ibid.: 49). He therefore recommended that the aim of the Western world should be to "promote greater cooperation and unity within its own civilization …; to limit the expansion of the military strength of Confucian and Islamic states …; to strengthen international institutions that reflect and legitimate Western interests and values and to promote the involvement of non-Western states in those institutions" (ibid.: 48–9). As could be expected, this militant call to uphold the hegemony of Western Christian civilization has been immensely influential in fuelling the flames fanned by extremist religious political leaders who play on the notion of a "war of religion" (among them Osama Bin Laden and Islamic State). Simultaneously, the strong reaction against Huntington's thesis in both academic circles and UNESCO generated renewed interest in questions of cultural diversity, as will become clear in later chapters.

Lourdes Arizpe, as President, inaugurates the International Congress on the Cultural and Biological Dimensions of Global Change held in Mexico City, 1993. *Source* Personal photo collection of the author

2.5.2 Multiculturalism and Cultural Pluralism

At the same time as Huntington was setting out his stand, in a fascinating counterpoint, critical anthropologists were intellectually redefining culture as a "site of contestation". As Stuart Hall argued, "We are living through the proliferation of the sites of power and antagonism in modern society" (Hall 1987: 20). As a result, cultural studies shifted its attention to urban, multicultural, multiracial neo-cultures that were becoming the most pervasive form of culture in a world where millions of people were migrating to cities.

A different picture emerges in relation to autochthonous peoples. The movement of Zapatistas in Mexico became internationally emblematic as a political movement which put forward the demands of indigenous peoples against discrimination within the Mexican state and worldwide. International organizations had already given prominence to issues concerning indigenous peoples: the World Bank with its projects for dams and economic development in indigenous regions (World Bank 1982), the International Labour Organization with respect to self-determination, and UNESCO in terms of the conservation of indigenous cultures and arts.

It is important to also mention other terms that were tried out to explain the increasing mingling of cultures worldwide. In the quest for culturally pluralistic models in the 1990s, the concept of 'transnationalism' was used to describe the diasporic creation of cultural communities (Ghosh 1989; Gonzalez and McCommon 1994; Lomnitz 1998). Massive outmigration both disrupts and renovates the social fabric of rural communities. It also accelerates cultural change in urban groups, especially among the poor.[1] As Kearney wrote, "The intellectual challenge posed by this deterritorialization of cultures and the creation of new transnational cultural spaces cannot be overemphasized" (Kearney 1995: 558–559). The new patterns of "segmented assimilation" (Zhou 1997), separation of community and culture through an alternative "demotic discourse" (Baumann 1996: 190–197), and the re-ethnicization of many immigrant communities in developed countries "point towards a diversification of cultural attitudes since the first generation of immigrants, in what may be called "minority cultures of mobility" (Neckerman et al. 1999, cited in Smith et al. 2001: 16).

2.5.3 Culture in a Postcolonial World

At the end of the 1970s, Edward Said's book *Orientalism* had opened the black box of cultural discourse by defining the intellectual construct of 'Orientalism' as "the

[1]The first comprehensive discussion of this field of studies can be found in Basch et al. (1994). A useful reader is Vertovec/Cohen (1999).

corporate institution for dealing with the Orient: dealing with it by making statements about it, authorizing views of it, describing it, by teaching it, settling it, ruling over it" (Said 1979: 3). Influenced by Michel Foucault and other French postmodern writers, Said ushered in a new critical trend, closely related to literature, to apply regimes of discourse and representation to cultural development. In the eighties, several authors in developing countries—for example, V. Y. Mundimbe in *The Invention of Africa* and Chandra Mohanty—used the same intellectual tools to challenge the concepts and narratives of prevailing development thinking (Mundimbe 1988; Manzo 1991). Postcolonial studies also challenged such narratives, by reading them as colonial discourse, whose predominant strategic function, in the words of Homi Bhaba, was "the creation of a space for 'subject peoples' through the production of knowledge in terms of which surveillance is exercised" (Bhabha 1990: 75, cited in Escobar 1995: 9).

Arturo Escobar, in his 1995 book *Encountering Development: The Making and Unmaking of the Third World*, also stressed that "the corpus of rational techniques of the development discourse organizes both forms of knowledge and types of power applied to bring peoples in developing regions under control" (Escobar 1995: 217). He argued that "development is at the same time self-destructing and being unmade by social action, even as it continues to destroy people and nature" (ibid.). Such social action by local communities may develop more creative and autonomous practices conducive to renegotiating class, gender and ethnic relations at local and regional levels. He called for a post-development cultural politics that would critically engage with neutralizing the dominant economic discourse (Rahnema/Bawtree 2001).

By the nineties, the development of the interpretive and textual-analysis methodologies that had been initiated in anthropology led to a general intellectual questioning of the term 'culture' itself, especially among anthropologists. For many years, this concept had been criticized because it creates hierarchy, tends to homogenize cultural patterns and flattens levels of cultural understanding. Writers such as Christoph Brumann argued instead that, even in highly fluid situations, cultural patterns form valleys and mountains that create landscapes (or, as Arjun Appadurai would term them, 'ethnoscapes') which give people a sense of identity and a certain understanding of the world (Brumann 1999).

Maurice Godelier, Fredrik Barth, Lourdes Arizpe and others at the First Plenary Session of the International IUAES Congress, Mexico City, 1993. *Source* Personal photo collection of the author

2.6 Gender and the Cultural Construction of Identity

Gender began as a topical approach to women's oppression in the seventies, which then developed into a major philosophical and social concern that strongly encouraged the kinds of introspection and reflexivity that the term 'culture' has taken up as its task. From sexual politics through women's emancipation to feminism, gender became a worldwide movement which asked the questions that have become the norm in thinking about cosmopolitics.

Are all women oppressed in the same way? This question brought to the world stage the old and yet unresolved theme of universals. Are poor and culturally and racially defined groups of women able to act together to end women's oppression? As the search for an answer investigated the many realms of women's and men's lives and a majority of cultures in the world, it cracked open, in a creative way, the hitherto unquestioned categories of daily lives: cultural conventions create versions of femininity and masculinity that vary between cultures; in every culture the perception of women's and men's bodies varies and leads to embodied knowledge that builds social norms which can be changed; freedom for women means having not only a "room of one's own" but a "mind of one's own". Gender therefore became an inescapable perspective in thinking about development, governance and all other central topics in international transactions.

The evolution of this deeper level of cultural reflexivity becomes evident in the marked distinctions in the theoretical currents of feminist scholarly research. It must be said, however, that the first version of the 1992 mandate for the World Commission on Culture and Development left out the topic of women, with several of us then successfully campaigning for its insertion. Deniz Kandiyoti, in her Working Paper for the Commission's work, argued that "economic and cultural globalization has created paradoxical consequences for the analysis of gender issues in general and women's rights in particular" (Kandiyoti 1993: 1). On the one hand, she says, "there has been a powerful drive to define women as significant actors in the process of development" (ibid.). 'Ironically', on the other hand, "in a world becoming rapidly homogenized at the level of material and popular culture, the locus of 'cultural distinctiveness' became increasingly identified with the arena of gender relations and the appropriate conduct of women, who are often singled out as the bearers and signifiers of their 'culture'" (ibid.).

Kandiyoti then makes a very important claim for the subject of this book: "Thus, the categories of both 'culture' and 'gender' became politicized in new ways and attributions of cultural authenticity or alterity became part of the political armoury of diverse social movements and ideologies" (ibid.). As a consequence, she continues, "we are now faced with the serious challenge of developing feminist agendas which avoid the dual pitfalls of ethnocentricism and Western bias, on the one hand, and unprincipled forms of cultural relativism which deny women their basic human rights in the name of 'difference', on the other" (ibid.). Kandiyoti's insight was extraordinarily prescient as we find ourselves, in the second decade of the millennium, confronted by fierce opposition to women's rights from religious fundamentalists and by the ugly terms of the campaign in the 2016 US presidential election.

In the next few years, the question of gender and culture was examined by many international meetings and research projects, which it is beyond the scope of this book to summarize. The role of women in culture has been present in all discussions on the subject and was a major focal point in the discussions and setting up of both the 2003 UNESCO Convention for the Safeguarding of the Intangible Cultural Heritage and the 2005 UNESCO Convention on Protection and Promotion of the Diversity of Cultural Expressions.

2.7 An Ethnographic Experience, 1988

By the end of the 1980s, the debates on multicultural policies (as initiated by Canada), endogenous development among developing countries, self-managed development (*desarrollo autogestivo*) in Latin America and cultural autonomy for indigenous peoples in many other countries had established culture as a mainstream development issue and a way for governments to show their concerned for 'marginalized' peoples. The West insisted that capitalist development was the answer to such questions; some countries, led by France, championed state policies to promote creativity and cultural freedom; socialist countries displayed the extensive support of their governments for social unity based on shared cultural activities.

The ethnographic account which follows, which the IUAES Executive Committee asked me to write in 1992, is included here because it describes an important international event and because it places an anthropologist—the author of this book—in the context of a rising interest in culture and its impact on international politics.

It was still the highly optimistic summer of 1988 and we were attending the 12th World Congress of the International Union of Anthropological and Ethnographical Sciences (IUAES) in Zagreb, Yugoslavia, a country that was considered one of the best examples of how culturally diverse groups could live together in peace. Had we known of the tragedy that would befall them only three years later, we anthropologists would have stilled that slice of time into an ethnographic monument to a human ideal. There were, we could see, plenty of signs that pointed to difficulties but not to an imminent collapse of the reality behind the discourse. In the Museum of the Revolution, one went around and around in a circular, heavy, whitewashed building, looking at black and white photographs, mostly of Marshall Josep Tito, which seemed to materialize the fixed discourse that accompanied them as captions. It seemed so empty, so stark, especially to me, coming as I did from a country in which the Mexican Revolution had given rise to an extraordinary, colour-filled, anthropological and museological culture.

The meetings of some three thousand anthropologists were held at the university, with most sessions reporting on what at the time seemed to some of us impatient Third World anthropologists as ostrich-head-in-the-sand ethnographies and futureless theoretical speculation. I remember asking what good it did to pile up ethnography after ethnography, when we should be focusing on the cultural and social sustainability as capitalism spread, undermining all small-scale and agrarian

Poster for the first Mexican-Central American Symposium in Women Studies, with Lourdes Arizpe as Secretary General, held in Mexico City in 1977

societies, which were 'our' tribes or 'our' villages. I was part of a whole generation of incensed social scientists. An unspoken solidarity could be felt among Latin American, African, South Asian and left-wing European young scholars. We did take time, though, to admire the high scholarship of Central European anthropologists, which rarely made it to specialized publications. There was also a particular bond among young women anthropologists as we embarked on a triple mission: as Third World scholars, as women and as politically engaged activists. This euphoria, or at least it felt like it, was tempered by the certainty that we knew exactly how capitalism functioned, how anthropology should be improved and how to defend the indigenous peoples of the world.

Society for International Development: Women's Conference on Gender and Equity, organized by Wendy Harcourt, held in New Delhi, India, March 1988. *Source* Author's personal photo collection

To my great surprise, one of the Vice-Presidents of the International Union called me aside in the cafeteria and asked me to be a candidate for President of the Union. Shocked that the patriarchal structures which I thought were unmoveable suddenly had a crack, I said no. This amiable Dutch man then said: "But you see, you women want to be taken into account and when we ask you to participate, you refuse." So of course I could not refuse. I was then to organize the next World Congress, in Mexico in 1993.

Other signs of change in anthropology were also present. The only out-of-the-ordinary incident at the Congress was a session in which Filipino anthropologists denounced the hoax of the "primitive cave-dwellers" recently 'discovered' in their islands. Believe it or not, then President Marcos had sent his henchmen all the way to Zagreb to disrupt the session, so Yugoslav police had to intervene to stop them. More signs that anthropological issues were suddenly becoming important in terms of national identity, money-making schemes and tourism.

The signs outside were even more telling of the turning of a page in history. Slobodan Milosevic, then Chairman of the five-man rotating Presidency of Yugoslavia came to the formal dinner, and congratulated me on being elected as the first woman, the youngest-ever and the first anthropologist from a developing country as President of the IUAES.

We listened to Mr. Milosevic's discourse, expounding on the perfection of the Yugoslav system that allowed all 'nationalities' to live, not side-by-side, but in culturally mixed societies in the different regions. Yet, the signs of different identities were still kept, conceptualized as "folk cultures", as we were able to see the next day at the closing festival of the Congress. Along grey streets, grey square buildings, grey shops came horse-drawn carriages of elderly couples in 'folk' costumes from each of the regions. Then the streets became alive with the flute and guitar music of the bands that preceded the dance groups, swirling in white, red and black. The Serbians with their richly embroidered skirts for women and black jackets for men; the Slovenes showing their nearness to Europe in their Austrian- or Swiss-looking costumes, aprons and scarves for women, black trousers and green gilets for men; the Bosnians with their flowing white; and the Macedonians in their Greek-style men's white stockings and short, fluffy skirts. Then it seemed an anthropologist's paradise. Now, it seems cultural conflict split into political management and cultural representation through folklore. A scheme tied with such thin threads that it couldn't allow even a single Joke, as Milan Kundera's novel had shown. The severity in people's eyes, the indifference to others, the emptiness—any cause would seem preferable.

And the question we ask even now: why did things go wrong? As was described later, among many others, a young Norwegian anthropologist, Tone Bringa, was able to record the slow deterioration of cultural coexistence, *convivencia*, in a Bosnian village that had no plausible explanation except politics, religion and testosterone. Culture had reverted to politics once again.[2]

A year after this meeting, in 1989, the Soviet Union and its satellite socialist countries collapsed and a new era of world cultural transactions began. Anthropologists watched with ambigüity the slow disaggregation of these formerly unified societies as they returned to previous and sometimes ancient forms of language, and historical, religious and regional loyalties and with anguish the ensuing civil wars. The case that heralded this new era—and became its emblematic conflict—was the war dividing Serbians, Croats, Bosnians and other groups in what had been Yugoslavia. "Ethnic cleansing" emerged as a new outcome of cultural and

[2]The text here is exactly as it appeared in the original report.

religious conflicts in many unexpected countries and places. The genocide in Rwanda between Hutu and Tutsi also laid bare how thin the veneer of peace was between groups differentiated by years of pent-up feelings of injustice, humiliation, envy or marginalization.

2.8 The Cultural Dimensions of Global Change

When I proposed "The Cultural and Biological Dimensions of Global Change" as a theme for the 13th International Congress of Anthropological and Ethnological Sciences (ICAES), my colleagues on the Organizing Committee asked what this new concept meant. "It means", I answered, "that we are taking up sustainability from a cultural and biological perspective." As explained later (see Sect. 9.1), as President of the International Union of Anthropological and Ethnological Sciences (IUAES) and a member of the Executive Committee of the International Social Science Council (ISSC), I had been involved in the dialogue with the International Council of Scientific Unions (ICSU) to find new theoretical and methodological grounds for cooperation between social science and natural and exact science in advancing knowledge about Earth systems. The term "global change" was already in use in the International Geosphere–Biosphere Programme (IGBP) and the ISSC had proposed a programme on the "Human Dimensions of Global Change". Since the study of "Man and His Habitat" had been one of the early research themes of anthropology, the Congress could be an opportunity to bring such research forward and to encourage anthropologists to take up this challenge.

At the plenary session, Fredrik Barth stated that there was now a need for a critique of development in terms posited by a "full world condition", since there was a qualitative shift from an empty world condition, with new resources to exploit, to a full world condition (Arizpe 1996: 19–30). This was all the more pressing because the transformation of production and technology in the Global South has resulted in what at one time had been merely cultural differences increasingly becoming differences in wealth and poverty. He also explained that, in our present-day world, industrial activity has passed "a fateful threshold of scale in relation to the global ecosystem" (ibid.: 12). He argued that new institutions were needed to facilitate rational individual and collective decisions in a full world condition, as was a human morality that could guide those decisions in such a way as to secure a future for life on earth.

To meet the challenges of the intensification of transnational migrations, an outpouring of commodities and worldwide communications, Eric Wolf said that "We must re-examine our central concept of culture, among other things by analysing how culture shapes people's observation and perceptions of society and nature and expand the scale and scope of an anthropological understanding of today's world" (ibid.: 13). With so much of present-day cultural learning dependent on situation and context, he wondered how people were coping with cultural coherence. (How much more penetrating his insight now seems in the second decade of the new millennium.) He concluded that each ethnic group may have its characteristic "just cause", phrased in its particular portfolio of cultural signifiers. An anthropological perspective must locate that "just cause" within the context in which identity-makers and identity-seekers respond and in which they have to operate (ibid.).

Roberto Cardoso de Oliveira affirmed that ethnicity is "a privileged realm for empirical observation of moral and ethical factors". He cautioned that the concept of culture alone will not suffice for stating the morality issue properly: as the philosopher Ernst Tugendhat said: "It is unacceptable to deem something right and good because it was established by custom, without proving it to be right and good" (ibid.: 13). He suggested that a way out of this dilemma is to distinguish between culture and norms, to enable advocating a code of ethics of responsibility which could be applied worldwide.

For some people, globalization is seen as a risk because it may mean further Westernization. Maurice Godelier aptly asked whether the West is a mirror or the mirage of the evolution of humankind. As the West became the permanent centre of the continuous expansion of the market economy, it drove differential effects on tribal and ethnic communities around the world. Godelier made the distinction between tribal societies such as those of New Guinea, which may belong to the same ethnic community yet make war on one another, and the European type of ethnic group in which political sovereignty belongs to the state. In the latter case, the same ethnic group can belong to several states while being dominant in one and dominated in another. A major point worth highlighting was his idea—stemming from the fact that "practically no society in the world today can develop without incorporating something that comes from the West, whether it be tools, weapons, techniques, ideas or social relationships"—that "the defence of 'traditional' patterns of thought and behaviour, or those considered to be such, can only be achieved paradoxically through the integration of the communities that exist within global structures which, at the same time, undermine their existence" (Godelier 1996: 70).

Without repeating what is included in *The Cultural Dimensions of Global Change: An Anthropological Approach*, which includes many of the papers presented at the ICAES, it is important to mention briefly here the topics and key points taken up by other anthropologists who participated in the plenary sessions: "people's reality", which may show a brutalization of relationships among families, aggravating gender disparity and gender inferiority as inequality rises in globalization (Karl-Eric Knutsson); people's new perceptions of their 'locality' as embedded in the new 'globality' of the world market, drawing on interviews in the Lacandon rainforest (Lourdes Arizpe); the "feminization of labour" as growing numbers of women enter the professions, clerical positions and the public sector (Helen Safa); the ethnicity that emerged with particular belligerence in post-Soviet countries in the areas of deep social change and radical reform (Valery Tishkov); the problem in the discourse on the environment and development caused by economists' inability to value what the environment is worth in anything other than monetary terms (Michael Redclift); the reworking of local cultural visions through the "electronic Trojan horse" of global television (Philip Carl Salzman); the success of development programmes in Africa as increasingly dependent on sensitivity to ethnic diversity, to intracultural variation and even to the culture of modern bureaucracies (Paul Nchoji Nkwi); and the need for cultural sciences to join with other social and biological sciences in cooperative endeavours if there is to be any real success in facing the challenges of understanding and coping with global change (Paul Baker).

Lourdes Arizpe studying the perceptions of social deforestation in the Lacandon rainforest in Chiapas, Mexico, 1992. *Source* Photo was taken by Veronica Behn-Eschenburg and is from the author's personal photo collection

This chapter has shown how, in the four decades following the end of the Second World War, the concept of 'culture' became a potent political tool as countries and international organizations came to terms with a world where there was an urgent need to establish global relationships along lines that would be acceptable to all the stakeholders involved. It was in this context that UNESCO turned its attention in the 1990s to the task of investigating the role of culture in development and how that knowledge could assist in formulating new guidelines and policies.

References

Arizpe, L. (1989). On cultural and social sustainability. *Development*, 1, 5–11.

Arizpe, L. (Ed.) (1996). *The cultural dimensions of global change: an anthropological approach*. Paris: UNESCO Publishing.

Bhabha, H. K. (1990). The other question: difference, discrimination and the discourse of colonialism. In: R. Ferguson et al. (Eds.) *Out there: marginalization and contemporary cultures* (pp. 71–88). Cambridge, MA: MIT Press.

Basch, L., Glick Schiller, N., & Szanton-Blanc, C. (1994). *Nations unbound: transnational projects, postcolonial predicaments, and deterritorialized nation-states*. Langhorne, PA: Gordon & Breach.

Baumann, G. (1996). *Contesting culture: discourses of identity in multi-ethnic London*. Cambridge: Cambridge University Press.

Brumann, C. (1999). Writing for culture: why a successful concept should not be discarded. *Current Anthropology*, supplement on culture, 40(S1), S1–S27.

Cernea, M. (1995). *Primero la gente: variables sociológicas en el desarrollo rural*. México: Fondo de Cultura Económica.

Escobar, A. (1995). *Encountering development: the making and unmaking of the Third World*. Princeton: Princeton University Press.

Ghosh, A. (1989). The Diaspora in Indian Culture. *Public Culture*, 2(1): 73–78.

Gluckman, M. (Ed.) (1964). *Closed systems and open minds: the limits of naïvety in social anthropology*. Chicago: Aldine Pub. Co.

Godelier, M. (1996). "Is the West the mirror or the mirage of the evolution of humankind?" In: Arizpe, L. (Ed.), *The cultural dimensions of global change: an anthropological approach* (pp. 63–75). Paris: UNESCO Publishing.

Gonzalez, N. L. & McCommon, C. S. (1994). *Conflict, migration and the expressions of ethnicity*. Boulder, CO: Westview.

Hall, S. (1987). Gramsci and us. *Marxism Today*, June, 16–21.

Huntington, S. (1993). The clash of civilizations. *Foreign Affairs*, 72(3), 22–49.

Kandiyoti, D. (1993). Gender, culture and development. Paper prepared for the World Commission on Culture and Development, UNESCO.

Keitner, C. (1999). Pluralism past, present and future: the evolution of UNESCO's approach to cultural diversity: an informal survey. Paper prepared under the supervision of Katérina Stenou, Director of the Division of Cultural Pluralism, UNESCO.

Kearney, M. (1995). The local and the global: the anthropology of globalization and transnationalism. *Annual Review of Anthropology*, 24, 547–565.

Lewis, O. (1959). *Five families: Mexican case studies in the culture of poverty*. New York: New American Library.

Lins Ribeiro, G. (2000). *Cultura e política no mundo contemporâneo*. Brasilia: Universidad Nacional de Brasilia.

Lomnitz, C. (1998). *Modernidad Indiana*. Mexico: Fondo de Cultura Económica.
Manzo, K. (1991). Modernist discourse and the crisis of development theory. *Studies in Comparative Development*, 26(2), 3–36.
Mondiacult (1982). *Mexico City declaration on cultural policies for development*. Mexico City: Secretaria de Educación Publica.
Mundimbe, V. Y. (1988). *The invention of Africa*. Bloomington: Indiana University Press.
Nash, J. (2001). *Mayan visions*. New York: Routledge.
Rahnema, M. with Bawtree, V. (2001). *The post-development reader*. London: Zed Books.
Rist, G. (1994a). Introduction. In: G. Rist (Ed.), *La culture: otage du développement?* (pp. 7–13). Paris: L'Harmattan.
Rist, G. (1994b). Des sphinx, des licornes et autres chimeres … Trois approches des relations entre culture et "developpement" face aux pratiques sociales. In: G. Rist (Ed.), *La culture: otage du développement?* (pp. 49–68). Paris: L'Harmattan.
Said, E. (1979). *Orientalism*. New York: Vintage Books.
Sen, A. (1999). *Reason before identity*. Oxford: Oxford University Press.
Smith, R. C., Cordero-Guzmán, H. R. & Grosfoguel, R. (2001). Introduction: migration, transnationalization, and ethnic and racial dynamics in a changing New York. In: R. C. Smith, H. R. Cordero-Guzmán & R. Grosfoguel (Eds.), *Migration, transnationalization, and race in a changing New York* (pp. 1–32). Philadelphia: Temple University Press.
UNESCO (1947a). *Commission on Social Science: second session of the General Conference in Mexico City, November–December 1947*. Document 2 C/K/9. Paris: UNESCO.
UNESCO (1947b). *Diario de la Conferencia General de la UNESCO, 7 de noviembre de 1947*. Paris: UNESCO.
UNESCO (1947c). *The programme for UNESCO in 1948: proposed at the second session of the General Conference in Mexico City, November–December 1947*. Paris: UNESCO.
UNESCO (1969a). *Cultural policy: a preliminary study*. Paris: UNESCO.
UNESCO (1969b). *A study of cultural policy in the United States*. Paris: UNESCO.
UNESCO (1970). *Some aspects of French cultural policy*. Paris: UNESCO.
UNESCO (1974). *Canadian cultural policy*. Paris: UNESCO.
UNESCO (1977). *Medium-term plan for 1977–82*. Paris: UNESCO.
UNESCO (1980). *Cultural policy in Yugoslavia: self-management and culture*. Paris: UNESCO.
UNESCO (1982). *World conference on cultural policies*. Paris: UNESCO.
UNESCO (1998). *World culture report 1998: culture, creativity and markets*. Paris: UNESCO.
UNESCO (2000). *World culture report 2000: cultural diversity, conflict and pluralism*. Paris: UNESCO.
Valentine, C. A. (1967). Culture and poverty: critique and counter-proposals. Chicago: University of Chicago Press.
Vertovec, S. & Cohen, R. (Eds.) (1999). *Migration, diasporas and transnationalism*. London: International Library of Studies on Migration.
Wang, G., Dissanayake, W. & Newton, B. (1983). *A cultural mediation approach to development and change*. Nagoya: UNCRD.
World Bank (1982). *Bank, tribal peoples and economic development*. Washington DC: World Bank.
Zhou, M. (1997). Segmented assimilation: issues, controversies, and recent research on the new second generation. *International Migration Review*, 31(4), 975–1008.

With President Ernesto Zedillo of Mexico and Mr. Federico Mayor Director-General of UNESCO, Paris, 1966. *Source* Photo by Sayeh Msadek

With Javier Perez de Cuellar, President of the World Commission on Culture and Development, Paris, 1996. *Source* From the personal photo collection of the author

Chapter 3
Internationalizing Culture

With the fall of the Berlin Wall in 1989 and the optimism generated by the possibilities of creating a democratic world, many people turned to culture in search of new avenues for political reorganization and development. As I repeatedly stated at UNESCO, culture became the last frontier in development programmes, after the environment, population and women had been highlighted as major international concerns. Conflicts in which culture was invoked were on the rise, mainly in developing countries, bringing a new agenda to international meetings. While the UN Decade on Culture and Development had stimulated interest in this field, the celebration of hundreds of cultural events in many countries had generated impatience with the inconsistent international narratives about culture.

To develop new guidelines in this field, the Group of 77 (which included many developing nations who had been promoting endogenous development for decades) and the Nordic countries, with the full support of Federico Mayor, Director-General of UNESCO, presented a resolution that was adopted in the report of Commission IV of the UNESCO General Conference on 6 November 1991.[1] Soon after, a resolution was presented to the General Assembly of the United Nations to establish "an independent World Commission on Culture and Development" to prepare a report and proposals "for both urgent and long-term actions to meet cultural needs in the context of development"; this resolution was adopted on 19 December 1991.[2] It would be a policy-oriented report based on the collection and analysis of information from all regions and from diverse sources. It would engage the general public and assist those responsible for formulating and implementing cultural and development policies at both national and international levels, including the different multilateral banks and UN organizations and programmes.

---

[1]Resolution adopted in the report of Commission IV of the UNESCO General Conference at the 26th plenary meeting, 6 November 1991, document 26C/3.4.

[2]Resolution adopted by the General Assembly of the United Nations, 19 December 1991, report A/46/645/Add.4.

L. Arizpe Schlosser, *Culture, International Transactions and the Anthropocene*,
The Anthropocene: Politik—Economics—Society—Science 17,
https://doi.org/10.1007/978-3-642-41602-6_3

The UN World Commission on Culture and Development (WCCD) that resulted from the December 1991 resolution was chaired by the former UN Secretary-General Javier Pérez de Cuéllar. In my professional capacity as an anthropologist I was a member of the Commission; later on, after I had been appointed as UNESCO Assistant Director-General for Culture, Mr. Pérez de Cuéllar requested that I be in charge of the Secretariat for the Commission. The two Directors of the Secretariat were Jerôme Bindé and Yudhishthir Raj Isar.

The Commission adopted an active international approach, holding nine regional meetings around the world. At each one, local experts, policymakers, social scientists, artists, cultural policy and development experts, and NGO activists presented their own concerns and ideas. Drawing from these contributions, and from commissioned papers, the Commission tested its own questions and explored a series of "lines of inquiry", consolidating some, abandoning others and opening up paths not originally envisaged.

In his welcoming speech to the Commission, Federico Mayor asked whether we could be mere passive observers in the silent opposition between societies of abundance and societies of poverty. He emphasized that the relations between culture and development had never been examined in a coordinated way at the global level—and that this would be the task of the Commission. As Chair of the WCCD, Mr. Pérez de Cuéllar stated in his opening address to the Commission that its work would take place in a world undergoing a scientific, technological, economic and great political and cultural transformation. Could culture, or another form of development, succeed in building defences against the contagion of the anti-culture spread by xenophobia and ethnic cleansing? He reminded us that our view of others is a reflection of ourselves, and that hate is, above all, self-hate.

3.1 Framing the Questions

On 17 March 1993, the first day of the first meeting of the WCCD, the members of the Commission were ushered into a small, dark room at UNESCO in Paris. There were twelve of us in addition to Mr. Pérez de Cuéllar, all highly motivated, attracted by the aura of a World Commission and impressed by each other's credentials. These first impressions were soon joined by concern as we were handed the mandate and lines of inquiry of the Commission but given very limited time to discuss it. Vigorous discussions ensued but there was only time to make a few minor alterations to the document before it was presented to the press.

The members of the original Commission, together with members invited to join it during the consultations, were:

- Claude Ake, development economist, Director of the Centre for Advanced Social Science at the University of Port Harcourt, Nigeria;
- Lourdes Arizpe, anthropologist, President of the International Union of Anthropological and Ethnological Sciences, Director of the Institute of Anthropological Studies at the National University of Mexico;
- Yoro Fall, historian, Head of the History Department at the Cheikh Anta Diop University in Dakar and Inspector General of National Education, Senegal;
- Kurt Furgler, jurist, Vice-President of the Club of Rome, former President of the Swiss Confederation;
- Celso Furtado, economist, former Minister of Planning and of Culture in Brazil, member of the South Commission;
- Niki Goulandris, Vice-President of the Goulandris Museum, Greece;
- Mahbub ul Haq, economist, former Minister of Finance and of Planning and Development in Pakistan and founder of the UNDP Human Development Report;
- Elizabeth Jelin, Argentinian sociologist, Senior Researcher at the University of Buenos Aires and at the National Council of Scientific and Technical Research, Buenos Aires;
- Ole Henrik Magga, linguist, President of the Såmi Peoples of Norway;
- Nikita Mikhalkov, film-maker, winner of the Golden Lion at the Venice Film Festival in 1991 for his film *Urga*;
- Tchie Nakane, anthropologist, Vice-President of the National Japanese Commission to UNESCO, Director of the Shibusawa Foundation for Ethnological Studies;
- Leila Takla, jurist, member of the Egyptian Higher Council for Culture, former President of the Committee for Foreign Relations in the Egyptian Parliament;
- Elie Wiesel, writer of Romanian origin, winner of the Nobel Peace Prize 1986, Professor of Human Sciences at Boston University.

Five honorary members also contributed to the work of the Commission: Prince Talal ibn Abdul Aziz al Saoud, of Saudi Arabia; Aung San Suu Kyi, winner of the Nobel Prize for Peace 1991, of Myanmar: the French anthropologist Claude Lévi-Strauss; the Belgian scientist Ilya Prigogine, winner of the Nobel Prize for Chemistry, 1977; and Derek Walcott, the St Lucian poet, winner of the Nobel Prize for Literature, 1992, and Professor of English Literature at Boston University.

The Secretariat of the Commission, initially headed by the French scholar Jérôme Bindé and then by Yudhishthir Raj Isar, was responsible for creating the Commission's Report. Several young, very intelligent and extremely hard-working researchers—Jean-Ives Le Saux of France, Guiomar Alonso Cano of Spain and Vladimir Skok from Canada, as well as Malick M'Baye of Senegal, were put in charge of collating all the ideas and materials generated from the Commission's regional consultations around the world and from the more than 250 papers which the Secretariat commissioned or received.

3.2 The Commission's Mandate

At this first meeting of the WCCD, Mahbub ul Haq made the case that future conflicts would not be between nations but between peoples and nations. But he warned, "we can't send soldiers to keep peace between peoples; you have to have socio-economic development".[3] He mentioned the contemporary conflicts in Somalia, Liberia and Sri Lanka. In the following years, many more such conflicts emerged in Kosovo, Macedonia, Chechnya, Nigeria, Liberia, Indonesia, India, Israel, Colombia and elsewhere. Twenty years on, such conflicts are innumerable, and have led to bloody wars and atrocious terrorist attacks.

Ul Haq specifically called attention to the impending problems in national and global governance, in environmental security and human rights. He said that development should be of, for and by the people. As one of the founders of the Human Development Report, he was well placed to speak on this theme. The Human Development Report has done more than any other UN initiative to construct tools with which to gauge this kind of development. He proposed that the mandate of the Commission be amended: we should establish as an assumption for our work the universality of human aspirations, while recognizing different routes to development that respect the diversity of cultures and languages. He spoke of the intrinsic value and sovereignty of the human spirit, which should never be stifled by barriers of race, religion, power or creed.[4]

Kurt Furgler wanted to emphasize respect for persons, the will to share, social justice and spiritual values in the work of the Commission. He cited Jean Monet, who said, after the Second World War, "que la guerre ne soit pas seulement impensable mais impossible" ("let war be not only unthinkable but also impossible").

Leila Takla suggested that the questions we must answer were where culture went wrong and how we use culture to fulfil our goals. She felt that exclusion must figure prominently in the concerns of the Commission: the need to take into account differences and the importance of emphasizing a global community.

Celso Furtado stressed that the Commission should focus on the possible and not on the ideal. He insisted that the starting point should be the crises in the world at that time: hunger, unemployment and exclusion, debt, the deterioration of prices for producers in developing countries and the loss of capacity for macro-economic decision-making in a world of monetary and financial internationalization.

Nikis Goulandris then raised the issue of the intrinsic relationship of culture to environmental concerns and proposed that we should discuss the 'intercultural'—that which joins the cultures of the world in a fundamental unity. Claude Ake intervened to ask why it was that wars of culture had increased. Cultures, he explained, were not developed or underdeveloped. Clearly ethnic conflicts arose as a result of non-development. Yet he warned that there also existed a culture of

[3]All citations in this meeting come from my personal notes of the discussion.

[4]It was with great sadness that we learnt of his premature death in 1996, soon after the presentation of the WCCD report.

hatred, of discrimination, of fascism so it was not a question of bringing in culture only in a "good way" in the Report.

Elie Wiesel contended that humanity had never risen so high as during the twentieth century but also that it had never been as inhuman. In his mind, the role of culture was that of a mirror. Along the same lines, Nikita Mikhalkov, speaking through an interpreter, declared that his proposals were not only about culture but about being human. He expressed his concern about the loss of livelihoods and about homeless children. We should reflect, he said, on the "destroyed cultures" in the ex-USSR, and speak of their renaissance that we need to set in motion.

For my part, I proposed that the emerging trends towards globalization be analysed by looking towards future trends, especially with regard to terms of exchange between countries, the depletion of natural resources and the rising expectations of people, particularly women and young people, all over the world. I also suggested that in the Commission's mandate priority be given to women's movements, ethnic strife, religious fundamentalism, social violence and science.

Comments were also received from some of the honorary members of the Commission. Ilya Prigogine praised UNESCO's initiative as reflecting the importance of the human element in the current relationship between culture and development. The explosive development of science, he said, was radically changing our vision of the world from one of conflicting certainties to one of interrogation, leading to greater unease not only about science and technology but also about our model of society. Perhaps highlighting the positive aspects of technology would allow us to conserve our cultural heritage while ensuring the increasing participation by society in culture. Most importantly, he asserted that "It seems to me difficult, however, to accept hierarchical models in which science would be subordinated to ethics, or, inversely, of science being considered as a vehicle for a superior truth overarching any social or ethical consideration." The end of omniscience, of determinism and of the intrusion of the observer in the description of the universe under study, required a new perception of Nature and of our world. On the one hand, we must overcome the lack of certitude of the late twentieth century, but, on the other, we must retain the idea of risk, of the incertitude that frames any tomorrow but which also implies the possibility of progress. He ended by stating that "such a profound change in our conceptions requires a dialogue between the different spheres that represent culture".

3.3 Amending the Lines of Inquiry

The "suggested areas of consideration" for the Commission set by the UNESCO General Conference were originally (in the order in which they were approved):

1. the progressive loss of indigenous knowledge, traditions and languages;
2. the more widespread use of local languages in literature and newspapers;
3. the growing interchange between different cultures;

4. aspirations to an improved quality of life;
5. participation by various sectors of the population in cultural life;
6. the contribution of women and men to culture and development, both in their everyday activities and as artists and artisans;
7. conditions for creativity and productivity in the arts;
8. relationships between culture and governance (the 'culture of democracy');
9. links between environment and culture;
10. cultural factors as key variables in overall development planning and implementation, taking into account the ethical challenges to scientific advancement and economic growth.

At the end of the first three days of meetings, the general topics in the mandate of the commission had been extended to twelve, through our insistence that the environment should be a separate topic, while special attention should be paid to women and culture, and to cultural heritage. They were then approved as follows:

1. culture, creation and creativity and the cultural sector;
2. culture, education, human resources and development;
3. information, communication and cultural industries;
4. the influence of cultural development on individual and collective well-being;
5. links between culture and development models;
6. science, technology, economics, culture and development;
7. development, culture and the environment;
8. development, culture and population;
9. universality, specificity, culture and development ("world culture");
10. cultural exchanges, intercultural relations and development;
11. development and the culture of democracy, human rights and peace, the culture of citizenship.

An additional point was:

12. the devising of strategies and machinery for assisting decision-making and international cooperation in development and cultural policies; plan of action and agenda; implementation, financing and follow-up mechanisms; proposals for reform.

At the second meeting of the Commission (which took place in Stockholm, Sweden, in July 1993), the lines of inquiry were compressed into just seven points, in which all previous topics were subsumed, with the following order of priority:

1. links between culture and development models: world culture, multicultural societies, diversity of relationships between state, society and social actors, and so on;
2. cultural development: the cultural sector as such and its various contributions to development, heritage, creativity, culture and economics and so on;
3. development, culture, population, environment and management;

4. development, culture, education, science and technology, economics and the impact of new technologies on cultural development;
5. development and the culture of democracy, ethics, human rights and peace;
6. culture, development, the communication society and cultural industries;
7. cultural exchanges, intercultural relations and development, new strategies for multicultural development, enrichment resulting from the stepping up of cross-cultural exchanges.

A comparison of the three different lists is a showcase for international political procedure, demonstrating how different actors (from countries, institutions and political movements to intellectuals and artists) wanted different priorities and emphases in the Commission's lines of inquiry. The locations in which such different views were transacted were the UNESCO General Conference, the UN General Assembly, the Secretariat of the WCCD, the first meeting of the Commission in Paris and the second meeting in Stockholm. By that time, consensus had been reached, thereby providing the mandate and lines of inquiry of the Commission with a political legitimacy that gave it wide acceptance internationally, both politically and intellectually.

This does not mean that some member states, political groups and social scientists were not opposed to the whole enterprise of debating culture and development or cultural policies, or, in other cases, were indifferent. But this negotiation had already taken place in the discussions about creating the Commission in the first place, within both the United Nations and UNESCO. The way that such processes work is that a group of member states takes the initiative, negotiates agreements with others to support a resolution, and then monitors the development of the initiative as a United Nations project.

The academic legitimacy of the mandate and lines of inquiry was partly assured by the presence of distinguished researchers in the Commission as both core and honorary members, reinforced later on by the incorporation of two further outstanding researchers, Elizabeth Jelin, a sociologist from Argentina, and Yoro Fall, a Senegalese philosopher. However, full academic legitimacy would have required a different procedure, one that was much more analytical, detailed and rigorous in its search for truth but this would have run counter to the political nature of the Commission, whose objective was to arrive at a consensus concerning principles and guidelines acceptable to the most powerful countries and cultural movements in the world.

The next stage of the process was to arrange the regional meetings for discussion of the lines of inquiry, at which local experts, social scientists, policymakers, artists, cultural policy and development experts, and NGO activists were invited to share their views. The first of these was held in Paris immediately after the establishment of the mandate.

3.4 Paris Consultation, March 1993

3.4.1 As Democracy Spreads, Does Governability Decline?

On 20 March 1993, the day after the mandate of the World Commission had been released to the press, the first of its open consultations was held in UNESCO. The participants were ushered into Room 10, a wood-panelled room with a high ceiling, the interpreters in their silent darkened rooms to one side and a small garden with a fountain visible through a floor-to-ceiling window. There is something about UNESCO rooms that engenders awe. Perhaps it is the combination of spaciousness and labyrinths, as if the Le Corbusier architecture could contain the words and thoughts of all the peoples of the world. Or, more likely, it is because the walls are impregnated with so many years of debate and questioning and trying to find human solutions to human problems, and this gives a human scale to the grandeur of the original ambitions. Today, Room 10, with its large oval table which has had to expand many times to bring in more representatives of delegations, hosts the meetings of the Executive Board, with its fifty-eight ambassadors, held every six months to review the work of the organization. I did not know it myself at the time, but it would be in that room that, as Assistant Director-General for Culture, I would soon be fighting epic battles over ideas and programmes, including three sessions to explain and defend the WCCD report itself.

At this first meeting, the French economist Ignacy Sachs fired from the hip by asking whether development was possible without destruction. We need, he said, a history of the axis of biodiversity and social diversity: that is, we need to develop a grid of ecosystems and culture as a basic framework. The main goal of development—social justice—can only be achieved, as Michel Serres had argued, by complementing a *natural contract* with a new *social contract* among humans. For the future a "civilization of the biomass", as an Indian thinker termed it, is needed to bring in new knowledge. A different range of strategies for development are needed in the North and in the South. The credibility of the North depends on reducing consumption by changing the style of development. Meanwhile, in the South, strategies must be found to innovate without losing cultural continuity. Mahbub ul Haq agreed that, if the South followed the development model of the North, it would need ten times the energy and two hundred times the resources that the Earth has. He then asked what would create new patterns of consciousness.

The Norwegian Minister of Culture, Åse Kleveland, reiterated the importance of environmental concerns, mentioning the lesson of the Brundtland Commission in fostering a new understanding of the interlinkages between environment and development. This led their country to support the Commission, since "cultural factors indeed enter the environment–development relationship" (Kleveland 1993). She also mentioned the population question, as well as the paradox that "just as the culture of democracy seems to spread, the governability of our societies appears to decline" (ibid.: 4). Another point she made was that of world income distribution: the richest receiving most of the world's income and the poorest only a minimal part, was "clearly unacceptable, whatever our cultural background. It proves how distant reality is from our ideals" (ibid.). Cultural industries, the new information technologies, the traditional division of labour between women and men, urban planning, architecture and industrial design, and the cultural needs of children and youth were also on her list of priorities. Ms. Kleveland asked for goals and proposals based on "cultural universalia" but with 'culture-specific' options for their operationalization.

3.4.2 *Political Reason Against Tribes and the Market*

Alain Touraine, the renowned French sociologist, announced that the grand scheme centring on the idea that the more reason was applied, the more the individual was liberated, had now been marginalized as a conception of economy and culture for two reasons. On the one hand, the concept of a substantive, objective rationality, as espoused by the sociologist Max Horkheimer, had been replaced by instrumental reason geared to needs and not towards building a rational society. On the other, the centrality of technology, and of psychological and cultural diversity, had given way to consumption. Thus, the world of the market has been separated from the world of values, of culture. While the global market is in the hands of the North, identities are strongest in the South. However, he argued that multiculturalism is unacceptable because it only leads to civil war, thus thwarting the unity of the world. What is needed is the creation of a mediating space of fundamental rights, not based on the market or on tribal identities but on political reason.

Responding to a question by Mahbub ul Haq on how to reverse these trends, Touraine answered that turning cultures into museums was not the answer. Traditional cultures, he said, are marginal and residual and, given the choice, people will choose cosmopolitan civilization. The choice must be one of intellectual innovation: not to see people as bearers of cultures but as actors of one culture that must have the liberty to change. This is modern culture and the search for creativity.

To my own question of why we should not speak of a global polis, Touraine replied that the reason was to avoid confusing globalization with political dialogue. He declared that he was not only a member of a family, of a nation and of United Nations organizations, but also a subject with rights in a broader context.

In the discussion, Åse Kleveland brought up the transition from oral to visual culture through information technologies. Leila Takla, reacting to mention of Christian culture, explained that she, as a Christian Egyptian Copt, felt close to many other religions because so many elements of one religion are similar to others.

3.4.3 *Culture Is Part of Democracy Since It Gives Freedom*

The philosopher Jacques Derrida warned against the new culturalism that may disregard the importance of techno-economic processes. He said that he was struck by the lack of reflection on the history of the concept of culture. Western literature, he held, is linked to democracy because writers have total liberty to write; he mentioned the case of Salman Rushdie as an example of non-freedom. On the linguistic hegemony of Anglo-English, as he called it, he said that it must not be opposed on account of "national identity". It must be considered good, yet it will not thrive unless such hegemony is different from itself, in the face of the Other.

He later explained his concept more fully. The project of a universal culture would speak about respect for the Other, which does not suppose a meta-language but a translation, understood as the language of the Other, not as a language that can be appropriated. He went on to insist that philosophy should be taught all over the world, since sovereignty, the market and the right to intervene all come from philosophy. Such teaching must not globalize the Western episteme but create an ethics of the relation towards the Other and the thoughts of the Other.

3.4.4 Culture as a Double-Edged Sword

Not mincing her words, Edith Sizoo, coordinator of the North–South Cultures and Development Network, began by stating that four decades of development had not been successful. She argued that culture is behind separatism, racism, machismo, the burning of widows and the mutilation of sexual organs. If that is the case, it can be seen as dangerous for people and for the Earth. However, culture may also be seen as a source of knowledge and cohesion. The question is which of the two understandings is to be taken into account in development.

She suggested that we need an approach of "starting from the other end", from the reality, which is that people think and act according to their culture. She gave two examples. Her first was from Senegal, where an income-generating project for sewing or carpentry might end up with no money. This situation would arise because, in the local culture, a poor person was not one without money but a social orphan, one without social relations. So the money might be used for an uncle's house or to treat a nephew's illness, since this gave them the most security. Her second example was the Brazilian government's project with the World Health Organization, whose clinics for information on AIDS had had no response. So they convened two groups, one with the priestesses of Candomblé (a local popular religious sect) and the other with prostitutes, a course of action strongly opposed by the country's bishops. The approach, however, was very successful. She ended by saying that a key problem was that actions and interventions are generally carried out according to a power relationship which determines whether things will change.

Catherine Lalumière, then Secretary General to the Council of Europe, began by stating that culture is not a question exclusively for the state but for all of society. It was a miracle that all the member states of the United Nations in 1948 approved the Universal Declaration of Human Rights, but we know that not all have followed them. In more specific terms, she explained that there is one European culture but also cultural identities to which people are attached. A mistake was made in building Europe in thinking that the market would later on lead to a European identity and values; monetary union will never bring cultural union. Europe must be built through the spirit and the heart.

3.4.5 The Loss of Compass in Understanding the World

The Japanese economist Michio Morishima spoke of Japan's success, which has frequently been attributed to its cultural values. In Japanese enterprises, he explained, the highest level of managers are bureaucrats, in the sense that they are not employed by the firm but are members of it: hence their loyalty to the firm, since they work for a career and not for wages. In contrast, in the West entrepreneurs want to maximize their profits and workers their wages. But he could detect a weakness: the Japanese high-level bureaucracy was becoming self-reproducing and university graduates were looking down on administrative workers.

Marc Ferro followed up this argument, saying that there is an increasing bureaucratic unification of the world. This is coming about through a unification of the church, the police, state functionaries and other institutions, which reinforces the dominant centre so that the excluded are driven further out to the periphery. This is one of the two major phenomena that Ferro identified as important at that time. The second phenomenon was increasing scientific and cultural uniformity, for example, in medicine and in the media. He cited a study just carried out that showed that the same information was being disseminated in Paris, Moscow and Buenos Aires. This had a negative effect in that there were fewer and fewer sources of counter-culture.

Earlier in his contribution he had said that more and more facts in history were known but now we understood less and less. In the West we tend to consider the media responsible for this "loss of compass" ("perte de boussole"). It is said that those who inform are not informed. Neither the Iranian revolution, nor the events in the USSR after Gorbachev, nor the wars in the former Yugoslavia can be understood within the formerly prevailing ideologies. Nationalisms are now overwhelming internationalisms. Ferro argued that democratic discourse should be set above internationalism, justice and human rights.

After listening to so many speakers, we had a wealth of ideas to begin considering. We all congratulated each other, with Federico Mayor beaming at the success of events. Had I known at the time how difficult it is to negotiate the creation of a programme, let alone a World Commission in an international agency, I would have congratulated everyone even more heartily.

Yet the difficulty of the enterprise had become apparent at this first meeting. Whatever the institutional or member state intentions, the different intellectual priorities of members of the Commission and the snowballing effect of political interest on a concept—culture—which had so many diverse meanings laid bare the challenge of creating an international agenda for culture while trying to find a consensus between opposing and perhaps irreconcilable views. An underlying concern exposed the same dualism as in the early 1950s: that culture could be an asset for development but also an obstacle to modernization. This dilemma had now expanded to a wider arena with the fall of the socialist bloc.

Subsequent consultations by the Commission further confirmed both trends: the bewildering variety of views on culture and the increasing importance of culture in national and world affairs. As different countries offered to host the meetings, the next consultations would take us to Stockholm in Sweden and to San José in Costa Rica.

The Report of the World Commission on Culture and Development was presented to the UNESCO General Conference in November, 1995. *Source* UNESCO

3.5 Stockholm, June 1993

3.5.1 Towards an Agenda 21 for Culture?

The WCCD Regional Consultation for Europe focused on cultural policies in the context of development. The Nordic countries had been strong supporters of the work of the World Decade on Culture and Development and of the creation of the Commission—they wanted culture to have the same kind of impact on world opinion that the Brundtland Commission had had for environmental concerns. Their priority in the meeting in Stockholm in the summer of 1993 was to gear the Commission towards practical, policy-oriented recommendations on cultural policies. As Carl-Johan Kleberg, one of the thinkers who originally proposed the creation of the Commission, put it, the aim of the meeting was to analyse "what the cultural responsibility of the state, the regions and the municipalities mean" (Kleberg 1993). He also warned of "the general lack of follow-up mechanisms for international declarations, which therefore remain empty words" (ibid.: 4).

Among the main points of his presentation, Kleberg noted that in most European countries the cultural responsibilities of the state were carried out through legislation, state institutions, the award of financial grants and attempts to influence public opinion (though he pointed out that in general there was a reluctance to use legislation in the cultural field, at least compared with education and hospital care). He laid particular emphasis on regional responsibility, noting that very often there is competition between different bodies working at the regional level and that the relationship to municipalities was unclear. In Germany and Scandinavia, and in other West European countries, the responsibility of the municipalities is looked upon as the cornerstone of cultural policy measures. In many European countries the municipalities had increased their cultural spending in the 1970s and 1980s, but the economic downturn in the mid-1990s was leading to cuts. Where to cut and how much was often the question.

Kleberg said that "It is impossible to make a clear picture of where we are [in cultural policies] even in one single country … everywhere there seem to be changes going on at all levels" (ibid.: 3). In the light of these changes, the question was "what responsibility is left for the State? …. Support to central national museums and operas cannot be considered as the main expression of a national identity (ibid). What would be the reality, he asked, if a possible agenda for culture led to the discovery that governments did not have the instruments for carrying out the necessary measures?

In another presentation together with Hans Sanders, Kleberg mentioned several European initiatives that could contribute to the report of the Commission. Foremost among them were the National Cultural Policy Reviews for each country, the Council of Europe's programme on the cultural dimension of culture, the Cultural Information and Research Liaison in Europe (CIRCLE) Report on Participation in Cultural Life in Europe, several initiatives on the Status of the Artist and the CIRCLE Conference on Cultural Rights.

As an example of a country review, Rod Fisher, President of CIRCLE, presented the case of France (CIRCLE 1993). When examining the conditions for creative activities, the evaluation was very positive. The famous doubling of the cultural budget which Jack Lang had achieved in 1981 had been used to improve conditions for the visual arts and architecture, for experiments in theatre and music, and to establish centres of creativity. Through decentralization a large number of new institutions had been established across the country. Once Jack Lang had raised the status of cultural policy, self-confident local politicians thought that they could find the necessary resources themselves. Available surveys, however, did not appear to indicate a more democratic distribution of visitors to cultural institutions and activities; rather, they showed that the number of people going to theatres, museums and concert halls was neither decreasing nor increasing by much.

In the CIRCLE Report, the first section, "The global economy and the end of the 'golden age' of culture?", highlighted the fundamental change in cultural life in Europe (Skok 1993). Presented by Vladimir Skok, it began by saying that Europeans were staying at home more, gathering around their increasingly sophisticated "electronic hearth". In one paragraph it summarized the set of cultural changes occurring in Europe:

> Established cultural habits are being challenged by a new, less clear pattern shaped by changing demographics—especially age and ethnic heritage, economic realities, [...] trade liberalization, competition and convergence of communications technology. The result is a scattered and complex post-modern cultural life with increased choices available to the individual. Confounding cultural policy makers and diverting traditional revenue flows, these changes are gathering steam, promising to even further propel discontinuities in European life and challenging the basic tenets of the role of the state in cultural intervention (ibid.: 1).

The report stated that the international cultural marketplace was becoming the major determinant for the public, marginalizing state policy. At the same time, low attendance rates for the performing arts had remained static or dropped by up to half in certain countries, unsettling traditional cultural disciplines. In Central and Eastern Europe, where state subsidies had dried up, the prediction of a severe drop in traditional cultural participation appeared inevitable.

While in the previous three decades, cultural policy-makers in western Europe had believed that the official role of the state was to bring culture to the masses, after the 1980s things had clearly changed. More people now enjoyed technologically enhanced audiovisual products and services, which supporters of traditional culture tended "to deplore as superficial, consumer-oriented mass culture which disregards national and local traditions and values" (ibid.: 3). The report went as far as to term this development "A form of Cultural Darwinism, inadvertently spurred by the democratization of choice and dominated by the home entertainment technology [which] will probably result in even more fragmented and sporadic participation patterns". As a result, "the individual European has—forever—become independent of official culture" (ibid.).

The report contended that there was a new opportunity to incorporate technology into culture. The focus of cultural policy should move from the intermediate demand of producers to the final demand of the consumer, as supplied by artists and designers. The report concluded by asking that "If the 'golden age' of traditional cultural participation is over throughout Europe, are we sure that the de facto cultural policy which values supply, excellence and cutting-edge innovation is the appropriate strategy for this fragmented but dynamic era?" (ibid.: 10). The answer was: "not in its present form".

Finally, the report argued that the modernist approach to European cultural policy that predominated from 1920 to 1970 needed to be rethought. That is, in a postmodern landscape, characterized by cultural diversity (due to immigration trends and the dominance of media communications technologies, the globally driven market and the fragmenting of mass culture activities), in order to successfully balance the influence of mass culture, increased public attention should be paid to promoting cultural diversity at regional and local levels and to supporting the decentralization of cultural responsibilities through the principle of subsidiarity. The report also emphasized the need for further research (utilizing comprehensive surveys, time-use studies and psychographic and socio-demographic profiling, and market research) in order to ascertain which cultural policy strategies are working and why.

3.5.2 *Diversifying the Sources of Funding for Cultural Programmes*

The trends noted above were having a direct repercussion on the financing of cultural policy in western European countries at the time of the Stockholm meeting. The Finnish social scientist Ritva Mitchell explained that the modes of financing were diversifying and that there was a new distribution of financing responsibilities, with concomitant changes in decision-making (Mitchell 1993). Three principles of financing had been established:

1. The principle of *compensation* appeared in measures which compensated artists and cultural workers for the public use of their works and creativity.
2. The principle of *required performance* pertained mainly to the management of cultural institutions, meaning that only the ability to function on a certain level of efficiency would guarantee public financing or its continuation.
3. The principle of *inducing* meant that public funds—usually central or federal government funds—were used to induce more regions and municipalities to finance projects or to generate private funding and sponsorship for specific projects (ibid.: 4).

The powerful impact of technological development took two forms: the introduction of new copyright and neighbouring rights legislation, and rearranging and reorganizing the financing of national cultural industries. Mitchell added that audiovisual industries seemed to follow the path of the traditional film industry in evolving into a complex network of taxes, funds and foundations which reallocated the internal flows of capital within the branches. New original schemes were emerging through which those involved were trying to establish new associations of investors interested in culture industries. Another response to such trends was seen in attempts to guarantee markets for national cultural products in the new television channels, or via the introduction of national quota systems, or both.

Mitchell further explained that the expanding arts world and cultural sector as a whole had been forced to resort to private financing during the 1980s. Conversely, private sponsors and patrons of the arts had discovered the possibilities that the arts offer for advertising and enhancing corporate image: "Thus, patronage and business sponsorship of the arts seem to be rather independent of the other trends and mores of mixed or private funding of the arts and culture" (ibid.: 5).

Regarding the post-Maastricht debate on the right of national governments to subsidize the arts and culture, Mitchell mentioned Article 128 of that Treaty, termed "the cultural paragraph", and the related change to Article 92(d) of the European Commission Treaty, which "as an astute observer has pointed out … affirmed in guarded terms the pluralism of European culture and the individual responsibility of member states for cultural policy (on the basis of subsidiarity)" (ibid.: 8). She ended her discussion by citing Peter Duelund's prediction of the future dominance of the "facilitating role" of the state as a means of orientating the national cultural policies in Europe and by raising the caveat that the contribution of artists and writers cannot be measured in simple terms of economic value-added shares—and that sheer market forces seldom enhance the innovative aesthetic aspects of culture.

3.5.3 The Paradox of Cultural Industries

François Rouet, later the French Minister for Culture, argued that there was a major paradox in cultural life in Europe at that time. He declared that cultural industries, especially the audiovisual industries, "contribute to a massive dissemination of cultural goods, while the sectors that are most aided by public institutions (visual arts, performing arts …) do not seem to respond to the hopes for democratization of culture that had been placed in them" (Rouet 1993). That being so, the cultural industries must not disappear from the field of perception of cultural decision-makers.

He noted that a close correlation between economic growth and the growth of cultural industries in Europe since the 1930s had been demonstrated in a review by the French Commissariat Général du Plan. Their paper showed that television and video viewing was the fastest-growing cultural habit in western European countries. It also demonstrated that Spain had had the highest level of growth of cultural industries in western Europe between 1980 and 1991. The contribution of these industries to GNP was highest in the U.K. (3.2%), followed by Spain (3.0%), Germany (2.6%), France (2.3%) and Italy (2.0%) (Commissariat Général du Plan 1993).

Another factor specific to cultural industries is that, because they produce cultural content, the value of their products is unstable since it is based on the recognition of their cultural value and leads to high levels of failure in the markets. This means that specific investment and management strategies are needed in these industries, mentioned by Rouet but falling outside the scope of the discussion in this book.

Rouet noted another never-ending contradiction. Cultural industries have a double orientation, like Janus, towards the potentialities of creating new products on the one hand, and towards responding to what different publics want on the other. This means that editors and producers need a certain level of autonomy, firstly towards creators and cultural workers, and secondly towards the mechanisms of distribution and commercialization. Rouet explained that this is one of the reasons why cultural workers have a low level of pay and why networks in the cultural sector are only weakly integrated.

He concluded that, given these constraints on cultural industries, they can only survive by maintaining a fine, yet unstable, balance in specifying their products, organizing their production and defining their relations with distribution and commercialization networks. He proposed that government aid for such industries should be directed at supporting the capacity of cultural industries for self-regulation.

3.5.4 European Identity and Multiculturalism

The French political sociologist Mattei Dogan argued that a decline in nationalism implies a weakening of primordial ties, but not the disappearance of national identity. Discussion in western Europe revolved around the issue that "people no longer believe in anything" (Dogan 1993). In 1987, the British people were asked: "In what kind of society are we living?". "In a society in which everyone is only out for himself" came the reply from eighty-eight per cent of respondents. Dogan explained that available data on public opinion suggested that individuals tended to view their personal development as an opportunity not to be achieved necessarily within the national community but in spite of it. Individualism, he stated, was gnawing away at nationalism and social solidarity but it should not be confounded with egotism. In Europe in recent decades nationalism and religion had declined together, while individualism was generated by the pluralistic character of advanced democracies and the multiple roles played by citizens in complex post-industrial societies.

The CIRCLE Report called attention to the widening ethnic rifts that were increasingly shaping European life. It noted that debates regarding multiculturalism dominated political and intellectual discussions, although they had not substantially influenced directional shifts in cultural policy. But it stated that the immigration issue was only part of the European identity crisis: "The cultural issue is that local, national or regional European identity needs to be pointed in a direction which so far lacks a clear political dimension … where, in a modern democracy, ethnicity is subordinate to the principle of equality" (Skok 1993: 10).

Interestingly, the CIRCLE documents reported that European youth was more ethnically diverse and more comfortable with racial, sexual and political diversity than any previous generation. Young people were being encouraged to feel part of a new, trans-border Europe. They were also the first generation to become empowered technophiles. Carl-Johan Kleberg suggested that the changing ethnic profile of most countries in Europe made it necessary to develop a special dimension to cultural policies, to guarantee that groups of immigrants keep contact with their origins at the same time as they are helped to integrate themselves into the country where they choose to live. This, he said, could help ease the tensions between immigrants and host populations that had recently created serious problems in many countries.

The consequences of economic integration for Europe's lesser-used languages were taken up by François Grin, an economist and political scientist who was later deputy director of the European Centre for Minority Issues. He concluded that such integration would create more favourable conditions for certain languages such as Basque, Breton, Catalan, Corsican and Galician, but would cause additional difficulties for languages such as Irish and Occitan. English would develop as a European lingua franca, owing less to the importance of the English-speaking countries within the European Union than to those outside it. Several European-majority languages would decline relative to others, among them Portuguese, Danish, Greek, Dutch and possibly Italian. French and Spanish could also serve as local lingua francas, while German might increase the number of its speakers.

3.5.5 Cultural Rights as Human Rights

Several participants in the CIRCLE European Round Table on Human Rights and Cultural Policies drew attention to the conclusions of its final document. It listed eight major concerns:

1. The need for reassertion and re-examination of values which should shape policies in the fields of human rights and cultural development, including respect for human dignity, equal rights, observance of the principle of non-discrimination, and tolerance and mutual respect between cultures, as well as participation, freedom of choice and artistic freedom.
2. The need to further consider the relationship between the rights of people as individuals and in community with others, and the collective rights of groups and people as a whole (of particular significance at the time since 1993 was the Year of Indigenous Peoples).

3. The recognition that democracy and respect for human rights cannot thrive without broad cultural practice within societies, giving rise to the need for reaffirmation of the value and importance of creativity within societies.
4. This needed to be accompanied by a strengthening of the legal protection of the principle of non-discrimination and of policies to ensure equality of access and opportunity to individuals and groups—not least the invisible majority of women.
5. Greater attention should be paid to the risk of creating a "Fortress Europe" and to acknowledging and respecting diversity and 'otherness'.
6. Further legal protection was clearly required at global, European and national levels to strengthen the protection of cultural rights.
7. Such strengthening of the legal protection of cultural rights was particularly important at the time, as cultural democracy was being eroded in some countries.
8. The position of the weakest and most disadvantaged members of society must be the central focus of attempts to define and develop the right to participate in cultural life.

Further recommendations were made in this document, which could be summed up by one of its final sentences: "Cultural policies should promote affirmative action to ensure equality of access and equal opportunities for everyone to participate in cultural life" (CIRCLE 1993: 7).

Soon after the WCCD meeting in Stockholm, the Vienna Declaration on Human Rights elevated cultural rights in terms of the right to development, and strengthened the human rights of women and children. According to the legal scholar Gudmundur Alfredsson, however, the Declaration would not succeed in maintaining or improving existing international standards of minority rights and the rights of indigenous peoples. He reported that, despite repeated efforts, the UN had never been able to reach formal agreement on definitions of the terms 'minorities' and "indigenous peoples". Nevertheless, he quoted a figure of 5,000–7,000 minority groups, with maybe a billion members in total; of these, about 200–300 million could be classed as indigenous peoples (Alfredsson 1993).

In Alfredsson's view, "culture cannot be an isolated phenomenon for individual consumption; it is inevitably tied in with the collective ability to make local decisions for the groups, to have serious influence on the relevant policy- and decision-making in the country concerned and to be heard at the international level" (ibid.: 8). Consequently, he advocated special measures in law to uphold group identities, cultures, languages and traditions by means of political and economic rights relating to self-government, land and resources; and that minority and indigenous peoples be incorporated into national and international plans and projects for development, the environment and economic and social cooperation.

With Kofi Annan, former Secretary-General of the United Nations and Federico Mayor, Director-General of UNESCO, at UNESCO, Paris, 1997. *Source* Personal photo collection of the author

3.6 San José, Costa Rica, February 1994

3.6.1 Culture, Production and Equity

The World Commission's Latin American consultation took place in San José, Costa Rica, on the site of an old munitions factory that had been transformed into a gleaming new cultural centre. Nikita Mikhalkov wryly commented how refreshing it was to see such a transformation at a time when, in Russia, cultural centres were being turned into arms factories.

On the first morning, Javier Pérez de Cuéllar and Jérôme Bindé were rushed off to the national theatre for the inauguration of our meeting but no transportation was provided for the other members of the Commission, so we sat fuming in the hotel for a long while. We walked into the national theatre just as Mr. Pérez de Cuéllar was finishing his speech, with one of the members of our Commission loudly apologizing but saying that we had never intended this to happen. We were then placed on the stage of the theatre, with the eminent intellectuals invited to the meeting sitting as an audience. Everyone was greatly confused as to who was who. During the intermission, one of the intellectuals, a friend of mine, politely asked whether we, the members of the Commission, could be introduced so that they would know who we were.

The debate was opened by Fernando Calderón, at that time head of the Latin American Social Science Council, who explained that Celso Furtado had begun work on sociological considerations of economic development at the Economic Commission of Latin America and the Caribbean (ECLAC) in the 1970s but that there had been no continuity in developing this line of thought. Now a new programme was emphasizing "authentic competitivity", with great importance placed on knowledge and education.

Gert Rosental, director of ECLAC, explained that civil society in the region was now demanding jobs, equity and democracy simultaneously. It was therefore essential to think in holistic terms, in terms of resonance. The core idea of the ECLAC Programme "Productive Transformation with Equity" was that in a globalized, competitive economy it could not be accepted that Latin America and the Caribbean would participate only through comparative advantages; rather there needed to be a process that included human capacity-building and creativity, which could only be fostered through equity.

Calderón added that many different groups were involved in development in the region so the question he asked was how to make all of them come together for modernization. These groups were defined by identities linked to memory and history and the desire to continue to be different. Yet a culture still existed of negating the Other, as well as of internalizing being Indian. "These are intercultural tissues", he stated, "that are the foundations of societies." He went on to say that many authors were now arguing that the conflicts that predominated in the sphere of production had been transferred to the sphere of culture. ECLAC was putting forward a proposal to overcome the dialectics of exclusion by linking individual identity, belonging to a community and productive modernization.

Lourdes Yero from the Universidad Central de Venezuela put in a comment that attempts were being made to leave behind ahistorical views and to see culture as a historical process: first, as implantation and forging of nation states and, second, as a "civilizational process", as Darcy Ribeiro, the Brazilian anthropologist, had called it. Only in this way would it be possible to analyse multiplicity in historical development.

At that point, I asked a question of all the scholars present: if Latin America is known for its discourse of fraternity, equality and solidarity, as Jean Franco had argued, why is it the "continent of inequality", as Pedro Vuskovicz rightly called it? The answer given was that this was precisely why ECLAC was emphasizing equity in the development of the region. I also asked whether cultural fragmentation could be held together by a strong state. Uneasy answers pointed to the need to think about cultural citizenship and cultural policies to find common ground for all peoples in a nation.

3.6.2 *Subjectivity and Integration*

The next speaker, Felipe MacGregor of the Universidad del Pacifico in Peru, emphasized the subjective dimension of culture which highlights admiration as a way of saving ourselves from the mundane. Human beings are more than mere actors on the stage of the world. There is transcendence, and it is human aspirations that drive culture. Culture means to admire, to recognize beauty, to adore God.

From Caribbean shores, Rex Nettleford, former Minister of Culture for Jamaica, was concerned, instead, that people were becoming alienated as tastes were engendered by devices that were not within everyone's reach. He quoted Roberto Retamar, who wrote, "we [in Latin America] are echoes of others. If we only look at ourselves we will realize we are like everybody else so we can't be universal if we cannot be particular". He explained that his area is called the "Crossroads of the Americas" because everyone there is the product of myriad encounters. In the Caribbean, people are all hyphenated beings, Anglophone-Afro-Caribbeans. The Caribbean is a laboratory of creole languages, and of differing political systems and approaches to public decision-making. Such rainbow sovereignties, the crossroads mentality of the region's peoples, may well lead to overlapping alliances, bringing into the twenty-first century new forms of functional cooperation with appropriate institutional frameworks which are now evident in other parts of the world. Life at the crossroads teaches pragmatism.

Carlos Lago, of Chile, spoke soberly: "We are realizing we are at the end of a dream: that economic growth would lead to integrated societies, that there could be modernization of one sector with another being excluded." Previously, it had been thought that the solution was education, but we have found that it is not enough because the quality of education is very different even though everyone may get an education. This is why the most important challenge for the Commission, he said, is that earlier paradigms have been exhausted.

How do we take up this challenge? he asked. If we want to do away with dualism we have to assign resources in a discriminating way with greater investment in poor people. We need decentralization, otherwise wealth will become concentrated in the areas with greater economic efficiency. Democracy is necessary to establish agreement on the rules. The market will have the most important role but it will not redistribute wealth, so it is necessary to invest in people through affirmative investment. There are limits to concerted action: for example, we now see growth without jobs or without equality. He ended by saying that the Commission should help build the paradigm for the twenty-first century.

3.6.3 *Confrontations, Reconciliations and Utopias*

Francisco Baraona, a Costa Rican scholar, brought in a new theme when he said that culture must be linked to the definition of positive and negative peace:

"Sometimes culture develops through confrontation." To create a new paradigm "we must realize that culture has been created through violence, through conflict". In fifty years, Costa Rica had built a democratic system but without having eradicated conflicts. Although there was poverty, there were formulas to escape a situation of violence. "In Costa Rica we don't cut off heads, we file away the floor" ("En Costa Rica no se cortan cabezas, se serruchan pisos"). It is important, therefore, to have techniques for conflict resolution, to bring an end to violence.

Following this, Niki Goulandris went on to evoke ancient Greece, where culture was called education—and this did not exclude foreigners. The first notions of democracy, solidarity, technology and the cosmos were established. There was the myth of Prometheus, and myths are sometimes needed. We must go back to relating politics to ethics.

José Joaquín Brunner, later to become Minister of Culture for Chile, began his remarks by saying that, in discussing culture and development, there is the risk of idealizing culture and seeing development as problematic, as having conflicts. It is then thought that culture will bring reconciliation and utopias. In fact, he said, it is the other way around. In today's world, it is culture that creates conflict, because it has to do with the meaning of modernity and because it is being cut through by the market. The relationship of culture and the market must be seen as more complex than when it is presented only as a dimension of development. Development is a dimension of culture. Culture is strongly resistant to a vision of it that is only administrative and managerial. There is no creator, but gods for whom we are willing to die. Hence we cannot arrive at a nineteenth-century-type generalization, of one culture that would emerge to reconcile and integrate all cultures in the twenty-first century.

3.6.4 Building a Democratic Multiculturality

Néstor García Canclini, an Argentinian anthropologist, began by saying that Latin America was emerging from a spiritual as well as a bureaucratic vision of culture. Previously, the emphasis had been almost exclusively on to the arts in cultural policies, which must now encompass multiculturality and cultural industries. Two different concepts of multiculturality must be distinguished, he said: on the one hand, multi-ethnicity; on the other, the segmentation of cultures in industrial societies. Multi-ethnicity can no longer be seen as an obstacle to modernity since modern technologies are no longer considered the only possible ones. It is just as legitimate to have other cultures. Some Indian groups no longer resist integration, yet can they retain their identities in the present context? García Canclini's answer was Yes. A few Indian groups have already inserted themselves into exports and commercialization. The Zapatistas speak of developing their communities. However, the discussion of multi-ethnicity does not end with discussing the future of multiculturality: we then have to discuss globalization. In globalization there is a tension between communitarian relationships and cosmopolitan relationships. In cities culture is polyglot, a multicultural superposition.

He then argued that free trade in globalization has had three different kinds of effects. First, it has created a historical heritage and popular culture circuit based on the traditional knowledge of traditional groups. The confrontation here is between the logic of self-sufficiency and that of maximizing profits in the market. The second effect is the mass-media circuit, which is leading towards greater dependence on US companies. And the third is the creation of selective networks through cable TV, satellite dishes and parabolic antennas, which will affect scientific and technological development. The challenge, then, is to build a democratic multiculturality that does not fall into either the neo-liberal confusion of the market and civil society or the unrealistic fundamentalisms that exalt tradition, such as Sendero Luminoso (the violent guerrilla group in Peru).

Two interjections came from invited observers to the meeting. Alan Tonelson of the Institute of Strategic Studies in Washington DC argued that the market gives greater freedom and that everything nations have done until today was to protect the individual from the market. A second interruption came from a young scholar from Surinam, who asked testily "Why aren't we cited as an example of peaceful cultural conviviality?" He pointed out that in his country twenty-five per cent of the population are descendants of slaves having an African culture of the eighteenth century. There is a large population of Hindus, who have their own schools, movie theatres and marriage laws. There are people from Java and Indonesia and many European (especially Dutch) immigrants. For this reason there is no national culture, but he argued that the population lives together peacefully. It seemed to me then that he had made an important point, which is the inherent bias in policy discussions that always focus on conflicts and rarely examine the many examples of places where things are going well.

This prompted Roberto DaMatta, a Brazilian anthropologist, to ask what the sources of national identity are. Rituals, he answered. Can other alternatives sources for that identity be found? He explained that the components of classical bourgeoisie have already been lost. In Brazil this is certainly true but instead there is popular music, the *Carnaval*, that gives a sense of place and identity. He went on to argue that culture cannot be measured because it has to do with meaning. The problem is that we still tend to view multiculturality from the standpoint of the classical bourgeoisie. He declared: "Because I can no longer stand to think of Brazil in terms of absences (*ausencias*), because we don't have such a bourgeoisie saying we have such a culture, I believe we have to rethink ourselves not as having absences but permanences (*permanencias*)." He concluded by saying that there are three or four contemporary cultural matrices, and we cannot remain in one unchanging form: we are all in transit.

As a conclusion, Rex Nettleford interjected to say "So our language has to change."

3.6.5 A Culture of Violence

In the discussion, increasing violence in the region was highlighted. Fernando Calderon spoke of the "culture of political violence" in the twentieth century that had coexisted both with the idea of progress and with that of a sovereignty which had always been incomplete. He made a point of noting that violence had also increased within the family. Now there was a convergence of human rights movements, 'popular' solidarity and mobilizations (for example, in Brazil against hunger), religious groups and movements for political ethics, democracy and transparent elections. A culture of peace should be built through concerted agreements, however, not deals between elites.

Elizabeth Jelin, who was later invited to become a member of the World Commission, gave a very clear warning that at international meetings there is a positive value given to self-determination but that this could come into conflict with the values of universal human rights when they clash with the values of local cultures, as in racism. She added that culture and development should also look at the problem of migration, and especially the growth of cities where new cultures were being created.

At this point Leila Takla took the floor to argue that there are three levels to improving the quality of life through development: first, the cultural constraints to development; second, the use of culture for development; third, the development of culture. After the military wars for land and the financial wars for capital, she predicted that we would now witness cultural wars. In the former Yugoslavia there was a cultural war. Russia was being invaded by American culture. So the new imperialism was based on culture.

3.6.6 Cultural Policies

Several issues regarding cultural policies were identified by José Joaquín Brunner. The first was how to exercise public control of cultural industries, given that governments are prevented from regulating cultural industries since this can no longer be done through laws and rulings. There is an internationalization of capital, and a convergence between television, telecommunications and informatics. All this is important because of the impact it is having on identities, schools and communities. He then argued that sociology had traditionally discussed capitalist development as possible in relation to certain forms of cultural ethos, mainly those of Max Weber and Protestant asceticism. In Latin America there had been a culture of the Church and the cultures of Indians, but both were now superseded by a Carnavalesque television culture. This created a contradiction, with ethical bases apparently needed for development.

García Canclini reacted to Brunner's comment by saying that the disjunction between participation, development and Carnavalesque culture is not general but by sectors. On the one hand, there is television for the majority of the population, with a hegemony of entertainment and shows; it includes very little drawn from traditional and local cultures. On the other hand, there is the demand for a culture of excellence in universities and companies. These are polarizing trends and lead us to think of the need for not one but several cultural policies. In Europe, he continued, there was a movement towards what had been a discredited idea, that of establishing quotas for television and transmissions. Spain had just decreed that thirty per cent of films screened in localities of more than 25,000 inhabitants should be European. Could UNESCO promote quotas? But, he said, it is not a question of protecting an anachronistic nationalism but of keeping a balance. It seems that we are still stuck with a Gutenberg concept of culture.[5]

Several other participants then joined the discussion. Some of the comments made referred to the fact that private industry was increasingly involved in cultural programmes: the Central Bank of Colombia, which houses an impressive collection of archaeological gold artefacts, was cited as a case in point. The Minister of Culture of Argentina had referred to the importance of funding: it all pivots on whether society is willing to pay for culture, and the truth is that it is not. Finally, attention was drawn to President Lacalle of Uruguay's proposal for a common market of knowledge for Latin America.

The Latin American and Caribbean consultation confirmed that a new generation of scholars were rethinking the terms of culture and development in the region. The difference from the positions taken by intellectuals in the 1970s, described in the previous chapter, certainly provides a contrast. The nineteenth-century preoccupation with "forging nations" and integrating Indians into national societies had been left behind, as had romanticism towards traditions and idealization of culture as a universal problem-solver. Instead, the multiplicity of cultures held the central space in the discussion. Multiculturality related both to indigenous peoples and to new cultures being created through modernization and cultural industries. This multiculturality cannot be understood as a polarization between ethnic groups in an era of optical networks, global finance and telecommunications.

There was an impatient dismissal of previous schemes that saw Latin America and the Caribbean as always an 'echo' or an 'absence' of European cultures. Multiculturality meant adding, not subtracting, so that people in the region are all "hyphenated beings". So the question was asked: how do we build a multiculturality which does not fall into either the romanticism of going back to traditions or the neo-liberalism of abandoning everything to the market?

Confrontations are an integral part of transformations, so culture may be seen as a technique for conflict resolution. However, cultural policies should not foster a new protectionism that may hinder new creativity in the cultural industries. The idea that was constantly reiterated was that there is a great variety of elite and subordinate cultures, of paths of modernization and of models of multiculturality.

---

[5]He was referring here to Marshall McLuhan's influential 1962 text, *The Gutenberg Galaxy: The Making of Typographic Man*, which analysed the effects of mass media on European culture and human consciousness.

3.7 Marshall Sahlins: The Cultural History of 'Culture'

In 1994, the distinguished American anthropologist Marshall Sahlins was invited to participate in the work of the WCCD. He drew up an "Outline of a proposed report, as suggested at the World Commission on Culture and Development meeting of 2 July 1994", which included a section titled "Why 'Culture and Development'?" (Sahlins 1994a), in which he referred to "the failures of 'economic development' in the late twentieth century … the dismal condition of the South, structural unemployment in the North, violence linked to exclusion, etc." (ibid.: 2). He described these "world-historical changes that have brought us to the present impasses in development and the present crises in culture" as 'Culturalism', the new cultural self-consciousness. There are positive values and negative expressions of such cultural identities, and an urgent need to cope with the latter.

In the conclusion to the following section, on world-historical changes, Sahlins pressed the need for a "global culture of cultures" and for the establishment of such a culture's values. This was followed by a section on "The Relationships between Culture and Development", which considered "models and results of development projects and policies that have integrated culture—or failed to do so—in concrete historical cases" (ibid.: 3). Its first subsection noted: "International C [culture] and D [development] projects: it would be good to have an autocritique of UNESCO" (ibid.); subsequent points mention "A critique or appreciation of specific international agencies/institutions", including the IMF, the World Bank and foreign aid from developed countries to less-developed countries, from a cultural perspective, as well as projects of national states with regard to enclaved ethnic communities and projects of cultural ministries in developing countries (ibid.). In the following subsection Sahlins raised the issue of interethnic violence in the case of the former Yugoslavia, India and Rwanda, and of conditions leading to religious fundamentalism and New Age cults (ibid.).

Section 3.4 of Sahlins's proposal covered "special problems of global extent in C [culture] and D [development]", and addressed issues of "gender and development", "education, youth and development", "cultural policies", "development, culture and the eco-system" and "the wretched of the Earth", this last referring to extreme poverty and the agonies of the fourth world (ibid.: 5).

In the final section, Sahlins considered "cultural enhancement through and as development", noting with regard to "Determination and use of appropriate social as collective-entrepreneurial units" that "We should proclaim the end of association of development and individualism. There are collective origins of credit, capital accumulation, production and marketing that can be used in many societies" (ibid.: 5); and that "Different kinds of culture order have different kinds of orientations to markets and production" (ibid.).

He then moved on to "political empowerment and protection". On the question of "development of local cultural and political autonomy" he argued the need for "Indigenization of control of the institutions of mediation between cultural communities and national states or international forces. Self-regulation in all areas of life, to the greatest extent possible, would be desirable" (ibid.: 6). This would be enhanced by "Supply of the means of communication—computers, faxes, video equipment, broadcasting equipment, telephonic improvement—as well as the education that would make [development of local cultural and political autonomy] and other development strategies possible" (ibid.: 6). In addition, he reiterated that "Local cultural communities within national states need access to a still higher political body—with higher ideals—than their national government. They need appeal to international bodies (The Hague, United Nations, etc.). How can this be effected?" (ibid.: 6).

In conclusion, Sahlins proposed "The Cultural Contract of WCCD":

> We need a world charter on C [culture] and D [development]: the establishment of "A World Culture of Cultures". This would be our call for an agreement between "peoples and cultures" on the principles of universal mutual respect and coexistence, for the promotion of universal equity and cultural diversity and for the constitution of specific forms of development consistent with those principles.

He specified that the names of the agreements sought would be "A charter of cultural ethics", "An agreement of development responsibilities", "An education project in multiculturalism for all countries" and "An agreement on positive cultural action (on the model of 'Affirmative Action' in the United States)".

As could be surmised, Professor Sahlin's proposal to the Commission raised many eyebrows and engaged many complex debates to his proposal in terms of international political transactions on culture. As a result of intellectual counter-positions, Sahlins' resigned from the Commission. In his letter of 14 July 1994, in answer to Mr. Pérez de Cuéllar's response to his proposal, Professor Sahlins stated that

> As you also say, the Commission has adopted the preliminary outline of the World Report, presented by you in Costa Rica and amended in the recent Paris meeting. For myself, the problem is clearly that I have entered too late into the work of the Commission, without any chance of participating in the development of the document now before it .... Consequently, I must, with regret, resign from the Commission as of today (Sahlins 1994b).

I remember that I deeply regretted Professor Sahlins's resignation since, as an anthropologist, I could very well see the importance of his possible contribution to the work of the WCCD. At the same time, I had gradually become cognizant of the intricate mechanisms of world political influence and, as the Assistant Director-General of UNESCO, I could now understand the opposition which some of his specific proposals triggered in the Commission and in UNESCO. I telephoned him trying to change his decision but he answered that the window of opportunity which had been opened in terms of his participation in the Commission had now been closed.

3.8 Pressures, Choices and Trade-Offs

As the work of the Commission progressed, disjunctions occurred as time pressures grew, different perspectives were locked into debate, and UNESCO member states took particular interest in the WCCD report. The Report was to be presented at the General Conference of UNESCO in 1995. Overall, the daunting complexity of the task and the choices and trade-offs that had to be made became more difficult to manage by the day. It was only Javier Pérez de Cuéllar's supremely refined diplomatic skills that kept the boat on an even keel in a stormy sea of ideas.

By November 1994, a new Executive Secretary, Raj Isar, had joined the Secretariat. His diplomatic skills and thorough knowledge of international cultural programmes was an asset in accelerating work towards producing the Commission's report. Meanwhile, the next consultation took us across the world, to a welcoming, culturally enthusiastic country—the Philippines.

Bearers of culture: Elderly women who have danced as *Chinelos* for thirty-five years in Yautepec, Morelos, Mexico, 2005. *Source* Photo by Lourdes Arizpe

References

Alfredsson, G. (1993). On the principles of nation-building for minorities and indigenous peoples. Document submitted at the Seminar in Kautokeino, Norway, 1993.

CIRCLE (Cultural Information and Research Liaison in Europe) (1993). Final draft conclusions and recommendations of the European Round Table on Human Rights and Cultural Policies in a Changing Europe, Helsinki 30 April–2 May, 1993. Document presented at the World Commission on Culture and Development consultation, Stockholm, 28 June 1993.

Commissariat Général du Plan (1993). L'evolution des industries culturelles en Europe (1980–1991). Paper presented at the World Commission on Culture and Development consultation, Stockholm, 28 June 1993.

Dogan, M. (1993). Comparing the decline of nationalisms in Western Europe: the generational dynamic. Paper presented at the World Commission on Culture and Development consultation, Stockholm, 28 June 1993.

Kleberg, C.-J. (1993). What cultural policies and for which development? Paper presented at the World Commission on Culture and Development consultation, Stockholm, 28 June 1993.

Kleveland, A. (1993). Speech at the Public Hearings of the World Commission on Culture and Development, UNESCO, Paris, 20 March 1993.

Mayor, F. (1993). Latin America's contributions to the new international community. Address to the United Nations Educational Scientific and Cultural Organization, 26 April 1993. Report DG/93/28. http://unesdoc.unesco.org/images/0009/000959/095900Eb.pdf, accessed 3 October 2016.

Mitchell, R. (1993). European cultural policies in transition: trends and problems in financing the arts and culture in the Europe of the 1990s. Paper presented at the World Commission on Culture and Development consultation, Stockholm, 29 June 1993.

Rouet, F. (1993). Les aides publiques aux industries culturelles en Europe. Paper presented at the World Commission on Culture and Development consultation, Stockholm, 28 June 1993.

Sahlins, M. (1994a). Letter of Resignation to the World Commission on Culture and Development. Copy of letter sent to Mrs Lourdes Arizpe by Jérôme Bindé, Executive Secretary of the World Commission on Culture and Development.

Sahlins, M. (1994b). Outline of proposed report, as suggested at the World Commission on Culture and Development meeting of 2 July 1994. Paper sent to the Secretariat of the World Commission on Culture and Development.

Skok, V. (1993). Cultural participation in Europe: comparative trends, strategies and challenges. Paper presented at the World Commission on Culture and Development consultation, Stockholm, 28 June 1993.

UNESCO staff consultations for the World Commission on Culture and Development, (1995). *Source* From the personal photo collection of the author

The Dance of the Tlacololeros, a fieldwork project on intangible cultural heritage, at Zumpango, State of Guerrero. *Source* Photo was taken by Lourdes Arizpe

Chapter 4
Recognizing Cultures

4.1 Manila, November 1994

The Asia-Pacific Consultation for the World Commission was held in Manila, the capital of The Philippines, in November 1994. The city itself is a tribute to blended cultures. Colonial churches and ramparts—and names—are Spanish; many of the people use Tagalog or other native languages in their day-to-day speech; and public affairs are conducted in English. Yet Filipinos are distinct as a cultural group, in their soft approach, in the ever-present smile, in their sincerity. They brought a choir to the consultation, which sang its way through practically all the cultures of the world.

Our meetings took place in one of the most important hotels in Manila, and were witness to an incident that was very telling of cultural perceptions. While the temperature outside was sunny and very pleasant, hovering around 30 °C, inside the meeting room in the hotel, where the temperature was kept at around 12 °C, we were freezing. After many unsuccessful attempts to have the temperature raised, with profuse explanations from the management as to why it could not be done—because it would have raised the temperature in all other meeting rooms—most of us opted to bring our European winter coats to the meeting. Colds for many of us were one of the results of this particular meeting, which was extremely rich in all other aspects.

4.1.1 Is There an Asian Civilization?

The culture of ancient civilizations continues to influence Philippine development. Professor Yao Jiehou of the Chinese Academy of Social Sciences began his presentation by arguing that culture is an entire system of spiritual and social values. Social transformation must focus on an increase in human intelligence, with

L. Arizpe Schlosser, *Culture, International Transactions and the Anthropocene*,
The Anthropocene: Politik—Economics—Society—Science 17,
https://doi.org/10.1007/978-3-642-41602-6_4

utilitarian and functional aims, but must guard against producing one-dimensional people. A market system always implies a cultural system: China maintained a market economy with social justice. He explained that the reforms in China sought to build a modernized, democratic socialist system. This needed a powerful national ethos, patriarchism, a vision of the destiny of the country and solidarity, combined with modernizing values. In other words, it needed a new cultural system since the country has fifty-six nationalities. Confucianism had had a deep influence—in terms of filial piety, fraternity, purity and honesty—but it had to be evaluated. In China, it had mixed with indigenous cultures, with Taoism promoting diligence in work, but it had also mixed with Buddhism, which in China was a new adaptation of Buddhist culture brought from India.

Kapila Vatsayanan, a leading scholar of Indian dance, art and architecture, went further, asking whether there was one Asian civilization or whether it was made up of a group of civilizations linked to African and Meso-American cultures. The only contrast among such civilizations was with post-Renaissance Europe and this was the difference of thinking—in terms of Man and Nature and not Man *in* Nature. This made for a very different understanding of what the self is, of what a human being is in the matrix of the natural world. She argued that pluralism is in the genes, which makes it necessary to build an eco-cultural model. Since the Second World War there had been a number of -isms used to define ourselves and this had taken place on a global, not regional, scale. Our problem today was the interpenetration of philosophy, political ideology and economics.

Are Asian values uniquely Asian? This was the first question posed by the Hawaiian sociologist Stephen Yeh. He wished to begin by saying that what is cultural in cultural policy contexts will turn out to be political or economic. Culture is used as a residual category, so this is what we have to theorize about. Asian values involve respect for authority, consensus decision-making, valuing community over individuals, emphasis on education, frugality and a strong work ethic with teamwork and competitiveness. The Asian intention was to get the best of the West and keep the best of the East. In considering whether there was a coming cultural crisis, he answered Yes, and that it was a crisis linked to poverty, population and the environment. But the second possible crisis was of "the poverty of culture": the poor needed to have access to things which would give them more culture. In well-off countries capitalist consumption produced egotistical individualism but there needed to be a shift from *having* more to *being* more.

Sang Bok Han, an anthropologist from Korea, explained that Korea did not have the West's capital accumulation or school of rationality but it did have its own development ethics: human resources and collaboration of government with business. He insisted that development was achieved through authoritarianism, collectivism and connectivism, as well as through close personal ties and supremacy in world achievements. Syncretism encouraged open-minded attitudes towards adopting new technologies.

Meeting Yasser Arafat, who gave his full support to cultural projects for the Middle East at UNESCO, Paris, 1995. *Source* The photo was taken by Sayeh Msadek and taken from the personal photo collection of the author

4.1.2 Globalization or Indigenization

Stephen Hill, an Australian specialist on technology, argued that culture is at the forefront of equality and success in competition in the market economy. It gives new advantages to players in a global world. The disempowering force of global culture is what remains behind in the shadows and does not allow people to reproduce technologies or to continue to produce. The problem of local cultures integrating into global cultures is that the knowledge is not integrated into the products they buy, making it very difficult to maintain or to reproduce machines or goods: for example, American or Japanese cars bought on Pacific islands where no repairs can be made.

In the entire chain from local village to global product, the critical link that destroys is that local innovations are of such low quality that they do not work. The acceleration of technology in fact acts to the advantage of small businesses. So corporations are breaking themselves up into smaller units, a Japanese corporation dividing into what they call 'amoebas'. In this way they are able to link up to local cultures and non-codified knowledge: what count are the webs of people and how they relate to each other. This means that a specific goal is how to develop partnerships for knowledge between local and global users and producers: that is, to build knowledge-rich local cultures by bridging the two levels of culture.

In the ensuing discussion, Commissioner Keith Griffin said that he would go to the extreme of saying that nothing is indigenous and all cultures are in flux, so that we are dealing with a spectrum, not a polarity. The sense of the national, in particular, becomes eroded as people become identified in other ways. The economist Bernardo Villegas, of the Philippines, warned that people living in dire poverty cannot participate in cultural activities, so we should avoid romanticizing local cultures. Stanley Tambiah, a Sri Lankan anthropologist, argued that the problem must be understood as a "politics of difference" that erects fences to win entitlements and benefits of various sorts. Its resolution must come through a "politics of accommodation". In response to this, Amara Ponsapich, an anthropologist from Thailand, asked whether accommodation implies co-optation.

4.1.3 Civil Society, Not Nationalism

Several Working Groups presented reports at the Asia-Pacific meeting. The main points presented in the report of the Working Group on Culture and Development were that emphasis should be placed on civil society rather than nationalism, and that comparative studies in the field of social science were needed to discover the relationships between values such as the work ethic and development. Development needed to be regarded as that of the whole individual rather than as that of the acquisitive individual. The elements for successful development have been drawn from rural to urban and tribal to agrarian movements, from the growth of cities, from the cementing roles of family values, and from the fact that the "body system of cultures" has been able to accept some parts of alien cultures and has rejected others. However, a 'phenomenal' rise of violence in the Asian region was noted: religious, ethnic, against women, intergenerational and sexual. Violence has become instrumental in solving conflicts, and so has become an integral part of development. Among its recommendations the group proposed including local knowledge (unstructured, informal and belonging to the communities) in school curricula, understanding culture as respect for the environment, and promoting bigger financial allocations to the conservation of cultural heritage in the region.

The Working Group on Cultural Pluralism made recommendations pertaining to the advocacy of the rights of all peoples to express and share an individual and communal sense of place and of identity. Equality of treatment and of opportunities could be achieved only through dynamic, participatory and democratic processes. National policy frameworks were needed for the continuation of the diverse linguistic patrimonies and other languages, as well as for the continuation and participation of indigenous peoples in intercultural exchanges within national and international contexts. This working group also advocated that governments should dismantle and prohibit monopolies in media to avoid cultural imperialism. The mass media should disseminate information and programming about multicultural groups, through special broadcast and narrowcast services for minority groups as part of the national cultural agenda of each country.

Culture and the arts reflect the differences between the haves and the have nots, according to the opening statement of the report of the Working Group on Cultural Expressions and Art. More intercultural dialogues were needed in the form of festivals, cultural exchanges and the preservation of past heritage through its being creatively used for the present. Educational policies for the arts should be applied in schools. Mass media should be harnessed as a vehicle for familiarizing people with indigenous cultures.

The Manila consultation provided some clear-cut messages: that development should not only mean having more but being more; the importance of intelligence and of spirituality; the threat of mass media conveying cultural imperialism; and the need to be cautious of the use of a "politics of difference" to win entitlements.

4.2 The Arab States, January 1995: How Did the West Become the 'World'?

The Regional Consultation for the Arab States was held in Muscat, Oman, on 10–11 January 1995, with a welcoming address from HH Sayyid Faisal Bin Ali Al-Sa'id, Minister of National Heritage and Culture, and presentations by Mr. Khalid Bin Talib Al-Zakawani and Dr. Al-Yaqdhan Bin Talib Al-Hinai of Oman.

Dr. Bin Talib Al-Hinai began his contribution by suggesting that development and culture were not mutually exclusive, and that the development process need not be at the expense of cultural values and norms: "Culture ought not to be perceived as a hindrance for development as many academics and practitioners, especially in the West, would like to believe" (Bin Talib Al-Hinai 1995: 1). Economic progress, he continued, helps raise the material standard of living, and this provides nutrition for the body, while culture in all its forms raises the moral standard of living and provides nutrition for the soul.

He cited two development strategies that had recently shown the importance of taking into consideration the cultural dimension. With respect to the Asian development experience, "it is agreed among observers that the culture in that part of the world has contributed greatly to their successes" (ibid.). In contrast, the failure of centrally planned economies could partly be attributed to their attempt to enforce a rigid economic system that did not recognize cultural diversity and that envisaged individuals as mere factors of production. "In Oman, we are proud to have recognized this crucial issue since the beginning of the Renaissance era, as the interaction between culture and development has been considered as an integral part of our development strategy" (ibid.: 3).

The information age, he continued, "is transforming the way we do things, invading our national borders, our communities and even our privacy in our own homes, sometime uninvited … we are bombarded daily by numerous cultural value systems and different forms and views of art" (ibid.). The big challenge, then, was simply how to preserve cultural identity and at the same time join the rest of the world in enjoying the fruits of technological advances. He pointed out that "Naturally the easy way out is isolation. But this is the wrong choice as it means death. Our challenge is to preserve what is good and beneficial from our past and tap into what progress has to offer and select what is appropriate for us" (ibid.).

Dr. Anouar Abdel-Malek from Egypt drew on a larger, historical discourse by opening his address with the question "How did the West, and its civilization, become 'the world'?" (Abdel-Malek 1995). He explained that the age of maritime discoveries leading to the age of revolutions—scientific, industrial, bourgeois, democratic was also the age when historical surplus value converged in the hands of the rising bourgeoisies of Europe, the end result of the domination of the Three Continents: the looting of Central and South America; the human haemorrhage inflicted by slavery upon Africa; the marginalization of the central Islamic area from the Crusades onwards; the later domination of Asia by Europe. From the Renaissance to hegemonic imperialism, such was the path that led the West—mainly Europe, but also the US since the mid-twentieth century—to its position of centrality and hegemony in modern times. Thus the need arose to seek new conceptions of the 'Other', to develop a new vision of the two sides of the river: a 'universalism' that posited the unity of humankind and the ideals of liberty, equality and fraternity.

At the time of 'universalism', he continued, dependent nations and societies stressed their 'alterity', but universalism had no place for otherness, in the concrete world of Realpolitik. Global reach—combining advanced technologies with nuclear weaponry and the hegemonic mass media—raised the challenge to a much higher level. This time, the very existence of distinct identities was at stake, resulting in the move from the mainly political vision of identity to the cultural (rather the national-cultural) positioning of the problem around the key concept of 'specificity'.

Dr. Abdel-Malek analysed world transformation in three domains, acting at different tempos and different levels of acuity and efficacy, with momentous consequences. He argued that: "While military might can maintain hegemony, and the control of major sources of energy and raw materials, it can neither develop new areas into valuable allies, nor break the resolve of peoples, let alone nations. … The center of gravity of power is shifting: from the Atlantic Euro-North American world, to East Asia" and to the Pacific (ibid.: 8). He also mentioned the new role of the state as "the regulator-protector of all advanced market economies with the shield of cultural identity during the interlude of the unipolar hegemony" (ibid.: 9). He explained "the resurgence of transcendence" as the rise of world religions, from liberation theology to integrism, around the national-populist fundamentalist mainstream, parallel to the revitalizing of liberalism, pragmatism and positivism, coupled with the negative mindset and nihilism in the West.

Dr. Abdel-Malek identified "the actors of human becomingness" in the next few decades as nations, major regional ensembles, and "civilizations (or civilizational circles/molds) defined as of a distinct vision of time, of the human condition in the world" (ibid.: 11). He went on to explain that from the Bandung Conference in 1955 (when representatives of twenty-nine African and Asian countries met to promote economic and cultural cooperation) throughout the 1960s and especially during the 1970s, alternative schools of thought and action in non-Western centres proceeded to develop "a panoply of civilizational projects". At the centre of this development was the Orient, resurgent: "Confucianism, oft[en] linked, and/or opposed to 'political Islam', opens the path … to the new civilizational project now in its formative stage" (ibid.: 14). The basic movement was, however, towards convergence with the new Europe in a new Eurasian heartland, with Asian–African Islam as its main link. Such was the 'ethos' of ensuring leading forces of development in a new world at peace.

A number of working groups again presented reports at the Regional Consultation in Oman. The report of the Working Group on Value Systems in the Arab World and Development was presented by one of the members of the Commission, Dr. Yoro Fall. One of the first points the group made was that the existence of crises in the Arab world was deeply linked to the imposition of a Western model of development, although a distinction was made between the Western *system* and Western *nations* (Working Group on Value Systems 1995). The Western system was the more visible to Arab peoples, and had created forces of rejection—the group preferred the term 'rejection' rather than 'fundamentalism'—as well as of acceptance and aspiration towards the Western world. Nonetheless, the Arab system of values was distinct in its history, its philosophy and its ethical foundations—in the last because religion could not be separated from social life, and Arab values were Muslim values, although there were nuances. The group was conscious that only fifteen per cent of Muslims were Arabs and that there were non-Muslim groups living in Arab societies—mainly Christian and Jewish communities. This was reflected in the value of tolerance in the Arab system.

The group explained that the difficulty for Muslim Arabs in accepting democracy and human rights arose from the core value possessed by this demographic group that there is diversity in the Muslim world, leading to different environmental, geographical and political forms of life in different Arab countries. The presence of immigrant workers in Oman, for example, was being dealt with through gradual development of a parliamentary system, and access to education and culture.

Correspondingly, the Working Group on Cultural Expression and the Arts: Energizing Forces for Development in the Arab States contested the Western-centric discourse which identified Arab culture as one culture (Working Group on Cultural Expression 1995). Mention was made of the variety of norms, lifestyles, roles of women according to countries, religious minorities and other social aspects. Participants also expressed concern about the identification of Arab culture, especially in the West, with aberrant or reformable practices (genital mutilation, forced labour of young children, and so on).

Regarding the tension between globalization and traditional cultures in Arab countries, Arab culture was generally perceived as 'self-sufficient'—a fact that was seen as an asset in the resistance to cultural 'invasion'. There was some debate, however, regarding its 'openness' and its capacity to face a world of growing interdependence (both global and regional). In discussing the idea of change, one participant emphasized the poor record of the region on human rights and the need to create opportunities for change, especially in the area of cultural norms and education. There was also some disagreement about the place of religion. While one participant said that Arab culture was in great part Arab religion, a concern was expressed that this should not reflect on the adaptability of the region: "no region has not been influenced by religion". Moreover, it was agreed that cultural norms are just as vitally connected to family and tribal traditions as to religious ones.

Desirable goals were identified by the group as freedom of expression, individual creativity, recognition of individual contribution to the arts and to development, the encouragement of individual initiatives, the giving of status to artists, and the need for dialogue such as the Omani tradition of the 'Al Barza', an open public forum designed for consulting the people, hearing grievances and mediating differences. All such tensions rendered imperative the definition of models of *shared* development.

4.3 New York, February 1995

The World Commission's North American consultation, held in New York at the United Nations headquarters, where a stream of UN officials heartily greeted Mr. Pérez de Cuéllar at every turn of a corridor, was duly opened by Robert Fowler, the Canadian Ambassador to the UN. He informed the Commission that in 1960 Canada had adopted legislation forbidding cultural, sexual or racial discrimination. In 1988, the Multiculturalism Law was passed, giving explicit recognition to multiculturality in Canada. At present, he said, Canadian policy for artists promoted professional development, provided access to resources and cultural centres, and gave active support to dialogues on cultural diversity.

4.3.1 A Vacuum of Cultural Policy

John Fiske, Professor of Communications Arts at the University of Wisconsin, began his remarks by stating that in the United States there was a vacuum in cultural policy, "so we have a cultural battleground" in which there were social and political motives that opposed culture's becoming a mere market commodity.[1] Diaspora, migration and exodus were normal in globalized societies, he explained, so it was rare now to live in only one country. Everyone, even those who did not migrate or live with migrants, was affected by a culturally unstable society.

Fiske stated that the United States would soon have no dominant cultural majority. The proportion of European Americans was falling compared to

[1]As elsewhere, all unreferenced quotations are drawn from the notes I took.

Hispanics and blacks. In 1994, seventy per cent of Hispanics spoke Spanish at home in their households. Cities were becoming increasingly fragmented, with people living with their own cultural kind, as in east Los Angeles, which was wholly Mexican. He predicted that cities would divide into multicultural centres, so that cultures would be local and global, not national. Hi-tech industries in California would be joined by local cottage industries: radio for Iranians and Mexicans; black intellectuals producing audio and video for a narrow black audience.

Yet he also said that cultural hybridity was another characteristic of the United States: Mexicans mixing Spanish culture and the American way of life. It was important to note that the difficulty of providing identity increases the further one is from the centre of one's own culture: "A separatism is needed to provide an identity. This must change the historical habit of the dominant group controlling the representations of the subordinates."

At the same time, Fiske cautioned, a central set of values could no longer be effective and must give way to multiple points of consent which had to be negotiated and achieved locally. The negotiations to achieve identities could be carried out productively by those having a secure position; otherwise they would lead to contention. He argued that others do not have to assimilate to white America, but rather that white Americans must also assimilate to other cultures.

He estimated that radio for local cultures required millions for investment in infrastructure, but argued that this needed to be done to achieve a peaceful America. With the closure of black liberation radio he declared: "I see a trend to repress publicly legitimated culture. The silencing of the dissenting, the weak, is a totalitarian tendency: this dystopian future must be avoided."

4.3.2 "No Modern Government Can Shape Culture, Because Culture Shapes It"

In the first session, Manor Moore, of York University, Canada, pointed out that it was being said that global trade and finance were increasingly being challenged by local cultures. "I think it is the other way around," he stated. "Democracy is globalizing and challenging global trade." He continued by arguing that we must re-examine the suppositions being made that global culture is homogenizing, that self-assertiveness and defence of cultural rights are the same, and that trade and economics are there to create cultures. One fallacy was that economic growth will do away with cultural disparities. Another was that if you increase the number of consumers of culture, you create more wealth: "In my view, [this will only happen] if you have more producers of culture."

His contention was that "No modern government can shape culture, because culture shapes it." One of today's issues was the question of who owns culture, who holds copyright: the clan, the distributor, the state, the world? Answers needed to be found not in culture, not in economic or political models or in psychology or neurology, but in individual integration, in multiple allegiances. Having access to culture was not the same thing as taking part in the conversation. Moore ended by declaring that "The tensest issue in culture is that in a world where each culture considers that it is as good as the rest, where is the standard of the best of the good?"

On behalf of artists, Carol Phillips, of the Banff Cultural Center, recommended that cultural creativity be highlighted in the Commission's report. She noted that in North America cultures were in flux but that reactionary forces were trying to stop this change. Art was being denied as exploration, experiment and invention. However, there had been a rediscovery of rituals, especially women's rituals, and also of nature.

In the discussion, Mahbub ul Haq, a founder member of the Commission, took issue by arguing that no government could shape culture but that they could influence cultural development: "If you think education is expensive, try ignorance", he said. Fiske replied that identities would no longer be national, and that the nation state would no longer be the frame for identities. Commissioner Griffin jumped in to say that the Commission's report should tackle issues of culture and power head-on: the "Othering of subordinate groups", gender discrimination and racial discrimination. Nathan Gardels, the editor of *New Perspectives Quarterly*, countered that in Los Angeles the dominant culture had disappeared and that communities were now attempting to create an anti-environment: "It is not reactionary; it is positioning themselves in a multicultural reality."

Marc Pachter, Consultant to the Secretary of the Smithsonian Institution, then brought in a contentious issue: "I've also wondered," he said, "if US culture has become the incubus for a global culture even if it is a salad and not a melting-pot." (In the next session, Gardels proposed instead the image of a 'collage' as a way of prefiguring the global culture in Los Angeles.) He continued: "I'm amazed at the pre-scientism of these debates: New York in 1900 is Los Angeles today. Hybridization has occurred before, for example, in American musicals."

Commissioner Fall took this cue to ask whether this could not be seen as a re-aestheticization of musical canons which are mingling? Immigrants take other music back to their countries and then bring new hybrids. Carol Phillips jumped up to say that art was conceived as the beautiful, but that music now is not only about harmony but about silences and intensity and harshness, so we have to leave behind the idea of aesthetics.

One of the observers at the meeting then joined the discussion to say "I'm interested in the cosmic culture which we need to have to negotiate in a global community." Moore closed the discussion by distinguishing between 'multiculture', in which we all live, and 'multiculturalism', which has been co-opted to direct things.

4.3.3 "Multiculturalism Has Become a Legitimation of Meritocracy"

This heading opened the presentation by Phil Lane, a member of Canada's First Nations, and a professor at Leithbridge University in Alberta, who began by speaking of four principles that link culture to development. The first, which applied not only to tribal peoples but to all peoples, was that development comes from within and is the basis for all self-determination of peoples. The second was that, if there is no vision, there is no development: as your vision is, so your power will be. The third—development of the individual—must go hand in hand with development of community and of nations. And the fourth was learning life-enhancing behaviour, responsibility and participation. Speaking slowly and reflectively he went on to say that the tribal peoples of the Americas could fill the gap in the dialogue between cultures. "If we hold on to these new ethics and spiritual development I believe we can save Mother Earth, for our life purpose is spiritual development."

In contrast, Harold Troper, of the Ontario Institute of Studies in Education, also from Canada, had a more sceptical view of multiculturalism in that country, which he said had become pejorative: "First it was a progressive notion, now not." In addressing why that had happened, he explained that multiculturalism is a collection of things: it is the reality of a street full of people with different cultures and backgrounds; it is a vision for participation; it is a government policy to manage cultural stresses; and, of late, it had become a code word for debates. Nevertheless, he asked, "is there a limit to the cultural carrying capacity? If government should stay out of the bedroom should it stay out of ethnic ghettoes?" He declared that "Canada is not a melting pot but a mosaic, while the United States seems a loose federation of shopping malls." In the end, multiculturalism had become a legitimation of meritocracy in an open society.

Harold Cruse, Professor of African American Studies at the University of Michigan, weighed in to say that "Culture is not important because it's all a question of race. Blacks are just left out of the thinking. Whites are calling the ideological and political tune and blacks are still dancing to it—dancing to it with more freedom of expression but they're still dancing to the same tune. Blacks still don't get it, that they have to stop trying to be Americans. Black scholars, particularly black historians, have got to begin challenging the whites' interpretations of blacks and slavery. It's all a question of race."

Thérése Paquet-Sévigny, Communication Professor of the University of Quebec at Montreal, Canada, declared: "I believe less and less in 'national cultures'. It is community cultures that will now go through the communication channels and it will be artists, artisans, databanks and software developers who will create the new culture, leading to a deep change in social relations." She suggested that informatics could bring high culture and popular culture closer and that information was already a value added to many products. However, the cost analysis of gains and losses starts from access: she indicated that this might lead to a greater divide between North and South and to more inequality. "How can we speak of democratization when, out of 3,000 languages spoken, only twenty go through the media?" she asked.

She finished by saying that "The role of states has been largely made redundant by globalization." Cultural protectionism was being used to mask the failures in national development. "Instead, we need new partnerships in cultural development, the flourishing of creations, a culture of humans, expansion of legislation and the subsidizing of works that are to circulate in the new networks," she added. The state must protect citizens against the sequestration of their artwork and rethink its role in forming citizens in this new information culture.

4.3.4 Art, Heritage and Moral Chaos

The second session was opened by Commissioner Kurt Furgler, who stated succinctly: "I have something: heritage. And I have the mission to add something: creation." He argued that this could not be done without a new ethics, to be fulfilled not only in words but in actions, and he ended with a flourish: "We must find a fit between the wonders of technology and the spiritual nature of human beings."

Mark Davidson Schuster, of the Massachusetts Institute of Technology, discussed the role of the arts in urban planning, pointing out that artists may argue that because the arts are fragile they have to be protected in a centre—a theatre—and so these uses get isolated. "However, instead of having art *in* cities, it should be arts *for* cities." He also referred to the importance of the economic impact of the arts.

Nathan Gardels called attention to the fact that "when we speak of cultural heritage in the United States, we speak of fifties films". He went on to say: "Postmodern thinking produced scepticism from mom to imam, so there is no authority: this is the subtext of Hollywood and the media. The message goes toward individual choice and economic freedom, the modernity of Disneyworld." He stressed that there was no distinction between freedom and libertinage; instead there was moral chaos, a soul mixed in irony: this was not nineteenth-century nihilism but a moral relativism that could be switched on and off like TV channels.

François LeBlanc, a Canadian architect, spoke of the generalized concern for the loss of small communities' heritage. To counter this trend, many new ways of conserving it were being invented: eco-museums, regional heritage parks, remodelling of main streets, and taking care of heritage in urban areas.

Comments in the discussion were very varied. In debating chaos theory it was noted that in the butterfly effect there was already a context that allowed for the effect itself to have effects. Biotechnology was mentioned as an emerging field that would alter many aspects of culture and the point was made that ethnic groups were being pulverized, for example, Latinos, Chicanos and Nuyoricans. At this point Harold Cruse reiterated his statement that "whatever the topic, the issue is race".

4.3.5 The Happy Paradox of the New Technologies: A Revolution of Power

Douglas Rushkoff (in his own words an author, social theorist, journalist and software developer) addressed Phil Lane when he said that "tribal Indians come from the Earth and I am in cyberspace but we are saying the same thing". The media have resisted Nature, he went on, "thinking that gravity makes something Ted Turner drops from atop trickle down to CNN on television. Now we have to do it through 'do-it-yourself technology', camcorders, participatory media spaces." He argued that people in participatory webs see the world differently: "Who you are doesn't matter; what you say, does. You don't recognize authority. The real fear we face is fear of empowerment. What everyone wants, that's the way it will be. We've been taught that we need leaders and popes to control the beast inside us, but you don't kill cultures, you kill people."

According to Marc Pachter from the Smithsonian, there was a happy paradox about the new technologies at present. In the history of standardization, there had been a concentration of authority and the growth of passivity in the reception of knowledge. But today, more and more information was being offered to more and more people, affecting their capacity to shape it according to their needs and comprehension. "The mass media," he said, "will be transformed into a medium, not a message"; in Marshall McLuhan's time the receptor was passive, but today that would no longer be the case. Pachter expressed his belief that oral reception of culture would replace texts and images, so that it would not pass through thought but through the senses. Sensory meanings had been devalued by abstract thinking. He ended by saying: "So I maybe see no limit to democratic access to info-tech. But industry will still depend on the narrative structure of knowledge they receive. It will take a tremendous philosophical leap to the new image and audio world. Museums are on the way out."

Commissioner ul Haq tackled the discussion of cyber-technologies head-on. Informatics, he affirmed, was not only a new technological revolution. It raised expectations, because even the poor have access to information on television; it might drive migrations; it might be a destructive force in economies, for example through globalized finance markets; it could lead to enormous disparity of opportunities: for example, the fight over intellectual property. So it was essential to talk of the informatics revolution as both a sociological revolution and a revolution of power.

I asked whether oral or even image communications would be able to sustain the fine lines of intellectual analysis that texts convey. I fully agreed that sensory meaning has been demeaned in certain cultures, but I could not see how it would be enhanced through the new technologies. "Sensory exchanges in human interactions mean tone of voice, touch, the softness of skin," I expanded. On another point, I argued that the tiered system of authority is part of the old Cartesian cosmopolitical scheme. We now have a new geometry—fractals—that may allow us to understand how horizontal groups are embedded in one another, while always repeating the same patterns.

Harold Troper cut through by saying "I have a feeling of déjà vu, of internal disharmony on cultural issues of blacks, Hispanics, et cetera". He declared that Canada had a different manner of dealing with these issues than the US because it had a safety net that the US did not.

The themes of the New York Consultation—multiculturalism, heritage and the new technologies—had drawn heavy fire in what were, at that time, blazing debates. The contrasting ways of dealing with multiculturalism in the US and in Canada were being held up to the world as examples of such policies. In both countries, multiculturalism was being dealt with by expanding democracy to incorporate individuals deserving of entitlement into the mainstream. This is why Phil Lane's comment—that multiculturalism had become a legitimation of meritocracy—was so important. But in the US this was done mainly through initiatives for affirmative action in universities, companies and communities; in Canada, by contrast, it was done mainly through government policies aimed at inducing such change. However, by 2003, such initiatives had been left aside in the US and even opposed by the George W. Bush administration; in any case, multiculturalism had been placed under the shadow of national security. By 2014, both the Prime Minister of Britain, David Cameron, and the German Chancellor, Angela Merkel, were to declare explicitly that "multiculturalism is dead".

Reviewing the WCCD New York consultation after a gap of two decades, it is clear that the optimism of the new technologies, even before the Internet was fully active, was at its peak among the North American participants. As Mahbub ul Haq correctly noted, they implied a revolution both in social interaction and in power. Accordingly, in subsequent international meetings on culture, the new technologies continued to be central to debates on the future of culture.

In the ensuing working meeting of the Commission these themes were discussed fully, with one member declaring that he thought that sociology was more important than technological determinism. Another member discussed the problematic questions of genotype and phenotype in racist theories and practices. A further member clearly stated: "we are not out to promote culture, but certain kinds of culture".

World Commission on Culture and Development meeting in Chiba Prefecture, Japan, 1995. *Source* From the personal photo collection of the author

4.4 Tokyo, 1995: Crossing Ethical and Cultural Thresholds

The Tokyo Working Meeting of the World Commission, to discuss the first complete draft of the report, was held on 8–9 June 1995. A public symposium took place the following day, with an inaugural lecture given by me and a panel discussion with several members of the Commission.

In my lecture I emphasized that the world is crossing a historical threshold in technology, economic interdependence, global change, population growth and communications, yet the corresponding ethical and cultural thresholds are only just beginning to be built through a new dialogue among cultures. I felt it important to reiterate our conviction that as people begin to assert their own cultural awareness in today's regional and global context, it is necessary to encourage intellectual and international cooperation to provide a new sense of direction that will ensure peace, sustainability and well-being for all the inhabitants of the world.

Throughout time and space, cultures have provided human beings with an identity, locality and sense of purpose that, unfolding through time, has shaped their individual, communal and national lives. Today, shifting cultural boundaries foster a feeling of uncertainty. The active intention conveyed by the original meaning of the term 'culture' must be restored. Culture first came from the verb "to cultivate". Now it is ever more necessary to cultivate human understanding and solidarity for peace. To cultivate is to develop, thus culture means developing minds and hearts through art and literature, through the appreciation of the cultural heritage bequeathed to us by past generations, and by living together harmoniously with cultural diversity at the same time.

The ethical and cultural challenges of our rapidly changing world also make it urgent that creativity be fostered in all fields of human endeavour. Philosophers and writers must be able to propose new meanings for the spiritual and intellectual development of humanity. Artists and performers must create new forms that reflect the aesthetic and emotional states (both individual and collective) induced by contemporary transformation. Our strategies need to promote the very status of artists and other creative individuals, as well as establish conditions favourable to the development of local cultural production.

Our futures will be increasingly shaped by the awareness of interdependence among cultures and societies as they face these global challenges. The flow of cultures through time has created boundaries that provide individuals and communities with multiple identities. It has also generated diverse sources of language, symbols and codes of behaviour as linear time crosses over major thresholds. We must encourage intercultural dialogue and actions by governments and by civil society that will lead to a better understanding and management of cultural pluralism.

Special attention must be given to the knowledge that each culture has contributed to the world's intellectual legacy. Such knowledge comprises valuable information on botany, agronomy and other areas of the natural environment, as well as on social areas such as traditional medicine and local histories. Indigenous groups and NGOs working in this field must help by collecting, systematizing and disseminating such knowledge, and combining it with modern technological knowledge.

In our own time, many processes have changed the cultural perceptions of women's and men's life cycles and social participation. Migration, urbanization and the fact that women are having fewer children, living longer and increasingly participating in the labour market are profoundly altering cultural patterns of gender relations. It is important, then, to give attention to inter-gender dialogues across cultures, particularly in the fields of cultural perception of maternity and paternity and of male and female identities. Much has changed since Lady Murasaki, from the Japanese imperial court, sat down with brush and writing case to compose what is still considered the greatest novel in her language, *The Tale of Genji*, six centuries before Mme de Lafayette and eight centuries before Emily Brontë.

I concluded my lecture by mentioning the most important UNESCO cultural programmes in the Asia-Pacific region, highlighting the importance of the revitalization and dissemination of cultural, traditional and popular forms of expression in constituting the backbone of many other activities in the region. Central among these were those which Japan and South Korea had pioneered to promote official recognition of people who possess outstanding skills connected with heritage and craft traditions.

4.5 The First Draft of the Report

In our panel discussion preceding the public symposium in Tokyo, we made a thorough study of the first draft of the WCCD report. We agreed that it should be short and to the point, aimed at policy-makers as well as NGOs, students and international civil servants, and should provide practical advice on policy lines. It should have a strong message, picking a clear position on each of the issues, and give emphasis to its core recommendations, which should be reduced from the fourteen in this draft to five or six.

Javier Pérez de Cuéllar stressed that the report should inspire cultural policies in every country. We then agreed that all key policy issues should be addressed. Missing policy issues, identified by several Commissioners, were: historical depth; power as part of social reality; social structures and institutions; the cultural vacuum in megacities; culture, power and freedom; the public sector; violence; the concentration of power in global media; and the oppression of artists. It was also important to bear in mind political problems that translate into ethnic conflicts, as in the historical dimension of the situation in Rwanda and Burundi, but also reflected in the conflict between the Flemings, the Walloons and the French in Belgium. The point was made that traditional cultures are not pure and harmonious and that the report should present the role of regeneration in cultures, especially where they were threatened. It was also noted that ethical problems should be considered as a question not only of values and morals but also of praxis.

It was suggested that we use a global formula: "culture is both the end and the means". General principles to be mentioned should be equity (both intergenerational and between genders), democracy and tolerance (some of the Commissioners did not like this term but no better equivalent could be found). Pluralism, women's freedom from patriarchy, creativity (as that which will constitute the heritage of the future) and competition in the media were also emphasized. In addition, mention was made of the balance between access to information and freedom of choice, on the one hand, and minimum regulation, on the other.

There was a growing consensus among the Commissioners as to the most important messages of the report, although differences had emerged sharply. These differences mostly related to historical perspective, social relations in terms of conflict, and questions of ethnic power. There was consensus that the report should not be traditionalist or conservationist but that it should give a dynamic view of cultures. As one Commissioner remarked, if every single sentence had to be dealt with through consensus we would have "ni chicha ni limonada" (neither beer nor lemonade) but just a bland, washed-out text. We wanted to ensure that a more analytical text be presented and so a new writer was appointed for the next draft,

Paul Streeten, a distinguished economist with long experience in development research. This draft would be assessed in the final meeting of the WCCD, to be held in Pretoria, South Africa, in three months' time.

4.6 Pretoria, South Africa, September 1995

4.6.1 "A Man Chiselling a Mask in His Backyard"

The WCCD consultation in Pretoria ran from 12 to 15 September 1995. It began with Commissioner Angeline Kamba's narrative of the previous consultations. She said that in Asia and the Pacific we had heard about culture, being and personal fulfilment; in Latin America about the coexistence of modernity and tradition; in the Arab States about the struggle for development but also about respect for spiritual and religious values; and in North America about technology, urban life, migration and exclusion.

Finally, in Africa, we should be concerned about war but we must remember the large steps that had been taken towards democracy: elections in Malawi and Mozambique, and the reconciliation, magnanimity and statesmanship of Nelson Mandela. She commented that we also knew that in Africa diversity could be a blessing but it would continue to be exploited for evil purposes: for example, African arts had been exploited for the benefit of others.

Commissioner Kamba was followed by Yoro Fall, who began by saying that the colonial period should not obscure the history of hominidization, which began in Africa, from where humans spread to the rest of the world. In the African present, he noted, democratization was an ending but also a beginning. Now that the majority of Africans were too young to remember colonial domination, now that Africa was not marginalized in art and music, there was a pluralistic elan in their development. Today, however, the homogenization of social relations was becoming a danger for society: it was therefore necessary to reinvent the ethical and cultural dimension of development, and to renew social cohesion and the networks of solidarity.

Next, John Cozne, Director of the Market Theatre in South Africa, declared that "it is time to close the bruises and wounds of apartheid with the intention of building a nation. It is time to address the wounds and needs of our people. There is not a single theatre or cultural centre in black people's towns … we are busy unlatching the political potential of the cultural plurality that was stunted by the political situation. Now access to and participation in art and culture are vital to the lives of individuals and communities." To look further ahead, cultural policies must be linked to other development policies. He finished by stating firmly that "we have put resistance in neutral but if anything happens with the national unity government, we can put it into first gear. We hope we will never have to do this."

In the ensuing discussion, Ms. Werewere Liking, a woman theatre director from Ivory Coast, noted that apartheid existed throughout the continent, because it was present in the mind: people were prevented from seeing themselves as equals. Following this lead, another South African pointed out the tendency to think that those who exhibit in galleries or are government officials of culture are the practitioners of culture. Apartheid had caused the segmentation of society: there were women's segments, a "lost generation" of children who had to abandon play to be part of the struggle, the problem of the cultural upbringing of white or black children, and the mentality of privileged children that had to be changed so that they did not think of themselves as better than peasant children. However, he believed that every person had the right to and possibility of creating culture: it was just as clearly being shaped by the man chiselling a mask in his backyard.

Another South African said that he wanted "to problematize the 'rainbow nation' as Mr. Mandela has called us: but in a rainbow the colours are separate. A counter-image is the idea of confluence when several tributaries converge in one river." He stressed that we have to think in terms of power and of the ethnicization of culture. In a session the next day, a black South African said that "culture did not divide us before apartheid, although tribalism did. I believe it is a matter of time for there to be a South African culture; we will be a people. I write in English and I am not ashamed. I have a black and white audience. We blacks will be able to shed differences while it will be more difficult for the Afrikaner." An important point in this session was raised by a Rwandan who gave the ominous caveat that there is a new thing in our communities: a culture of impunity.

4.6.2 Kinship Provides Checks and Balances to Ensure Accountability

Olivia Muchena, the Zimbabwean Minister of Agriculture, rocked the meeting by asking outright: what is African nationality? Her answer was that it is a holistic world view integrating the physical and metaphysical. Their common root is the land. However, through education we have adopted a compartmentalized world view. Indigenous views have been judged mystical and irrational, so they have been put aside, while the utilitarian view is considered useful and modern. She proposed the importance of transformational development: that is, a change of structure. She argued that, in this hi-tech era, transformational development is expected, but this takes many years. Countries are expected to set up democracies overnight, are expected to set up structures to which they are not accustomed. Time becomes a central issue, since not enough time is allowed for a learning process to take place.

Minister Muchena suggested that the Commission should consider the following issues: that Africa is not a homogeneous continent, but that there is an inclusive nature to African society through the centrality of the extended family. She stressed that kinship provides checks and balances to ensure accountability, and that a means needed to be found to raise this to levels of political and economic accountability. Finally, she put forward the question of how the essence of African cultures could be used for development. She ended with the words: "I was taught not to beg because I lost my dignity. We now live in a continent that has lost its dignity because it is always begging."

After Muchena's fiery words came the sobriety of Mamadou Dia's contribution. An economist at the World Bank, Mr. Dia put his point bluntly: "In Africa we have trained people. We have equipment so we have the supply side of institutional capacity, but on the demand side we have failed: in maintenance, in the legitimacy of institutions and in accountability." He argued that "Indigenous institutions are limited: they have women and age discrimination." Transplanted institutions were similarly problematic, so even with investment good results were not seen. This made the pendulum swing back to the other side, to the "fundamentalist traditional" institutions, to the romanticization of traditional ways of life. The need now was for the renovation of indigenous institutions and the converging of transplanted institutions with indigenous ones. These would generate transformation costs in order to achieve successful change: costs of leadership, local ownership, accountability and incentives.

Mr. Dia explained that it was essential for formal institutions to adapt and for there to be renovation of indigenous institutions that act as safety nets for people. He noted that twenty-one case studies on "reconciliation institution-building" had shown that the civil service was still an extension of the colonial administration set up for extraction purposes, so that a normal citizen could have no access to it unless they knew someone in power: "It is based on privilege, and they have no incentives to serve civil society." He concluded by saying that service delivery was most effective when it took into account indigenous institutions. This had also worked with HIV when the medical service was modified from clinical treatment to holistic healing. The most successful entrepreneurs in Africa were those who had reconciled corporate and societal culture and had built and maintained social capital by transferring resources back to the community. The necessary preconditions for this to work included building on local communities and institutions and having communications networks which were inherent in their operations.

In the discussion, Commissioner Griffin questioned whether institutions that were rich in people but poor in resources were truly indigenous or a result of the imbalance in power. He argued that East Asian development was based on a massive redistribution of resources and that this model could not be repeated in South Africa.

A white South African museum director said that he had toured African museums and had been astonished at how colonial they were. For example, in Africa intangible culture was much more important than tangible heritage but this was not reflected in the museums. Later, Ms. Shaje Tshiluila of Zaire also mentioned that, in contrast to Latin Americans, Africans attached great importance to the restitution of cultural objects. But what mattered was that Africans should concentrate on safeguarding what they have now. A participant from Burkina Faso brought up a different point: that, with deforestation, animals were disappearing so that the morality tales based on animals also disappeared. A way had to be found to reconstruct them.

The director of several community centres in Johannesburg strongly recommended that the World Commission should take on the media as part of its responsibilities, and that its report should look at traditional cultures, "since sixty-five per cent of Africans are becoming urbanized. In cities there are new ethics, new ways of behaving. It's not a question of the melting pot and cultural particularity; there must be a right balance, between multiculturalism and nation-building." He further noted that cultures are not static: they move and change. In America people are proud to say that they are Irish-American or African American.

4.6.3 Between Universal Knowledge and Lived Experience

Ms. Tshiluila called attention to the fact that cultural policies in Africa had a complex pattern, since some countries followed Roman law while others had private rights based on common law. She suggested that "What we need are regional African culture programmes". For many people, statuettes and masks were still considered talismans. This would not change until Africans acknowledged the connection between universal knowledge and their own lived experience. Africans still only saw their art as folklore and used it to sell or to support a life abroad.

A film director from Burkina Faso, Idrissa Ouedraogo, was concerned that most images now came from satellite communications, and that all young Africans were watching the *Santa Barbara* soap opera. Young Africans needed references to their daily lives but they tended to identify with sportsmen and extremists. Nevertheless, audiovisual means were a way of reaching even those who were excluded. So, instead of buying Rolls-Royces, she argued, we should buy satellites. However, she also argued that Africans had something else: solidarity and closeness. "In Europe they don't talk to me of sentiments or relations between people as we do here."

Samba Sarr of Senegal also deplored the African mimetism of the West that expanded people's perceived needs. He saw it as essential that Africans must develop through successive stages. Colonial intervention must be stopped, and this could be achieved through the help of cultural industries, through books, and through audiovisual media.

Damien Pwono of the Rockefeller Foundation concurred that culture was hampered by several factors in Africa. At the macro-level, this was the result of the legacy of the colonial economy, the lack of knowledge about local cultures, and the brain drain. At national level, culture had not been given priority in government policies. Additionally, there was a lack of funds and there was harassment of artists and creators. The low quality of research into culture did not help, nor did the prostitution of African artists who had become apologists for powerful figures. His recommendations were to harness the technological revolution and to professionalize the culture sector. Priority should be given to the environment, health and cultural activities.

In the discussion, Commissioner Kamba made a striking point. She said that the Afrikaners (that is, South African whites of European descent) were wanting to present themselves as a minority in UN terms, in order to gain protected status. She criticized the Organisation of African Unity (OAU) for having a non-interventionist policy, saying that this was not African, since the traditional stance was to intervene in family and community matters. A South African from the African National Congress countered that Mr. Mandela wanted to accommodate an extreme right-wing group that presented itself as a minority. The problem was that the moment that boundaries were established it would all go back to apartheid. Another participant defended the OAU by saying that it stood by its policy of non-intervention in the affairs of other countries, including humanitarian aid. This caused several Africans to disagree.

Samba Sarr then made a strong statement: "One cannot say that culture is dynamic and, therefore, one cannot say whether something is autochthonous or not. But one needs points of reference; there are signatures that are at the base of this transformation. We need databanks based on cultural areas: for example, Bantu culture. All the signatures must be internally referential, since the exogenous could destabilize the local community."

The Pretoria consultation was marked by the euphoria over the end of apartheid and the new Mandela government. However, there was concern about new destabilizing factors in Africa related to culture. These included tribalism, which could be countered by not thinking in terms of the melting pot or of cultural particularities but of a right balance between multiculturalism and nation-building. There was a loss of accountability tied to kinship, community responsibilities and even tales of animals that were disappearing as forests receded. The new cultures in transition in cities called for new ethics and new ways of behaving. Finally, the global media and audiovisual culture were locating young African minds in California and not in Ouagadougou or Pretoria. Participants agreed that cultural policies were urgently needed but they were well aware of the difficulties in terms of lack of finance, of professionalization and of political will to take up the challenges of conserving African heritage and providing support for artists. These were all important issues that we carried in our minds as the Commission considered the final draft of the report and prepared for its presentation to the UNESCO General Conference later that year.

References

Abdel-Malek, A. (1995). Culture and development in the making of the new world. Paper presented at the Regional Consultation for the Arab States, Muscat, Oman, 10–11 January 1995.

Bin Talib Al-Hinai, A. Y. (1995). Culture and development. Paper presented at the Regional Consultation for the Arab States, Muscat, Oman, 10–11 January 1995.

Working Group on Cultural Expression and the Arts (1995). Cultural expression and the arts: energizing forces for development in the Arab States. Report presented at the Regional Consultation for the Arab States, Muscat, Oman, 11 January 1995.

Working Group on Value Systems (1995). Systèmes de valeurs dans le monde arabe et développement. Report presented at the Regional Consultation for the Arab States, Muscat, Oman, 11 January 1995.

Chapter 5
Negotiating Cultures

After some thirty-one months of work, in November 1995 the WCCD presented its report, *Our Creative Diversity*, to both the UNESCO General Conference and the United Nations General Assembly. The central message emerging from it was that development embraces not only access to goods and services but also the opportunity to choose a full, satisfying, valuable and valued way of living together. Looking at development as a process that enhances the effective freedom of people everywhere to create cultural expressions and to exchange them broadens the widely accepted notion of human development. Culture's role is not to *serve* the ends but to *constitute* the social basis of the ends themselves. In other words, culture is not a means to material progress: it is the end and aim of 'development', when that is seen as the flourishing of human existence in all its forms and as a whole.

The Report included an "International Agenda", setting out ten action points for the international community based on the findings of the Commission. Several of these recommendations had immediate results. UNESCO began to publish the World Culture Reports. An Intergovernmental Conference on Cultural Policies for Development was planned to be held in Stockholm in 1998, emphasizing cultural policies, creativity, respect for cultural diversity and the use of new communications and information technologies for cultural programmes. And the UN Volunteer Services that helped with the conservation of cultural heritage sites received an enthusiastic response all over the world.

Our Creative Diversity was solemnly presented by Javier Pérez de Cuéllar and myself to the plenary session of 186 member state delegations at the UNESCO General Conference on 5 November 1995, after an exhausting final year of precisely negotiating the text with major international actors. Following its presentation, meetings to launch it were held in Amsterdam, the Netherlands, and in Ottawa, Canada. The Report was also discussed at UNESCO's Executive Board meetings

L. Arizpe Schlosser, *Culture, International Transactions and the Anthropocene*,
The Anthropocene: Politik—Economics—Society—Science 17,
https://doi.org/10.1007/978-3-642-41602-6_5

during 1996 and in the Culture Sector of UNESCO with all the officers for programmes related to the themes taken up in it. Other seminars were held in several regions, of which those convened in Santiago in Chile, Lillehammer in Norway and Birmingham in the UK are summarized in this chapter. As an example of one of the meetings of UNESCO National Commissions held to explore ideas related to culture, the sessions of the French National Commission are also briefly summarized.

Presenting the WCCD's Report *Our Creative Diversity* to the General Conference of UNESCO, Paris, 1995. *Source* The photo was taken by Viviane and is from the author's personal photo collection

5.1 The Official Launch Meetings

After the official launch of *Our Creative Diversity* took place in Amsterdam on 8 November 1995, it was presented in more than twenty countries, with reactions that varied from enormous enthusiasm—as that of the Vice-President of Bolivia, who said it was the Bible—to mild rejection in countries that objected to the imposition of what they considered to be Western values on the rest of the world.

A second launch was held in Ottawa, Canada, on 29 February 1996. One of the first comments made at this meeting was that a new legitimacy had recently been bestowed on culture as the third pillar of foreign policy. The Report was praised because it had given attention to sustainability, to women, to indigenous peoples and to cultural diversity. Robin Higham, a senior Canadian diplomat, said that he was very glad to see that the United Nations was a catalyst to help show how universal values link us together in our similarities and differences. However, aspects that had not been dealt with were cultural factors in aid and development projects, spirituality, the individual and culture, and the growth of civil society. The suggestion was made that a new perspective should be given in terms of a "cultural ergonomics".

Susan Scatti, a communications specialist, talked of the changing role of the state, where decision-making was being given back to the municipal level, especially for culture: films, broadcasting, museums. "We cannot abandon creators to the realities of markets", she said, arguing that we need development policies for cultural industries in all countries, not only developing countries. She also spoke of identity and the nation state. Canada had evolved a 'post-nationalism' to deal with plurality and the country's success was the balance between nationalism and ethnic allegiance. The challenges that she identified were: how to use the new technologies for diversity, how to foster diversity in the media, how to interest young people in cultural heritage and how to link cultural rights and gender equality.

The report of one discussion group then went on to ask: "How can we affirm universal rights now that we face a planetary challenge through global change? We are, after all, one species. How do we make these rights an effective force?" They were interested in debating the role of the United Nations, and of Canada itself. They questioned how many rights should be included in a global ethics (one of the central proposals of the Report), and argued that they should be kept to a minimum. The fundamental rights—to work, for a fair share, and for women—must be fought out at a political level. They felt that a global ethics would emerge gradually, as civil society grew, changing the rules of the game and bringing more accountability from governments and more gender-aware planning.

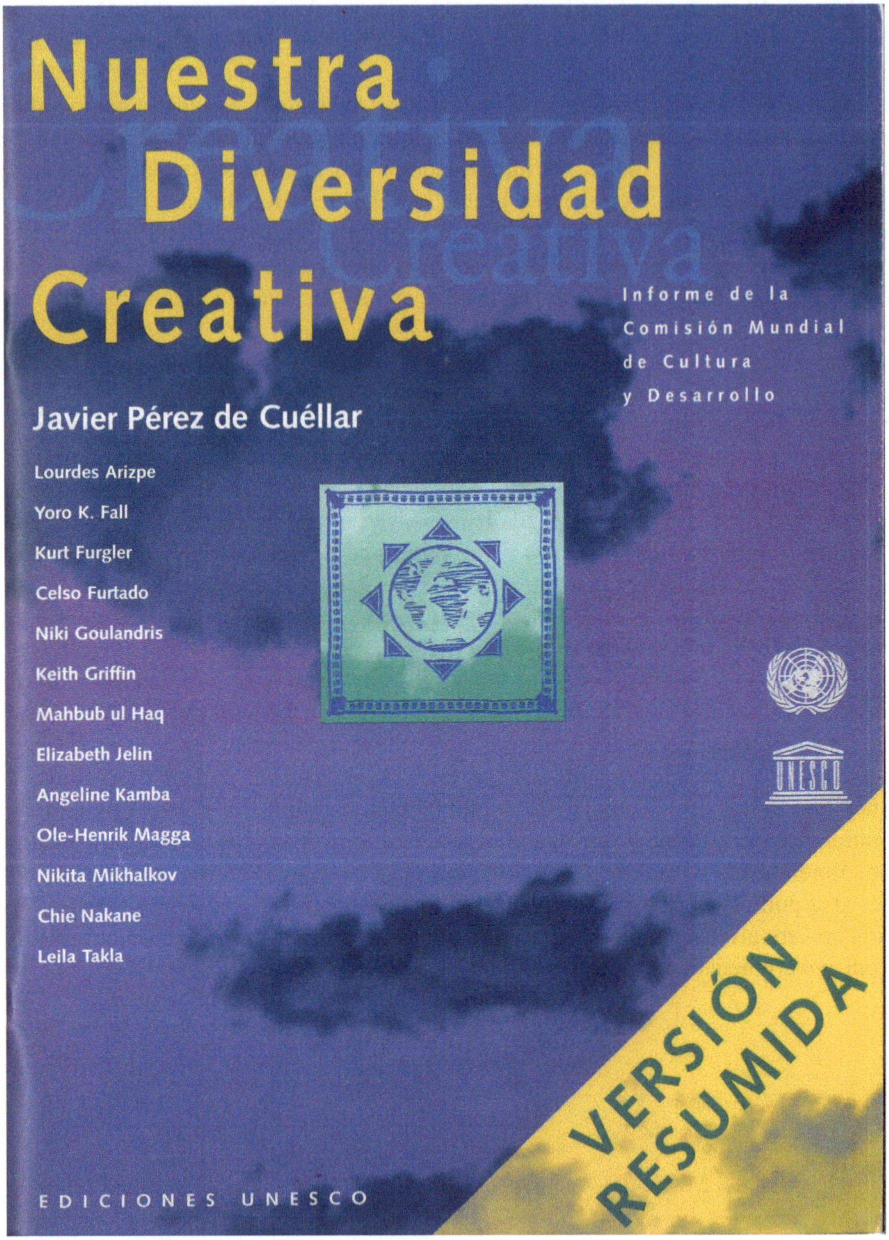

Nuestra Diversidad Creativa, Spanish translation of the Report of he World Commission on Culture and Development, UNESCO, Paris, 1995

5.2 Our Creative Diversity

What of the content of the Report that generated such animated discussion at the Canadian launch? Over the course of two and a half years, the World Commission had accumulated an abundance of information, viewpoints and analyses. It was explicitly mentioned in the working meetings that the Report could not be a treatise, nor a work of original research, nor a handbook on cultural affairs in the world. So the analysis was focused on a selected number of topics where the interactions between culture and development appeared especially strong and pertinent. In doing so the Commissioners aimed to address a diversified audience across the world, ranging from government officials and politicians to scholars, and artists, community activists, and field workers. The Report was to inform the world's societies and to guide its policy-makers. We wanted it to capture the attention of intellectual and artistic communities, as well as the general public.

Seminar to discuss *Our Creative Diversity* at the University of Birmingham, UK, 1996. *Source* From the author's personal photo collection

At the time of the presentation of the Report, I found it unbelievable that so much material could have been processed and discussed, and a consensus text arrived at, in so short a time. The delegations expressed their satisfaction at the presentation of the Report and requested that it be discussed at the next UNESCO Executive Board meeting, in the spring of 1996. As it happened, the Report's political and cultural import and the boldness of its recommendations for the international world led to its being discussed in three Executive Board meetings—an extremely rare occurrence. As Assistant Director-General for Culture at UNESCO, I had to respond to delegates on these occasions.

The documents drawn up by the Secretariat of the Commission, which are summarized in the following pages, stated that

> the argument advanced in this Report is that development embraces not only access to goods and services, but also the opportunity to choose a full, satisfying, valuable and valued way of living together, the flourishing of human existence in all its forms and as a whole. Even the goods and services stressed by the narrower, conventional views of development are valued because of what they contribute to our freedom to live the way we value. Culture, therefore, however important it may be as an instrument of development (or an obstacle to development), cannot ultimately be reduced to a subsidiary position as a mere promoter of (or an impediment to) economic growth. Culture's role is not exhausted as a servant of ends—though in a narrower sense of the concept this is one of its roles—but is the social basis of the ends themselves (WCCD 1996: 15).

The ten chapters of *Our Creative Diversity* convey this message from different perspectives, starting from a normative *global ethics*, and continuing with reflections on *pluralism*, *creativity*, *media*, *women*, *youth and children*, *cultural heritage*, *environment*, *cultural policies* and *research*. In its "International Agenda" the Commission had preferred to recommend a selective and illustrative short list of ten actions to "help energize and motivate people throughout the world" (ibid.: 18), at a time when so many international initiatives being launched on many different important issues made a set of limited options more realistic. These ten actions were:

1. An annual report on world culture and development
2. Preparation of new culturally sensitive development strategies
3. International mobilization of cultural heritage volunteers
4. An international plan for gender equality
5. Enhancing access, diversity and competition of the international media system
6. Media rights and self-regulation
7. The protection of cultural rights as human rights
8. Global ethics in global governance
9. A people-centred United Nations
10. Working towards a global summit on culture and development.

5.2.1 Global Ethics and Pluralism

As our futures will be increasingly shaped by the awareness of interdependence among cultures and societies, it is now essential to promote cultural conviviality through new sociopolitical agreements which should be negotiated in the framework of a *global ethics*. Cooperation between peoples with different interests can only flourish when they are all committed to certain shared principles. Rejecting the idea that ethics is somehow a Western idea, the Report maintained that some moral themes appear in nearly all cultural traditions: for example, the ethical impulse to alleviate suffering and the teaching that one should treat the other as one would want to be treated oneself. Together they form an indispensable point of reference and a pillar of support on which to build a global code of conduct. What we are looking at here is the development of a normative code that would effectively govern the peaceful understanding of people among themselves, in their relationship to states, and between states.

This type of code would contain the minimum ethical principles that any political community should observe, principles which rest on five pillars: human rights and responsibilities, democracy and the elements of a civil society, the protection of minorities, the commitment to peaceful conflict resolution and fair negotiation, and equity within and between generations and sustainability. If fundamental human rights are at the top of the list, it is because the Commission considered that the inherent value of human dignity is a wealth belonging to all peoples and societies, and must be respected in all places and circumstances.

It was in this spirit that the Commission urged the implementation of global ethics in global governance in the following three areas (International Agenda action 8):

- Greater democracy in global governance needed to be considered—including the decision-making process of G7. This was an inevitable development since democracy does not stop at national borders. At the same time, demands for democracy in international institutions were not very convincing if the country concerned did not practise democracy at home.
- Respect for human rights should not be expected only from poor nations. Rich nations must set an example, particularly by respecting the rights of their own minorities and immigrant populations.
- The rich nations must be prepared to open their economies and to undertake structural adjustments just as they were eager to press poor countries to do so.

The *principle of pluralism* is regarded in the Report as fundamental. Cultures are neither isolated nor static, but interact and evolve. Cultural diversity is inherent in any human society and this implies that both cooperation and confrontation within and across cultural boundaries have always been present in human history. In our contemporary world, however, shifts in cultural boundaries—at the local, national, regional and international level, as well as in communication—make it essential that new legislation, new policies and new actions dealing with cultural pluralism be established. The main message is that cultural pluralism is an enduring characteristic of contemporary societies, that ethnic identification is a normal and healthy response and that ethnicity triggers violent conflicts only when it is mobilized and manipulated to do so. Among the many policy approaches to diversity, the most durable way to accommodate differences is to create a sense of the nation as civic community, rooted in the values that can be shared by all ethnic components of the national society.

At the other end of the spectrum, we were aware that massive breaches of human rights have often been motivated by cultural considerations, when culturally persecuted individuals and groups have not found adequate recourse in the existing framework of human rights protection. With that in mind, the Commission strongly recommended that cultural rights need to be protected as human rights (International Agenda action 7) through the creation of an inventory of cultural rights that are not protected by existing international instruments. This would enable the world community to enumerate and clarify existing standards of international law concerning the protection of cultural rights. An International Code of Conduct on Culture should subsequently be drafted so as to provide the basis for consideration and action in cases of blatant violations of cultural rights. Such a code, or its provisions, could be made part of the "Draft Code of Crimes against the Peace and Security of Mankind", which was then under consideration. Ultimately, the possibility of the establishment of an independent International Office of an Ombudsperson for Cultural Rights needed to be considered, for the peaceful settlement of disputes related to cultural rights.

5.2.2 Creativity and the Media

Pluralism is only an empty word if those concerned cannot take democratic initiatives and manifest their creative imagination in tangible ways. Creativity should be recognized as a lever of development in relation to technology, new political forms, empowerment and artistic production. The development of new communication technologies and industries for cultural production has opened unprecedented avenues for the creation and diffusion of new forms of expression. If handled for the benefit of all of humanity, such technological advances offer unlimited possibilities for access to information and for cultural exchange that may foster understanding and democracy. The contribution of civic culture seen as a form of social capital based on the accumulated forms of cooperation and solidarity of local communities must be harnessed so as to help democratize societies in ways that are meaningful for local peoples. This means nurturing social creativity by allowing the possibility of producing new models, concepts, institutions and procedures—such as delegation, empowerment or citizenship rights—which are already helping people to adapt to rapid changes and to transform their realities. An example of new forms of political representation that the Commission proposed to reintegrate to the UN was its main attribute of being people-centred. Seventy years ago, the United Nations was created in the name of "We, the People". Today, however, it is primarily the member states' governments that run the world body. The Commission felt that the time had come for representatives of civil society—NGOs, private foundations, indigenous peoples and cultural minorities, to name but a few—to have a stronger voice in the UN when dealing with such issues as the global Code of Crimes against the Peace and Security of Mankind. The Commission also recommended:

- a review of the criteria for NGOs' accreditation to the UN so as to ensure the widest possible participation of civil society;
- the setting-up of a permanent World Forum composed of all NGOs accredited to the UN, which would offer its views on key global issues. The international community had already accepted the presence of parallel NGO forums at all major UN conferences and summits, and so it was only logical that this *ad hoc* presence became a permanent one;
- the establishment of a UN World People's Assembly. Composed of representatives directly elected by the people of all nations, this body would run in parallel with the existing UN General Assembly. Such a two-chamber system would ensure that the voice of the people was heard at all times.

Plural voices were also to be heard through the new media technologies. The Commission viewed technological developments and the creation of a truly transnational media environment as positive: they enlarged choice, created opportunities for diversity and promoted a freer flow of information globally. But it was also noted that problems had arisen similar to those that nations faced in the early days of broadcasting, namely concentration of ownership, homogeneous programming and difficulties in balancing freedom of expression with certain minimum standards. It was important to anticipate these problems and to act promptly before powerful vested interests were created. The Commission's starting point was that the airwaves, including space, should be seen as a collective asset, a global commons. The issue was then how this global commons should be managed. Should those who got in first and had the most money be able to use this collective asset free of charge and without regulation? Alternatively, had the time come to establish some kind of charge and form of regulation?

Nossa diversidade criadora, Portuguese translation of the Report of he World Commission on Culture and Development, UNESCO, Paris, 1995

The Commission sought a media system that would allow many voices to be heard and many different points of view to be expressed, making a genuine global dialogue possible. Our proposal was that global public resources could be used to fund diversity and pluralism in programming (International Agenda action 5). Alternative services could be created by charging for the use of the global commons, exactly as nationally there are broadcasting or licence fees. Revenues could be collected and allocated by an intergovernmental body. These funds could support the production of independent programmes which could be aired on commercial media. There would be resistance, since nobody likes to be charged for the use of resources that were previously free, nor welcomes greater competition. But the Commission underlined that any effort to generate diversity of content through the definition of a public media space was in the interest of the private sector, which was already investing heavily in international communications infrastructure. Creating demand for new programming and services might even encourage further investment in regions where such infrastructure development was slow.

Although national experience in promoting competition and diversity lies mainly with developed countries, the Commission was very much aware of the question of global equity. What could be done to open up the media and make sure that those media were not restricted to a global elite? Opening up implied deregulation: the state's relinquishing its monopoly and encouraging private investment. In many developing countries, this was likely to imply foreign investment, a price that the Commission felt should be paid to avoid marginalization. This process of opening up also required partnerships between international agencies, governments, the media and civil society.

Standards of decency, respect for others and self-restraint vary from country to country, and from one period to another. While all forms of censorship must be avoided, nowhere is freedom unqualified or allowed to operate regardless of consequences. What is true nationally should also be true internationally. The Commission therefore recommended that a global debate be promoted to create a better understanding and coordination of national efforts (International Agenda action 6). This might eventually lead to some form of self-regulation by media professionals that protected young audiences in particular from images of gratuitous violence, human degradation and sexual exploitation, while respecting freedom of expression. The Commission also recommended the creation of an international forum to reflect on media violence and pornography, whether in TV programmes, videos, or interactive games or services. International cooperation would draw on a range of national initiatives.

5.2.3 Women and Children

In our time, many processes have been changing the cultural perceptions of women's and men's life cycles and social participation. Migration, urbanization and the fact that women are having fewer children, living longer and increasingly participating in the labour market have profoundly altered cultural patterns of gender relations. The rights and needs of women and the interdependence between men and women are therefore important in redesigning their identities and roles in societies. The Commission felt that it was time to broaden the discussion and propose strategies to encompass fundamental changes in the structure of societies all over the world, and to develop agendas which would avoid the dual pitfalls of ethnocentrism and Western bias, on the one hand, and unprincipled forms of cultural relativism which deny women their basic human rights in the name of 'difference', on the other. There are two grounds on which women's rights can be asserted: that of equity and that of efficiency. Equity calls for an end to the status of disadvantaged groups discriminated against in education, income, consumption, status and access to power. Efficiency calls for the proper use of women's productive capacities. Better utilization of resources raises not only female but also male living standards. Redistribution of income, assets and power from men to women requires the consolidation of a political base, empowerment and appeals to fairness in men.

In this regard, *Our Creative Diversity* singled out four areas where policies could be devised and applied. First, it called for redistributive strategies to be devised and equity to be introduced as a key criterion for decisions on macroeconomic policy. Second, social services were to be implemented to meet basic needs and eliminate poverty. Third, modernization policies were to be designed in order to remove barriers to women's access to factors of production, new technologies, training, scientific education and the arts. The fourth area called for government action to introduce policies for the development of citizenship. While the Commission strongly endorsed the Platform of Action for women's empowerment adopted at the 1995 Beijing UN Conference on Women, it believed that political advocacy for women should be carried out constantly, not just periodically when international gatherings take place. With that in view, we recommended in particular the application of an International Plan for Gender Equality (International Agenda action 4), based on:

- the establishment of a permanent high-level UN agency for women's advancement to ensure the swift implementation of gender-equality policies;
- the setting of a concrete timetable for the countries that had not yet signed or ratified without reservations the Convention to Eliminate All Forms of Discrimination Against Women (CEDAW) to do so before the year 2000;
- the preparation by specialized UN agencies of an international plan to strengthen women's strategies for participation in the cultural, economic and political fields.

Turning to children and young people, the Report noted that roughly one-fifth of the world's population was in the 15–24 age group in 1995. No generation had ever been so large or so young. In developed countries, the number of young people was predicted to decline by the year 2000. In the developing world, however, they would soon make up at least eighty-five per cent of the population. At the same time, no generation in the history of humanity had ever been faced with such swift and sustained transformation—everything from work practices and gender roles to communication and geographical limits. This, combined with their lack of power, presented a special need to protect young people against exploitation and neglect. Not only was this their basic human right but it was also an important investment in the future.

5.2.4 Cultural Heritage, the Environment, Policies and Research

Tangible and intangible cultural resources embody the collective memory of communities across the world, for humanity is unique in its ability to gather inspiration and meaning from the past and use it to provide meaning for the present and the future. As wars, environmental disasters and the bulldozers of indifferent development put the natural and cultural heritage of humanity in greater danger than ever before in history, local communities—as well as international institutions —rose to this challenge by joining together in concerted efforts of safeguarding and revitalization. This heritage is made up not only of sites and monuments but also of a multitude of arts and crafts, objects, documents and manuscripts, oral traditions and expressive culture in all its forms, including the performance arts. The Commission felt strongly that safeguarding the creations of our ancestors had to go in hand in hand with fostering the creation of our contemporaries. Current interpretations and usage of cultural heritage still tended to be dominated by a combination of aesthetic and historicist criteria. These needed to be broadened so that each society might assess the nature and precariousness of its heritage resources and determine the contemporary use that it would make of them.

Observing the discrepancy between the ends and means of heritage conservation throughout the world, we recommended that international efforts be made to mobilize the goodwill of volunteers of all ages to work as "cultural heritage volunteers" under professional guidance and alongside professional staff (International Agenda action 3). It was argued that the task of organizing this new effort should be entrusted to the UN Volunteers and should start, if possible, during 1996. The effort should be directed to give priority to the preservation of the most threatened forms of human heritage: deteriorating monuments, artefacts, books, disappearing languages and traditional know-how in a variety of disciplines, to name just a few.

Cultures cannot survive if the environment on which they depend is laid to waste or impoverished. Humanity's relation to the natural environment had so far been seen predominantly in biophysical terms, but there was a growing recognition that societies themselves have created elaborate procedures to protect and manage their resources. These procedures are rooted in cultural values that need to be taken into account if sustainable and equitable human development is to become a reality. The Commission addressed the fact that soon, for the first time in history, the majority of the world's population would live in towns and cities. It was the beginning of a truly new era with new challenges to face.

Within the broader context of culture as the enhancement of people's capabilities, the Commission saw the need to reformulate cultural policies and consolidate their position as an integral part of development policy, planning and capacity-building. The ethical and cultural challenges faced also made it urgent that creativity be fostered in all fields of human endeavour. Philosophers and writers must be able to propose new meanings for the spiritual and intellectual development of humanity. Artists and performers must create new forms that reflect the aesthetic and emotional states, both individual and collective, induced by contemporary transformation. It was essential that strategies promoted the very status of artists and other creative individuals, as well as establishing conditions favourable to the development of local cultural production.

On research, we advocated interdisciplinary work that integrated variables from different disciplines. There was a need to generate and monitor new information about the links between culture and development and to be critical of misleading doctrines. As I had noted in my Tokyo lecture, special attention had to be given to the knowledge that each culture had contributed to the world's intellectual legacy, from botany, agronomy and other areas of the natural environment to traditional medicine and local histories. Support and financial was needed for collecting, systematizing and disseminating such knowledge, as well as in combining it with modern technologies.

The World Commission had incorporated as many analyses as possible into *Our Creative Diversity*. But the most important thing about the presentations of the Report was that they were able to galvanize thinking through "living knowledge" in the same way as, at precisely this same time, we began work at UNESCO to incorporate contemporary cultural practices in cultural heritage, and started referring to "living heritage". Thus, the transactions on culture that were under way created a template against which governments, civil society organizations, intellectuals and artists could position themselves. In the following sections the most interesting and relevant remarks that arose in the follow-up meetings are summarized, focusing on those that aimed to continue an open intellectual and political debate on culture and development.

5.3 The Appraisal of *Our Creative Diversity* at the UNESCO Executive Board

5.3.1 "Many Lanterns into the Future"

At the first meeting of the UNESCO Executive Board in the spring of 1996, a first round of member state reactions to *Our Creative Diversity* set the tone for subsequent discussions of its implications. As normally occurred with all items on the agenda, a close tally was first taken of the topics mentioned by member states. Both mentions or omission of topics could signal important political positionings of a government; the ensemble of those positions would then indicate alliances and therefore the strength of support or opposition to a given idea or project.

Most of the delegates to the Executive Board were high-level ambassadors or diplomats: UNESCO's representatives had been mainly intellectuals, academics, scientists or artists over the previous four decades, but since the 1990s, following a resolution by Japan, the majority of delegates had been direct representatives of governments. Some of them were distinguished intellectuals in their own right; others, however, were minor functionaries of embassies in Paris with little knowledge or understanding of the matters being debated. As a result, discussions in the governing bodies of UNESCO increasingly tended to give priority to the political implications of programmes and projects in terms of international relations, rather than to their substantive intellectual or scientific aspects. The weight in decision-making thus shifted the balance between government representation and civil society representation. UNESCO had been the only major agency of the UN system that had incorporated civil society representation through the UNESCO National Commissions and the UNESCO Clubs—precisely as the WCCD advocated in our Report. But, just when this recommendation was being made, an internal shift within UNESCO was heading in the opposite direction, towards greater decision-making power for government delegations.

The general reaction to *Our Creative Diversity* at this first meeting was strikingly positive. Many delegates said that the Report took us from having an instrumental view of culture to giving it a humanistic, integrative role. The moral and political foundations for looking at culture in the long term had now been set by the World Commission, while the ten recommendations for action could be used to energize governments and people. One of the delegates stated that "the Report has lit many lanterns into the future".[1]

Unanimous support was given to the Report's formulations on global ethics and on women's equality, with one comment made that it was vital that women demand participation based on their capacities and not only on their reproductive role, and another comment that it was critical to relate culture and gender since culture had not been included in the Beijing Plan of Action. The recommendation that UNESCO publish a World Culture Report was welcomed by all. A few delegates wondered whether the necessary statistics could be obtained on a worldwide basis, but it was pointed out that such a process would be important precisely because it would show the discrepancies in statistics among member states, which would therefore encourage the gathering of additional data. Sweden offered to host the Intergovernmental Conference on Cultural Policies proposed in the International Agenda.

Other sections of the Report drew mixed reactions. The proposals on cultural rights and democracy had "full approval" from some delegations but it was noted that "trying to define cultural rights brings problems of limitations of culture rights". One delegate made the point that not enough was said about culture and

[1]Quotations that are not otherwise identified are taken from my notes.

religion. An ambassador stated that the most valuable idea in the Report was that tolerance is a passive attitude but should become an active attitude. Another ambassador briefly noted that "biodiversity as cultural diversity must lead not to conflict but to harmonious work".

The most contentious issue, as we had witnessed since the drafting of the Report, arose from the sections on the media. One delegate stated "we share the concerns expressed [about the media] but this should not undermine the principle of the free flow of information. Measures to convince will be more effective than those that prohibit." The Association for Press Freedom, exceptionally granted the floor at this Executive Board, expressed their "deep concern about the sections on media".

What was missing in the Report? Art was mentioned several times as not receiving sufficient attention. Some delegates raised the need for a deeper analysis of 'underdevelopment'. And the relationship between cultural diversity and biodiversity was also declared to be an area that needed to be investigated more fully.

Delegates commended the Report because it could help countries modernize their cultural policies. They said that it was now necessary to derive local community initiatives, thus creating an Agenda 21 for culture.[2]

5.3.2 Will Culture Become Another Conditionality?

The Executive Board of UNESCO had exceptionally decided that the WCCD Report should be discussed on three different occasions. This gave clear proof that *Our Creative Diversity* had fired up a long overdue debate of central, and sensitive, concerns among member states. Now that the impact of the proposals was starting to sink in, however, misgivings about different proposals in the Report began to emerge.

The Report's recommendation for a "culturally sensitive" development was very well received yet it failed to illustrate the socio-economic conditions under which people lived and how the fanatical application of the market economy was making people's—especially women's—lives more difficult. At the second meeting, in contrast to the first, one delegate declared that society was losing control of media technologies and of their content, which consisted only of violence, global control and profit-seeking. Instead, media should be used by communities. Concern was expressed about the growing "monopolies of electronic images". Another delegate criticized the fact that many of these activities actually fostered the creation of more bureaucracies. Several ambassadors proposed that, before the Report could be published, it should be considered in more depth and that an outline of its contents should be drawn up.

[2]Agenda 21 was the result of the UN Conference on Environment and Development held in Rio de Janeiro in June 1992. It laid out an action plan for sustainable development.

One ambassador declared that cultural rights were outside the competence of UNESCO. Indeed, said another, culture may become another form of conditionality towards the "Third World". Furthermore, the recommendations related to women's rights were termed "overly ambitious projects". In my role as Assistant Director-General for Culture, I responded thus to the last of these comments: "We agree that strengthening institutions is better but there is an interface between human rights, cultural rights and women's rights in which women are not protected, as has just happened in Bosnia with rape used as a tool for ethnic domination."

5.4 Santiago de Chile, August 1997: *Our Creative Diversity* "Gave Me Back My Feelings and My Utopias"

Our Creative Diversity was presented at the meeting of the Economic Commission for Latin America and the Caribbean (ECLAC) in Santiago de Chile in August 1997, presided over by Javier Pérez de Cuéllar. The artist Claudio Girolamo, then Director of the Chilean government Commission on Culture, spoke enthusiastically about the Report: "Reading it opened up new vistas for me, it brought back my capacity of awe and it gave me back my feelings and my utopias." He went on, "It invites us to be subjects of our transformation, not objects; it brings us back to our origins as a species; and takes us beyond coexistence to *convivencia*.[3] As an artist I say that culture creates values that give alterity and even subversion." Ernesto Ottone, an economist at ECLAC, added that one positive point in *Our Creative Diversity* was that it captured the widespread concern about the diminishing of the sovereignty of states over economic policies.

But not all comments were positive. Manuel Antonio Garreton, a professor at the Catholic University of Chile, claimed that the Report built a normative vision that obscured the contradictions in reality. He argued that it did not address the question of how the actions of citizens, the structures of nation states, and social relations have all been subordinated to market forces. He asked, "Can there be a high level of cultural development with a high level of inequality? If culture is everything, then we are evading everything. What would distinguish cultural from economic development? The Report keeps wanting to give a substantive definition of development, as when it defines a global ethic, but it doesn't take up inequality in defining this global ethic."

Professor Garreton continued by saying that the 'revolutionary' contribution of *Our Creative Diversity* lay in proposing the democratization of world government and of the United Nations, and controlling the monopoly of the media at world

[3]This word has no equivalent in either English or French. More than "living together", which is conveyed simply with the Spanish term of *convivir*, the term *convivencia* means to share life experiences, that is, *vivencias*. In my own work, I have constructed the word 'conviviability' in English, to convey the sense that, if there is no ability to live together, there will never be sustainability. In other words, there will be no sustainability without conviviability.

level. Yet he argued that the concept of cultural diversity does not explain where the positions are from which an ethical citizenship can be exercised, in relation to communicational power and the environment.

I made the brief comment that now that we more or less agree on *what* has to be done in culture and development, we should turn our attention to the *how*: organizing, encouraging wide popular participation, financing meetings and projects, and continuing to develop theories linking culture and development.

5.5 Lillehammer, September 1997: "There Is No Darwinism of Culture"

At the meeting held in Lillehammer, Norway, titled "Culture and the Environment and Coping with Conflicting Goals and Cultural Diversity", the young Norwegian philosopher Andreas Follesdal declared that "There is no Darwinism of culture", in the sense that the best cultures in terms of dealing with the environment will be better suited to survival. He warned that learning may fail to have a long-term perspective, or may be deterred by factors imposed on the group from the outside, or unexpected natural phenomena may do away with that culture. He cited recent feminist research that had shown that, for women to be equal, they must share equally in power so that deep social changes could be achieved.

The more appropriate question to ask, he argued, is why culture is important for the individual, and what interests are at stake for the individual in cultural rights. He gave three answers: first, collective belonging must be regarded as valuable, because, secondly, cultural membership must provide the infrastructure for individuals to be able to work towards a good life; and thirdly, culture is to be valued in order to look towards the future, identifying paths and options. However, cultural changes should not be too abrupt, so that correct expectations could be formed, reducing the risk of false expectations.

Working, as the World Commission did, by defining development as the human capabilities and functions available to agents, following Amartya Sen's formulation, Follesdal ended by suggesting three areas for further research. The first was the measurement of how one's capabilities relate to the decrease of those of others. The second was that metric measurement was not enough because it did not capture the feminist argument that options must not only be there but that people must be aware that they are there. Third, research must be conducted into human rights as a means through which people protect their culture without giving them means to abuse their members, so that there is not a right simply to preserve cultures.

Several other speakers then went on to criticize the Report, each in harsher and harsher terms, saying that it did not go far enough in calling for respect for indigenous cultures, or for women; that it did not give sufficient historical depth to the issues; that recommending cultural rights created many legal and juridical problems and that it called for the creation of yet more bureaucracy. By the end of the session, it seemed to me that the speakers had not realized how difficult it had been to write this Report, in such compressed, complicated and combative circumstances, having averted the hijacking of it by several political interest groups breathing down our necks. During lunch, however, a friendly Norwegian approached me to say that it was their cultural custom to first demolish the report being presented and then reconstruct it together. And, indeed, that was what happened in the afternoon session, when they praised the consistency of the Report, the clarification of the main issues related to culture and development and the forward-looking outlook of its actions.

5.6 Birmingham, February 1998: The Power to Define and "Culture as Everyday Life"

The UK presentation of *Our Creative Diversity* was held at the Department of Cultural Studies and Sociology at the University of Birmingham, on 20 February 1998. The event also marked the moment when Britain rejoined UNESCO. Susan Wright, Senior Lecturer in Cultural Studies and organizer of the meeting, wrote the report of the discussions at the seminar. She said that the seminar was inspired by the Report's broadening of the concept of culture and its argument—as relevant to income-rich as to income-poor countries—that successful development would result in "a flourishing of human existence in all its forms and as a whole". *Our Creative Diversity* criticized the view of culture as exclusively heritage and art because this was "biased towards the elite, the monumental, the literate and the ceremonial". In contrast, it offered a view of culture as "ways of living together" and "everyday life". That is, culture was not just the province of the great and the good, but was produced through the ways in which people make meaning for themselves and express value through their daily activities.

Like the Report's authors, anthropologists and exponents of cultural studies had recently viewed culture as a "site of contestation" over the power to define key terms, including the meaning of culture itself. They were exploring and revealing the ways in which decision-makers use different meanings of culture politically and they analysed the effects on those who were marginalized and impoverished. Anthropologists were therefore studying who, in a flow of events, has the power to define what for whom, and with what outcomes. At its most authoritative, a particular meaning of a key concept or symbol becomes 'natural' or "common sense", "taken for granted" as if it were nothing to do with politics and power. At this stage, the idea can be called hegemonic. However, even the most hegemonic ideas can be contested, unsettled and changed. The question to ask was how those asserting the power to define prevent other ways of thinking about key concepts from being heard.

Wright argued that a good example was provided by how, after many years as a central concept in anthropology and cultural studies, culture as a way of life had at last gained authority through its inclusion in a UN document. *Our Creative Diversity* had unsettled the idea of "culture as heritage" which had become so dominant that it was no longer remembered that particular people had argued this definition onto the agenda in times past. Now that "culture as heritage" had been dislodged, it was revealed to be one equally politicized concept in what Nira Yuval-Davis had called a "battleground of meaning" (Yuval-Davis 1997: 41). However, taking the understanding of culture as "*one* way of life" allowed culture to be conceived as a bounded entity containing "a people who all have *one*, shared, way of life". Early anthropologists thought of the world as a mosaic of discrete cultures. Decision-makers had adopted this language: they used it when they referred to a population as having a "national culture", and when, often in the next breadth, they described the nations as made up of several 'cultures', each with a distinct "way of life".

In such cases, the rationality of governance was laid out as follows. The population is classified in terms of one element of their identity, such as ethnicity. The people defined in terms of that category are then assumed to share all other characteristics and to be homogeneous. They are also assumed to know each other and to be a group that interacts together. The authorities look for leaders of the 'group'—often unelected and self-ascribed—whom they can involve in discussions and through whom they can channel funds. This process has been shown to generate leaders and to enhance their authority over the people whom they, adopting the authority's discourse, claim as their ethnic group or community. It is their view of the group's 'culture' which prevails and is naturalized or claimed to be 'tradition'. People who are positioned differently within that category or group may be marginalized; their voices are not heard in the present and are silenced in history. For some this method of administering a plural society is empowering; for others, it is oppressive.

The second meaning of culture as "everyday life" conceives not of a plurality of discrete cultures but of a diversity of cross-cutting differences. In areas where there are polyglot, multi-ethnic urban populations or majority and minority voices, the WCCD Report noted that an open and pluralist society fosters creativity in art, technology and government. This idea had itself been the subject of political contestation in Britain in the 1980s and 1990s. Stuart Hall, one of the founders of cultural studies at Birmingham, argued, in a similar way, that migration and diaspora had generated populations with multifaceted differences. He enjoined his audience to rejoice in cross-cutting differences and fluid identities, and to celebrate the creativity inspired by such hybridity. Such ideas informed the work of anti-racists who sought to change the workings of state institutions or laws in the interests of treating all citizens equally.

The New Right in Britain had captured and inverted the meaning of this language of difference. They appropriated anthropologists' language about culture's being a way of life and used it to redefine 'nation' in terms of culture rather than political rights in order to make a distinction between the English and "foreigners holding British passports", as one former chair of the Conservative Party had put it. The New Right mobilized "culture as a way of life" to reinforce exclusion, using it as a euphemism for renewed racism, with profound implications for public policy and people's lives. One proposal for advancing the ideas developed in the Report would therefore be comparative research into the ways in which culture is mobilized politically, in order to analyse how its various meanings imply different forms of governance, with different effects on those who are marginalized and impoverished.

5.7 Response from the French National Commission: The Artist as Central to Culture

On 9 January 1998, the French National Commission for UNESCO held a meeting on cultural diversity at which speakers were divided for and against globalization. A few of them opposed 'mondialization' because it destroys the "root people" ("les peuples racines"), while other argued that globalization would then create spaces for multiple cultural allegiances ("les appartenances multiples"). "The universal is the local without walls" was how one participant described the way in which globalization was simultaneously a local process. However, in order to prevent the risk of domination and increasing uniformity through globalization and the new technologies, governments had to give greater attention to culture. One speaker remarked that "there is no government without culture and no culture without government". Instead, consensus emerged calling for exchanges which bring about the recognition of the Other and lead to greater cultural enrichment.

With reference to the dissemination of knowledge and culture in society and the risk of ever-increasing uniformity, it was noted that publishers and the press have always been in crisis, yet the 'written' is still the most important instrument for conserving culture, and thus for pluralism. The new communications and information technologies would, through the global networks, foster greater exchanges that would escape the censors and, therefore, promote pluralism and liberty.

A discussion ensued about the subversion of the social order through these technologies, since families, schools and other institutions are no longer the location for the transmission of knowledge. Instead, the Internet allows the individual to have direct interaction with others around the world. In the area of audiovisual production, however, the outlook was considered to be gloomier. Participants argued that cultural production from countries in Europe and in the Global South was being wiped out. Three proposals were suggested for countries of the South: firstly, to provide means through international organizations to preserve the memory of peoples; secondly, to train creators and journalists; and thirdly, to sustain the role of the public sector in the new communications and information technologies, since it remained as irreplaceable as it was when printing was created.

The consensus reached was that it is essential that the artist be placed at the centre of culture. This had been the general conclusion of one of the working groups of the meeting held by the French National Commission in its debate on the

WCCD Report in 1996. It is a restricted definition, different from that reached by other countries. Several comments were reiterated. These included emphasizing that there can be no cultural policy without the creators of culture. We are at a stage in which art is resistance, because it gives power to choose the Other. The great danger is that art become merely merchandise, which is why the state must provide spaces for experimentation and confrontation between the arts and the sciences, without expectations that this will fulfil a demand for social consumption or acceptance on the part of the public. There are, in fact, many publics. Art must therefore go beyond the laws of supply and demand.

Regarding the Report as a whole, several specialists pointed out that we must be careful of the destructive risks of the ethnicization of culture. A specific remark was made that one must see the fallacy of considering "culture as the continuation of war through other means". As an antidote, universalism should therefore be reinforced as a strong and unequivocal principle.

5.8 *In from the Margins*: The Report of the Council of Europe

The 1996 report *In from the Margins: A Contribution to the Debate on Culture and Development in Europe*, prepared by the Council of Europe, reviewed aspects of culture and development in that region, as one of the many statements and testimonies of countries and regions provided to the work of the World Commission. It also became one of the most influential documents as a counterpoint, both supporting and offering alternative points of view to those of *Our Creative Diversity*. There were, of course, political underpinnings to this initiative in terms of influencing the Commission's report, some of which were resolved, while others remained. As *In from the Margins* is easily accessible, in the following pages only the most salient points included in it will be presented.

The Report begins by stating that

> There is much in [*Our Creative Diversity*] which the task force finds it can endorse, in particular its commitment to the centrality of culture in development; its analysis from a global perspective is interesting and challenging. The two reports share common elements and a number of recommendations follow the same lines of thought. That said, the task force was concerned not to be unduly influenced by the World Commission's proposals. The commission had a world-wide brief and those of its concerns that are relevant to developing countries are not necessarily all shared by European countries (Council of Europe 1996: 16–17).

The task force also makes it clear that their conclusions complement those produced by the World Commission; where there is a convergence of ideas on action these are reflected in the "Follow-up"—the case for urgent action.

In the Executive Summary, the task force explains that "we endeavour to link our ideas on culture to three major aspects of development: their contribution to the sum of human knowledge and understanding; their economic and social outcomes;

and their potential for harnessing human resources as a medium for empowerment and entitlement" (ibid.: 18–19). In its view, the task of government is to re-position culture at the heart of decision-making.

Considerable attention is given in the Report to culture in the framework of individual liberties and the renewal of civil society. The authors identify the single most important social trend in Europe since the Second World War as individualization and a corresponding weakening of social structures, controls and values:

> It echoes the logic of market liberalism (most succinctly expressed by the UK premier, Mrs Thatcher, when she was quoted as claiming "There is no such thing as society"). The growth of individualism is consistent with the rationale of civil society, which as we will see, depends on a careful balance between the state and the stand-alone citizen (ibid.: 145).

Salvador Giner is quoted as stating that "The unhampered market allocates resources, honour, authority as well as goods and services, through a spontaneous and ultimately anonymous process of countless individual transactions among individuals and their associations" (ibid.: 150). The Report notes an escalation of advertisement-driven consumerism, the rise of political intolerance and increased feelings of insecurity and alienation as a result of the economic recession and structural unemployment (ibid.: 152). It ends by recommending that "Culture can make a major contribution to a tolerant and diverse community, partly through the intrinsic value of creative activity and partly through the independent voluntary associations to which it gives rise and which are the building blocks of civil society" (ibid.: 276).

In the section headed "Towards a New Policy Agenda", the Minister for Arts, Culture and Gaeltacht of Ireland, Michael D. Higgins, is quoted as saying:

> When I first attended the Council of Culture Ministers in 1993 … people were saying that when economic growth came back we could talk about cultural projects again. I said it is *precisely* when you have stagnation in the economy, when you can't create jobs in the old way, when people are being attacked by racism—*that* is the time when you should be investing in culture because you are then investing in tolerance, you're investing in diversity, you're investing in creativity and in imagination (ibid.: 265, emphasis added).

Such a statement highlights the dilemmas which, at the time of writing this book, in 2017, have become dangerously acute in so many countries.

In describing the four key principles which have guided cultural policy—support for cultural identity, for cultural diversity, for creativity and for participation in cultural life—the authors nevertheless entitle this section "Cultural Policy: An *Imprecise* Art" (ibid.: 265, emphasis added). Although in this new policy agenda they forthrightly assert that "There is no question that state-supported culture has been a powerful promoter of cultural identity", citing Finland as a case in point, they also assert that

> events have moved on. Today's new individualism means that many people move at will among a wide variety of identities, finding themselves more at ease with global culture than pleases national policy makers and, at the same time, but without any sense of contradiction, delighting in the satisfactions of locality and regionality (ibid.: 265–6).

Significantly, the statement is then made that "An exclusive and monolithic approach to cultural identity no longer reflects (if it ever really did) the values of modern European society, which is more accurately seen as a 'culture of cultures', both in terms of the composition of its population and its openness to international influences" (ibid.: 266).

Contrasting *In from the Margins* with *Our Creative Diversity*, we see that the former takes a more processual theoretical viewpoint, analysing culture as it influences governance, social consensus and social cohesion, and enhancing human resources, the impact of the communications revolution, creative employment and the dynamics of international exchange and co-operation. The task force also explicitly states that "We were asked to look for European solutions when there is not a single Europe. But perhaps the thinking at intergovernmental level is flawed. We *can* identify some broad trends, but the impact these trends will have will differ from country to country" (ibid.: 270, emphasis in original). Finally, a consideration that is important for understanding which transactions on culture are possible at the global level and which are unresolvable is reflected in the recognition by the task force that their Report "does not pretend to have solved the methodological problems inherent in its work" (ibid.: 21) nor the many unfinished terminological tasks involved in defining culture and cultural policy.

5.9 The Stockholm Intergovernmental Conference on Cultural Policies for Development, 1998

The critiques of and comments made on *Our Creative Diversity* generated an international process that had ramifications beyond the narrow discussion of culture and development. By 1998 it had been issued in the six official United Nations languages—English, French, Spanish, Arabic, Russian and Chinese—and it has since been translated into many more. While the ideas and the discourse it had put forth drew the attention of the academic community and civil society organizations, governments were keen to convert such ideas into actions. Naturally, the different actors in this process—UNESCO member states, national commissions, academic institutions—would then have to position themselves with respect to this new international standard. As cautiously expressed in the Executive Board meeting, however, *Our Creative Diversity* was not an official UNESCO document and therefore, insofar as UN intergovernmental agencies hold the authority to determine international normativity, this Report could not be considered normative. Member states interested in consolidating the principles and policy guidelines set in the Report then initiated the long, embattled road to making some of its ideas and proposals official by inserting them into UNESCO's Medium-Term Plan and Biennial Programme (C/5).

The next step to keep up the momentum of the international debates was to formulate an Action Plan endorsed by UNESCO, at the next Intergovernmental

Conference on Cultural Policies for Development. Accordingly, the Swedish government, one of the strongest supporters of the WCCD from the beginning, agreed to execute one of the recommendations of the Report by hosting this conference, which took place in Stockholm from 30 March to 2 April 1998. The conference brought together some 2,400 participants representing 149 governments, twenty-three international intergovernmental organizations (IGOs) and some 135 NGOs, as well as individual artists, scholars and experts.

As Assistant Director-General for Culture, I was directly responsible for the organization of the Conference. My role also meant that I was well aware of the rift that existed between governments' discourse on culture and that of the practitioners of culture themselves. Accordingly, I set up a tripartite organization for the Conference, juxtaposing three different sections. In the official Plenary sessions, government delegates read their speeches in the main auditorium; in the research Forum, researchers discussed themes; and in the Agora, IGOs, NGOs, associations, foundations, artists and intellectuals organized seminars and workshops with a view to highlighting ideas and new initiatives across a broad spectrum of policy-related areas.

As it happened, the arrangement created some tension during the Conference, since the Forum and the Agora drew bigger audiences than the Plenary sessions. The political importance of the Conference was in this official section because countries were in the process of negotiating international normativity and developing international conventions, programmes and procedures to consolidate a world space for agreements on culture. This was the most difficult process in operation. Both the Forum and the Agora, of course, were exuberant with participants pitching ideas, initiatives, projects, claims, complaints and audiovisual shows for the whole world to see and hear. All this content was intended to pull the concerns over culture and development into the future. In spite of the tensions, however, participants commented very positively on the fluid exchanges that took place between government officials, researchers and cultural practitioners.

Although such a dual structure is now usual for UN Conferences—for example, in conferences on population, women or the environment, where NGOs and civil society organizations have their own forums—these tend to draw participants from groups with specific concerns. In comparison, the theme of culture seems like an open call for everyone in society: artists, indigenous peoples, cultural activists and so on. This supports my contention that culture is the broadest dimension and, given its polysemic layers, the one that engages all people most profoundly in discussing the planetary human outlook.

There were ten themes in the Forum, each sponsored by different countries:

Forum I	A commitment to pluralism (Canada)
Forum II	Improving research for cultural policy (France)
Forum III	Cultural rights (Bolivia)
Forum IV	Mobilizing resources (Republic of South Korea)

Forum V	Culture, children and young people (South Africa)
Forum VI	International cooperation in cultural policy (Egypt)
Forum VII	Cultural creativity (UK and Jamaica)
Forum VIII	Culture and the new media technologies (Finland)
Forum IX	Cultural heritage (India)
Forum X	The role of the media in cultural policies (Philippines).

The range of topics discussed at the Agora was very varied. Ten main strands surfaced as key concerns: multiculturalism and the practice of cultural diversity; the cultural role of civil society; creativity in the new global scenario; freedom of expression and the silenced voices of journalists and artists; other silent voices: women, children and indigenous peoples; new research agendas; public service and the old and new media; public cultural institutions such as libraries and archives; international initiatives and forms of cultural expression of local interest groups; and towns and cities as spaces for cultural practice, innovation and analysis (UNESCO 1998: 45).

There were also a Business Forum, a conference on "Performance priorities for cultural policy-making" organized by the International Federation of Actors, and a meeting organized by the Swedish Joint Committee for Literary and Artistic Professionals entitled "Towards a cultural Agenda 21". This agenda was to be the most important reference for many subsequent UNESCO and UN programmes internationally, as well as for national policies on culture around the world.

5.9.1 "A World Culture Is in the Making"

Among the keynote speakers, Jan Pronk, then Minister for Overseas Development of the Netherlands, emphasized that "as large multinational centres of economic power gain control of the channels along which the transfer of culture takes place, the trend [towards a uniform global culture] is reinforced. A world culture is in the making, influenced partly—but by no means exclusively—by the West" (UNESCO 1998: 85). The Minister cautioned that there was a new cultural dispute: "a conflict between cultural diversity in an open society on the one hand, and cultural self-containment in closed communities, characterized by static conventional wisdoms, on the other". He spoke of a "sinister paradox" in the impact of commercialization on the realm of the media. "While technology permits the release of an inexhaustible stream of information, commercialization—fed by television ratings—limits the scope of that information. The news is presented through a magnifying glass. It looks bigger, but we see less" (ibid.).

He also referred to the upsurge of identity-related conflicts in the world, leading to "an uncontrollable stream of violence". He pointed out that "It seems to make little difference whether those in question represent a minority group or a dominant culture. The projected image of a culture under threat appears to be a highly effective rallying cry to mobilize people who would otherwise never have dreamed of butchering their neighbours" (ibid.: 86).

Mr. Pronk concluded by arguing that two conditions must be fulfilled if the open society were to have a chance of developing in every country on the planet: the fostering of global ethics, and freedom of expression in all nations. "That is democracy", he went on, and "Democracy is not a system, it is a culture. Without such a culture, there is no development" (ibid.).

In my own statement at the Closing Plenary, I highlighted the fact that all debates on culture and development converged on the need for freedom to create. This would be the only way to reduce the gap between those who have the resources to continue to conserve and re-create their cultures and languages and those who do not. Creative people should be given opportunities to generate the symbols of tomorrow through the work of their minds and their hands, particularly in developing countries. A culture of citizenship was also needed that would be conducive to the peaceful resolution of conflicts between culturally different peoples. Finally, I strongly advocated that it was necessary to explore the seemingly boundless possibilities that new forms of communication and expression were placing in our hands (ibid.: 89–92).

5.9.2 *The Stockholm Plan of Action*

The Conference ended with the recommendation that UNESCO member states adopt the Stockholm Plan of Action. Five major policy objectives were proposed. The first reiterated the general recommendation of the World Commission on Culture and Development to make cultural policy one of the key components of development strategies, by, among other things, advocating a broader vision of national cultural policy, with the participation of civil society, including the media. The second objective dealt with creativity and participation in cultural life: through respect and equal opportunities for the different components of the nation, placing emphasis on local initiatives; and through knowledge, by strengthening the cultural content of formal and non-formal education, in particular by encouraging the learning of one or more foreign languages.

The third objective called for reinforcing policy and practice to safeguard and enhance cultural heritage, both physical and intangible, moveable and immoveable, and to promote cultural industries. The fourth advocated cultural and linguistic diversity in the information sector, especially through the development and use of new technologies and communication services and the digitalization of information. The fifth objective focused on making more human and financial resources available for cultural development from governments, especially local authorities, business (through fiscal incentives) and international organizations (UNESCO 1998: 14–18).

5.10 Freedom to Create

The issue that may be highlighted as the most important emphasized at the Stockholm Conference was that of the emerging possibilities of the new information and communication technologies. In terms of political positions, the fault lines between groups of countries were clearly stated. On language conservation, for example, countries of the North advocated the saving of all languages and encouraging the learning of other languages, a policy which was readily opposed by African and Asian countries, who pointed out the impossibility of implementing such practices in their own countries, in many of which more than fifty native languages are spoken. On global ethics, some Middle Eastern, African and Asian countries also rejected the idea that the system be based only on Western values, arguing that there are a number of ways of arriving at such a global ethics.

On the basis of similar arguments, a minority of governments were opposed to cultural pluralism, arguing from two different political positions. On one side were governments that said such pluralism could destroy the political basis of republics; on the other side of the political spectrum were a few governments who claimed that there was only one true religion and therefore that there was no possibility of accepting plurality in either culture or religion. However, the majority of government delegates at the Conference embraced the commitment to pluralism advocated in *Our Creative Diversity*. On the topic of the media, aggressive arguments were put forward to enhance freedom of expression, while only a few speakers insisted on the need for the regulation of media professionals as recommended by the World Commission.

Two important facts emerged clearly from the Stockholm Conference. The first was that, in comparison to international intergovernmental conferences on economic policies, conferences on culture and development demonstrated a wider variety of positions held, both within and between countries. Of course, such positionings can only be made visible in conferences such as the one held in Stockholm, in which the tripartite structure of government, researchers and activists allowed such diversity to become apparent.

The second conclusion was that, as in all political discussions, governments sometimes have contradictory positions. On the one hand, this might happen because different ministries hold divergent positions, especially between the economic and cultural ministries. On the other hand, it is possible for government delegates to stridently advocate cultural pluralism in terms of international relations, yet be adamantly opposed to such a policy within their own country. In international conferences where only government delegations speak to each other and where negotiations are driven by diplomatic considerations, such contradictions are smoothed over in reaching consensus over resolutions or policy documents. In a conference such as that held in Stockholm, the presence of a diversity of actors in debates allowed such divergences to be brought out into the open. Such diversity and new ideas no doubt enrich the discussion, yet it must be said that sometimes they make it much more difficult to build consensus, since different governments and practitioners make claims on different political scales.

References

Council of Europe (1996). *In from the margins: a contribution to the debate on culture and development in Europe*. Strasbourg: Council of Europe Publishing.

UNESCO (1998). *Final report of the Intergovernmental Conference on Cultural Policies for Development, Stockholm, Sweden, 30 March – 2 April 1998*. Paris: UNESCO.

WCCD (World Commission on Culture and Development) (1996). *Our creative diversity: report of the World Commission on Culture and Development*. Paris: UNESCO.

Yuval-Davis, N. (1997). *Gender and nation*. London: SAGE Publications.

The Economics of Culture and Media: seminar on Culture in World Trade held at New York University, New York, March 2000. *Source* From the personal photo collection of the author

Chapter 6
Diversifying Cultures

Our Creative Diversity had set an agenda for the discussion of culture and development. This agenda had been contested and new topics and prospects had opened up in international debates about culture. The Stockholm Conference provided an open forum where governments, intellectuals and civil society had robustly debated the way forward in more practical terms in relation to cultural policies, the new information and communication technologies, and the arts. The first UNESCO World Culture Report (subtitled "Culture, Creativity and Markets", and published in 1998) sharpened the issues by bringing in the latest scientific research and statistical data and eliciting more complex discussions on the main issues of culture and human prospects. It also opened up the debate to new themes, including cultural pluralism, the economics of cultural heritage, culture and trade, and so on.

These events set benchmarks and activated different constituencies that had remained on the margins of the debate. As the debate grew more mainstream, arguments and positions began to coalesce around substantive questions about what would influence the economic and political course of globalization. Four concerns emerged as those drawing the greatest attention in the global arena: languages; the economics of culture, including trade and culture; the role of new media technologies; and "world culture" versus cultural diversity.

6.1 Languages

6.1.1 Linguistic Pluralism: The Zimbabwe Conference

In 2002, linguists were saddened to hear of the death of Elsie Vaalbooi, at the age of about 100, on 7 October. As noted in a report in *The Economist* ten days later, "She was one of the last speakers of n|u, a click-based dialect of the once-nomadic San

L. Arizpe Schlosser, *Culture, International Transactions and the Anthropocene*,
The Anthropocene: Politik—Economics—Society—Science 17,
https://doi.org/10.1007/978-3-642-41602-6_6

people. Only a handful of her ageing relatives can still use n|u, the last known example of !Ui, a family of San languages spoken across southern Africa by hunter-gatherers, possibly for the past 30,000 years" (*The Economist* 2002).

The debate on how countries should deal with a plurality of languages had been present on the agenda of UNESCO since the 1950s, and emerged as a major political issue after the decolonization of African and Asian countries. As the end of the Cold War revived the question of pluralism, the failure of many developing countries to formulate linguistic policies that could avert clashes over the use of languages or repression against languages put this issue on the international agenda once more. Equally importantly, the effects of globalization began to be felt in terms of the marginalizing or disappearance of languages and cultures.

In 1996 PEN International presented the "Universal Declaration on Linguistic Rights" to UNESCO. It strongly advocated that all languages be respected and that governments should take all necessary steps to prevent the loss of languages. It also declared that languages should be incorporated into education at all levels and that in literature and art, in particular, all languages should be allowed to be used. This document, hotly debated at the UNESCO Executive Board that year, drew strong opposition from the governments of many developing countries and cheers from many civil society and academic organizations. I remember that one of the definitive pronouncements on this debate came from the delegate from Indonesia, who carefully and respectfully called the attention of those present to the fact that Indonesia has more than 800 languages and its government would simply go bankrupt if even the minimum of those actions called for were to be implemented.

Partly as a result of that meeting, African delegations were able to insist on holding a conference to discuss this issue in Africa. Of an estimated 1,400 African languages, 500 are in decline, half of them facing imminent extinction. African delegations at UNESCO, as well as the Organisation of African Unity (OAU; now replaced by the African Union) had called for this meeting with the aim of "defining a realistic linguistic policy that could establish the statute and functions of languages existing in every country and define an appropriate strategy for each situation". Accordingly, the Zimbabwe Intergovernmental Conference on Linguistic Policies in Africa took place in Harare on 17–21 March 1997.

As I saw it, the language debate in Africa had arrived at an impasse. Debates on this topic in UNESCO meetings always flared up fiercely and burned out into two inflexible positions: those who advocated using vernacular African languages and throwing out all ex-colonial languages, especially English and French; and those who argued for the importance of keeping such languages as lingua francas while recognizing other African languages as national or official languages. In organizing the Zimbabwe conference, I found that the two African programme officers who would assist me held exactly these two opposing views, as did many of the African ambassadors and delegates to UNESCO.

After a protracted and intense debate at the Conference, and very careful lobbying with the delegations, the Harare Declaration was adopted, calling for a recognition of African ethno-linguistic pluralism in a framework of democracy and the peaceful coexistence of all peoples. As a guideline for the creation of general linguistic policies by governments, it suggested that "Every country's policy framework should be flexible enough to allow each community to use its language side-by-side with other languages while integrating with the wider society, within an empowering language policy that caters for communication at local, regional and international levels" (UNESCO 1997: 75). The Declaration called for every government to establish a linguistic policy based on an exact typology of the linguistic situation in their country and to make adequate preparations for language policies. The training of language practitioners in various professions and the production of teaching and learning resources, including those required for second-language teaching and learning, were also emphasized. At pan-African level, governments were to reactivate the Language Plan of Action for Africa, promulgated by the OAU in 1986, to implement the resolutions taken at the Accra Conference in 1996 to make Kiswahili one of the working languages of the OAU, and to produce didactic materials for schools in African languages (ibid.: 75–6).

Culture and Linguistic Pluralism: the Zimbabwe Intergovernmental Conference on linguistic policies in Africa, held in Harare, Zimbabwe in March 1997. *Source* From the author's personal photo collection

Reactions to the Declaration were plentiful. An African United Nations official who had participated in the Conference contended that "Too much emphasis on local languages has the potential of isolating a country not just from its subregional and regional neighbours, but perhaps more importantly, from international forums."

The case of Nigeria was cited as an example: even if one language was chosen among the multiplicity of them—say, the Yoruba language—it still has over 160 different dialects, some of them incomprehensible to other Yoruba speakers. Education in the mother tongue of any particular region of a country can be equally isolationist within the country itself. He ended by saying that, in any case, most new vocabulary coming into use, especially in the field of technology, was in English.

The Zimbabwe Intergovernmental Conference on Linguistic Policies in Africa was only one of the highlights of UNESCO programmes that focused on African sources of cultural evolution, and major policy issues. Doudou Diène, one of the African progam officers, carried out very important activities on the Slave Route that transported millions of Africans as slaves across the Atlantic. He was also head of the Silk Road programme which developed research and action projects linking cultures in Asia with those of the Middle East and Europe, generating a constant flow of ideas, scientific discoveries and art across Eurasia.

6.1.2 The Debates About Disappearing Languages

Discussions over the use of different languages are not only a prime political theme in Africa but have become so in all countries where a policy of multiculturalism has been implemented. The European Union has not solved the problem and has merely postponed it by declaring twenty-four official languages. Several European countries have deep cleavages on this matter: for example, the Basque country of France and Spain, and Belgium, to name only those where the issue has led to violence. Small European countries have all debated the future of their languages. In the Netherlands, for example, the eventual disappearance of Dutch has even come to be envisaged; nevertheless, education there is trilingual, with Dutch, English and any other international language mandatory in schools. The demise of the Soviet bloc also brought hope to many culturally oppressed people that they would be free to use and develop their own languages. One problem is that this would have to lead to mass geographical displacements, not the least because Russian speakers, who were fast becoming oppressed linguistic minorities in many transition countries, would have to be repatriated to Russia.

The same debates over linguistic pluralism were held in Mexico in the 1970s and in many Latin American countries thereafter, in relation to indigenous languages. With the rise of Indian militancy since the 1970s, their languages are no longer in decline. Now, in fact, most indigenous groups have opted for new names for their languages and ethnic groups, since these had previously been ascribed by the Spanish colonial administrators. Thus, the derogatory term of 'tarascos' (meaning 'mindless') has been replaced by the proud indigenous term 'purhépecha'. Also, instead of the erroneous use of 'Indians', imposed on the peoples of Latin America by Christopher Columbus, several new options are being considered:

"Meso-American cultures", "pueblos originarios" (original peoples), 'etnias' (ethnicities). The debate on bilingual and bicultural education has re-emerged in the new millennium in Bolivia, Peru and Ecuador—even more intensely, since it is now entangled with nationality. Evo Morales, president of Bolivia, set the pace, not only announcing that his is a government for indigenous peoples but boldly proposing a new philosophy of "Buen Vivir" (in my interpretation, as discussed in the next chapter, this expression translates as "Living Goodly", which is different from what I see as the more Western translation of "Living Well"; as so often, it is a question of culture).

In the United States in the 1990s the debate was fuelled by the growing numbers of Hispanics who were transforming cities such as Los Angeles, New York and Miami into bilingual zones. The test for bilingual education came in 1998 with the vote on Proposition 227 in California, in which, to the surprise of Mexicans, many Mexican residents in the US voted against having separate classes for Anglophones and Spanish speakers in schools. Their reasoning was that Spanish-led classes became second-rate classes that meant Latino students became an underclass unable to compete under equal conditions in labour and politics.

There is the risk of repeating such debates in a never-ending story. I believe that history, however, gives a few lessons that we can learn from. Along these lines, let us imagine what would have happened if the Roman Empire had established a Roman Language Academy to protect Latin against linguistic change and regional diversification. Clearly, neither French nor Spanish nor Romanian nor any of the other Romance languages which are now so zealously defended would have evolved. Strangely enough, unlike many other once imperial languages, Latin died out as a living language, in spite of a formidable literary, artistic and political cultural heritage. Or perhaps it would be more accurate to say that Latin continued to follow its natural course and became Italian.

The difficulty in defining the status of the languages spoken around the world, and their number, may now have become clearer. One could state as axiomatic that every generation inevitably transforms whichever language is transmitted to it through learnt education. This genetic inheritance is the last straw that bring the whole edifice of the debate over linguistic pluralism to its downfall. The reasoning goes like this: every individual has the capability of altering the grammar and vocabulary of the language they learn; the collective outcome of such a capability is that every generation changes the language or languages they learned. The outcome is that no language can be kept the same over the generations, although change can become more gradual and systematic through the regulation of linguistic changes by authoritative bodies (academies, governments or writers and artists). Such bodies can totally resist change or can shape it by accepting some changes and not others. An illustrative case is supplied by the Royal Academy of the Spanish Language of Spain which, until the 1970s, considered what Latin Americans spoke as second-class Spanish, until it was shown that some countries in the region (for example, Colombia) had better resisted change than Spain itself, with their people speaking a more ancient, 'correct' Castilian than the Spaniards themselves. Subsequently, the Academy's prestigious *Royal Dictionary of the Spanish Language* began to include many neologisms that had evolved in Latin America so as to reflect more accurately the development of the Spanish language in the modern world.

Cultural policies for language conservation in a globalized world must therefore start out from the assumption that people have the capability to constantly create language. It is impossible to halt linguistic change because people adapt their oral communications to the environment in which they are evolving. In any case, people wanting to conserve or even reconstruct their language must be given the freedom to create that will enable the safeguarding of their mode of existence.

6.2 Economics of Culture

6.2.1 Culture Counts: The World Bank Meeting in Florence

The issue of financing cultural policies and actions had emerged in different meetings and documents, as described in previous chapters. It became a major topic in international discussions about culture in several meetings of the World Bank. This was particularly true during James D. Wolfensohn's tenure as President of the Bank: the fact that he was an accomplished musician meant that issues related to culture moved as never before. On one memorable occasion, I remember walking with Mr. Wolfensohn and the UNESCO Director of the Division of Cultural Heritage, Mounir Bouchenaki, along the winding, narrow streets of Fez, admiring the restoration work that UNESCO was carrying out on the beautiful architecture of its buildings. The World Bank had just approved a loan for urban planning in the city and it was planned to cut through the Kasbah with a new, wide avenue. Mr. Bouchenaki and I were trying to convince Mr. Wolfensohn that a few of the existing narrow streets could be widened without altering the historical core of Fez. Mr. Wolfensohn agreed.

At that time, Ismail Serageldin had just been named to a new position as Vice-President for Special Programmes in the World Bank with a primary focus on the issues of culture and development. Mr. Serageldin—who was later to lead the project to re-create the Bibliotheca Alexandrina in Alexandria—spoke on "Investing in Culture: A View from the World Bank" at the Stockholm symposium (see Sect. 5.9) on 30 March 1998. He declared the complete support of the World Bank for an action programme on culture and development that recognized "the intrinsic worth of culture, not just what it generates in tourist revenues; the educational content of culture; the special case of historic cities; going beyond the "do no harm" posture; the need for a culture of participation; the importance of promoting the sense of community, social inclusion and social cohesion" (Serageldin 1998). The

World Bank, he stated, was therefore "willing to commit to the financing of operations in culture in all countries who seek such loans and credits from us" (ibid.).

Eighteen months later, the conference "Culture Counts: Financing, Resources, and the Economics of Culture in Sustainable Development", co-sponsored by the Government of Italy and the World Bank, in cooperation with UNESCO, was held in Florence on 4–7 October 1999. According to the published proceedings,

> The premise of the conference was that culture is crucial to advancing sustainable development. The objectives were to (1) promote the expansion of economic analysis in, and resources available for, culture in sustainable development programs; (2) expand the range of institutions and actors involved in culture with a development perspective; and (3) increase the instruments to be used for these programs (World Bank 2000: viii).

Hillary Clinton, who delivered the concluding address, stressed that globalization is neither good nor bad but it can threaten cherished traditions, it makes consumers but not citizens, and it may, for example, lead to the burning of daughters-in-law (she was referring to India) because their dowries have not brought enough to buy a colour television (Clinton 1999: 1). It also lays a thin veneer over human awareness, despite people's great need to assert their cultural identity and to set themselves apart from one another. Ignoring this basic human need is what marginalizes people and sows the seeds of fundamentalism and extremism. She added that, though many committed to alleviating poverty still see culture as a luxury, not only can culture help achieve their aims through the encouragement of tourism and the revitalization of traditional crafts, but it gives meaning to people's lives and binds them together (ibid.: 158–9).

In his opening address to the conference, Lamberto Dini, the Italian Minster of Foreign Affairs, mentioned the enormous, chaotic growth of cultural products and forms of acquiring knowledge, given the proliferation of communications. He hoped that the Conference might contribute to "defining sound parameters for a new strategy", which took into account the need for individual identity, the problems created by a concentration of wealth and knowledge, and the impact of new ways of knowing and producing (ibid.: 6). The President of the World Bank, James Wolfensohn, quoted the results of the 60,000 interviews that had recently been conducted with poor people from sixty countries. The overwhelming response was that these people wanted, in addition to a steady source of income, a "sense of well-being which is peace of mind: it is good health, care of the spirit, family and community, and happiness". He mentioned their need to have their cultures, of "simple human signs of solidarity". Conserving peoples' cultures was not an optional extra but an essential element of development (ibid.: 10).

In contrast to the opening session, the atmosphere hardened when David Landes of Harvard University stated that "not all [cultures] are equally suited to successful high productivity in a material sense. They may be spiritually equal, or even superior. They may get people to heaven faster. But they will not make faster cars." He argued that there were "toxic cultures … which handicap the people who cling to them … in their ability to compete in a modern world" (ibid.: 30). To my dismay, he took my own cultural region as an example, and declared that "There has been extensive machismo, and machismo is poison. … Now I don't have to tell you that if you exclude one-half of the population of the human race [i.e. women] from effective contribution production … you are losing a large share of potential productivity." He closed by emphasizing that "true autonomy comes only with a level of education that allows you to learn from others, change what they've told you, improve on what they've told you, and invent your own things" (ibid.).

Professor Landes's paper certainly galvanized people's attention. He was followed later in the same session by Juan Ignacio Vidarte, Director of the Guggenheim Museum in Bilbao, Spain, at the heart of the Basque country. He argued that cultural initiatives were "becoming part of economic development strategies as a factor to encourage cultural and business tourism, as a factor to promote or to bring about an increase in tertiary activities, and as a tool to influence business decisions; in general, as a factor to enhance the quality of life of the citizens of the areas" and that in these initiatives there was a underlying assumption that there is "an explicit and symbiotic relationship between the degree of cultural activity of an area and the economic development potential of that same region" (ibid.: 46). He went on to explain that the Guggenheim Museum in Bilbao had been a response to globalization. With the industrial recession of the early 1980s, there had been a need to reposition the city and the surrounding country. One of the eight priorities that had emerged was the cultural enhancement of the Basque region, and the creation of the Guggenheim was "an important tool in reshaping the personality of the city". Its success was evident in the fact that it was having an effect not only on the image of the region but also on the local economy. In its first year of operation, 1.4 million people visited the museum, producing a direct economic impact of US$ 210 million in the area (ibid.: 47).

There were a number of interesting questions from the floor. One speaker asked: Who decides what to preserve? If you do it on the basis of economic interests do you give the economists the power to decide? A delegate from India declared that "Landes's scheme wants to make us all Protestant Calvinists. So in the end it is Westernization". There was prolonged applause in response. A Jamaican participant also criticized the fact that all the speakers were white men, with no blacks or women, and continued "in this new dispensation of globalization some of us are not converts: we already knew that culture was the basis of society. Now the World Bank finds out because wars are increasing. All speakers are from a single economic area. What about those of us from poorer countries who are unable to speak on structural adjustment or level playing fields? We're not called human capital, cultural capital. What does the World Bank want of this conference? I say that we have a right to exist." Again, this contribution was met by sustained applause.

In his summing up, Ismael Serageldin, Vice-President of the World Bank and the organizer of the meeting, who at that time was a candidate for Director-General of UNESCO, said that he had seen the emergence of four themes in this Conference:

first, a setting had been established; second, the problematique had been analysed and explored; third, actors had been identified to create a coalition of the caring; and fourth, different approaches had been explored to find out what works and why. He distinguished between cultural relativism and cultural pluralism and he argued that the challenge remained to "quantify the unquantifiable and to value the invaluable" (ibid.: 162–71).

While the Florence conference was the result of the cultural vision of the World Bank's President, the theme faced strong opposition from the bankers within the Bank. A comment often made in UNESCO, and repeated by UNESCO programme officers in Florence was that the Bank was trying to build a new, streamlined view of culture and development, just as it had done previously for indigenous peoples, women and gender, education, and science. It was actually reproducing the United Nations within its walls, without the General Assembly or the UNESCO General Conference. The meeting in Florence showed clearly that UNESCO was building a global society based on cultural diversity; the World Bank, in contrast, was focusing on a global economy for the exchange of cultural goods and properties. UNESCO had great social capital and no economic capital, while the World Bank had great economic capital and very little social capital.

6.2.2 Culture in a World of Trade: New York University, 2000

As research and debates on culture and development rippled through the political world, the subject drew the attention of both international institutions and the governments of many countries. The culture ministries of Canada and France agreed to create international spaces to discuss such issues. The first event along these lines was the seminar sponsored by the Canadian Minister of Culture, Sheila Copps, on "The Prospects for Culture in a World of Trade", held at New York University on 3 March 2000. The basic terms of the debate were set at this meeting.

I opened the meeting by stating that diversity is the source of human capability for development: we think by associating different images; we identify by contrasting ways of living; we elect by choosing from an array of options; we grow by rebuilding our confidence again and again through dialogue. In this new beginning, to cope with the momentous challenges of sustainability, governance and conviviability in a global era, we need all the creativity we can summon around the world. I asked whether globalization would take us down a road of diminishing returns as uniformity dried up "our creative diversity" or, on the contrary, whether it would give us greater "freedom to create". The answer would weave the texture of the twenty-first century and beyond.

World trade in cultural goods had grown exponentially in recent years: between 1980 and 1991, world trade in printed matter, literature, music, visual arts, cinema, and photographic, radio and television equipment almost tripled from US$67 billion to US$200 billion (UNESCO 1998: 432). Cultural industries were providing increasing numbers of jobs and becoming an important source of revenue from international trade, as could be seen from statistics for Canada, the UK and the US (Statistics Canada 1997; Creative Industries Task Force 1998). Nevertheless, the scale of North America's role in global media, both as exporter and investor, remained unique, with US-based multinational companies (such as Time Warner/Turner Broadcasting, Walt Disney/Capital Cities/ABC and Viacom/Paramount) dominating cultural export markets. These huge cultural industries were taking over the traditional forms of production and circulation of culture, capitalizing on the promise of this emerging market, as demand for cultural products increased.

In response to the expansion of the trade in culture, there had been a strong polarization of government positions which expressed not only an ideological debate but also positioning in the market. As the herald of free market principles and with strong economic interests in exports in the culture sector, the US strategy was to place cultural and economic goods and services in the same playing field as other exports, as deserving of exactly the same treatment. Culture was viewed as an industry like any other, subject to market laws and benefiting from all the advantages of free trade. In close alignment with this view were countries such as Japan, the UK and Germany, which were the biggest exporters and importers of cultural goods.

At the opposite end of the spectrum was the French doctrine of "cultural exception", based on the principle that cultural goods have an intrinsic value on their own that it is essential to maintain and protect, not only for artistic production and diversity, but also for national identity and cultural sovereignty. Close to the French position stood Canada, with less emphasis on public subsidies but with supporting measures "to help domestic companies to remain an element of choice in their cultural markets". Canada also demanded special treatment for cultural products and services in trade forums in order to ensure respect for the cultural concerns of equality of access, diversity of content and the rights of the creator within the global information society.

Much could be learned from the recent experience of the failed Multilateral Agreement on Investment (MAI) discussed at the Organisation for Economic Cooperation and Development (OECD) (summarized in Alonso Cano 2001). Multilateral investment rules were aimed at enhancing investors' confidence by providing a stable framework of clear and transparent rules applicable to all. The MAI proposal would require governments to treat foreign and domestic investors equally and would create a set of global rules to replace a patchwork of 1,600 or so bilateral investment treaties. However, this was the tip of the iceberg; there was a much more complex issue connected to technological change, the liberalization of international trade and investment, the role of the state, public service, and the balance between public interest and market forces.

Some negotiating countries in the MAI believed that the agreement needed to mention national public interest, and fought for exceptions to protect culture, public health, the environment or social rights. Many movements from all around the world railed against the MAI as a treaty of corporate rights and government obligations which ignored the concerns of citizens and was a step backwards in international human rights. Eventually, the MAI was quietly shelved. However, it had set the stage for negotiations at the World Trade Organization meeting in Seattle in 1999. But there was no agreement even on the agenda to discuss the exception clause for cultural industries. A UNESCO document signed by thirty-five ministers of culture was presented, stating that goods with cultural content should be given special treatment in international agreements on trade and investment.

I concluded my address by arguing that trade in commodities with cultural content would continue to expand in the years to come. If such trade offered opportunities for culturally diverse groups to participate in the production, circulation and consumption of such products, globalization would become a vehicle for communication. If, on the other hand, globalization led to oligopolies that restricted possibilities for creativity among different peoples, resistance and opposition would ensue. It was therefore imperative to develop new strategies to link globalization to cultural diversity and creativity.

6.3 The Role of the Media

6.3.1 "The Expression of Cultural Diversity Is a Right"

My opening address in New York was followed by a talk from Bonnie Richardson, Vice President for Trade and Federal Affairs with the Motion Picture Association of America (MPAA), in which she set out to debunk three myths concerning culture: first, that Hollywood opposes all policies of government on culture; second, that globalization is against cultural diversity; and third, that world trade rules destroy cultural diversity. "The expression of cultural diversity is a right", she contended. "The Motion Picture Association of America has opposed some tools that governments use because they are ineffective or inappropriate or have effects on third parties, but it is favourable towards acceptable tools", according to my notes.[1] She specified that these would be particularly useful on intellectual property rights, on tax incentives and on subsidies to promote cultural expressions for things that governments believe are important such as the fine arts, ballet or learning a minority

[1]Quotations not otherwise ascribed in this chapter are drawn from my own notes.

language. She pointed out that Europe now has fewer languages than in medieval times yet it is much richer in cultural terms. She also added that digital technology had vastly reduced the cost of filming and editing, and that the Internet allowed much cheaper dissemination.

Ms Richardson insisted that the MPAA was not the enemy of cultural diversity. There is some truth to this: the production of Hollywood films has always involved multicultural themes, actors and directors; part of their films' global success comes from the global outlook created by this multicultural method of composition. Nevertheless, the cognitive structure of story-telling in Hollywood films tends to create a uniform mind-set among publics in countries around the world. Unfortunately, there is still no research that analyses the cultural impact of Hollywood films in relation to local or national cultures. The evidence is mixed, with some countries such as India and Egypt being able to produce films in large numbers because their national publics prefer the styles and narratives of their own films. Others, such as Chile, Switzerland and New Zealand import ninety-five per cent of their film-viewing from Hollywood, while most countries fall in a middle range that combines Hollywood, European films and a few films from other regions.

Sondra Myers, the Director of the President's Millennium Seminars: The University for a New Democratic Era at George Washington University, had different views. She believed that in humanities institutions in the United States it was very difficult to think of culture in terms of policy. There were "culture wars" in the US, which included the religious right's wanting to abolish all public support for culture. More dangerous were the culture wars among ethnic and religious groups. She also mentioned the cultural war between neighbours, now that Canada was wanting to keep its own culture. She suggested that globalization could be harnessed to appreciate and enrich culture through the contributions of others, locally and globally. She recommended the creation of a cabinet position for Cultural Affairs, to nurture culture and to improve cultural relations with other countries.

6.3.2 *"The Idea that Local Voices Be Kept Alive Is not Opposed to Free Trade"*

The next session dealt with corporate consolidation and global media empires. Mark Crispin Miller, Professor of Media Ecology in the Culture and Communication Department at New York University presented the findings of a survey on whether people knew about the give-away of the digital spectrum, an item of public property worth seven billion US dollars. The survey team were outraged at the idea. They believed that media companies should pay for the use of the spectrum, and that they should be required to offer other content such as non-commercial educational programming for children.

Edwin Baker, Professor of Law at the University of Pennsylvania, argued that the idea of keeping local voices alive was not in opposition to free trade. Culture should sometimes be treated differently but most arguments made in the US for this are elitist or are harmful to consumers. He described three types of market failures. The first assumes that the market correctly prices goods but products may be mispriced because of externalities so that the price does not cover the costs. The second occurs when international products replace local products. The third economic problem is that the market works only as it measures preferences, so there is no reason to think that market measure is more truthful. Thus the argument for free trade fails, since products may satisfy consumer preferences but not provide that which is good for society.

6.3.3 "The More We Are Wired, the More We Are Split Apart"

Technology has changed what companies do and how they work, said Marnie de Kerckhove, Executive Director of the North American Broadcasters Association, at the opening of the session on technological convergence and the future of regulation. Text was bound in time and space and people had time and place; regulations organized this world. Now broadcasting and the electronic media were providing material for consumers, and they were not bound by time and space. However, she stressed that history has demonstrated that new media do not do away with previous media. The same would happen with cable, direct broadcasting and so forth. The key for regulators was not to impose a general regulatory framework but rather to understand the new medium.

Yochai Benkler, Director of the Information Law Institute at New York University, countered that regulation is not neutral but can nudge providers in a certain direction. Technological convergence could open or close avenues. A closed system is one where passive consumers choose from options provided by professionals. Yet this can go together with active consumer discussion of the options that professionals could then provide. "Letting the users use it" would mean cable sending very low downstream products so that the users could manipulate them. Nevertheless, he concluded that regulatory instruments cannot have it both ways: they must choose either a closed, right-wing media or a democratic, open one.

6.3.4 "Give People a Choice of What They Want to See"

The issues were clearly summarized by Beth Novec, Director of International Programs of the Yale Law School Project on the Information Society. What do we mean by regulation: rules for technology, or cultural values, or advertising? So what is the future of culture as a regulator for new media, whose culture is it, what is it, and what is its impact on technology? Self-regulation is generally preferred to government regulation but how do we define it? It means giving people a choice on what they want to see. The advantages of self-regulation are that it is cheaper, it has fewer transactions when individuals can make choices directly and not through a central broadcaster. It is also a more effective solution where laws cannot keep pace. At the same time, you can build into the regulatory structure other interests, such as the protection of children versus civil liberty; technology can provide that option. But information overload, the inability to get to the truth, may also isolate individuals, and it becomes a problem of privacy. It may also lead to too much individualism, taking the power away from culture, from collective life: the more we are wired together, the more we are split apart—this is digital loneliness.

She argued that there are both positives and negatives: "Regulation redistributes intelligence from one set to another. We have to ask who the new actors are that have control over the technology." She finished by asking: what can consumers have from culture? We can have new norms, new interfaces and personalized technology, more interactivity, the setting of rules separate from collective values. By engaging technology to address cultural development, it will be possible to rejuvenate cultural production in the era of convergence.

6.3.5 *How to Create Winning Conditions for Creators and Consumers*

The Assistant Deputy Minister for Cultural Development in the Department of Canadian Heritage, Michael Wernick, described the Canadian model for cultural policies. It had been centred on the freedom of creators and freedom of cultural choice. The government had acted as provider, sponsor, rule-maker and referee, with the main objective being "to create winning conditions for creators and consumers". Canadian cultural policy was intended to support creators and small industries, but it was not protectionist because it accepted all cultural goods and choices.

All groups should be allowed the conditions to express and create their culture internally and externally. For this, Mr. Wernick continued, the challenges are horizontal and vertical integration, and the convergence of television, films, video and the Internet, but there is concern that decisions will be left in fewer and fewer hands. According to my notes, he closed the session by declaring that "Canada is an emphatic supporter of economic liberalization but rules are made by economists and lawyers. It needs to become more diverse. A new international instrument on cultural diversity is proposed by Canada, in which all countries will participate."

6.4 One "World Culture" or a "Global Organization of Diversities"?

At the beginning of the 1990s, as economic globalization began to make real the possibility of a single world market, the term "world culture" began to crop up in international discussions. I had often heard it in academic meetings in the US so when I first arrived in UNESCO in the summer of 1994, to take up the position of Assistant Director-General of Culture, I was surprised to find that my use of it hit a brick wall. This was the total rejection of the term both by the Group of 77, especially the Asians, Africans, and those from the Middle East, because it implied the expansion of Western culture and by representatives of Western Europe because it implied the expansion of American culture.

My own knowledge of the term had been drawn from contemporary discussions in academic research. In his development of the world-systems theory, Immanuel Wallerstein contended that global processes operated according to the single logic of global capitalism (Wallerstein 1974). Extending this theory to culture implied that one culture is in the making worldwide. Other authors argued, to the contrary, that globalization operates according to several different dynamics and, therefore, many different cultures are being constructed (Featherstone 1990b; Escobar 1995; Arizpe/Alonso 2002). Featherstone, for one, strongly stated that "the varieties of response to the globalization process clearly suggest that there is little prospect of a unified global culture, rather there are several global cultures in the plural" (Featherstone 1990a: 10). Some authors differentiate according to locally specified patterns of consumption (Sahlins 1985; Featherstone 1990b; Miller 1995; Barber/Waterman 1995), others by the fact that commodities, ideas, media images, technologies and so on circulate according to quite different criteria (Appadurai 1996; Hannerz 1992).

The view which seemed to bridge the two opposing views was that of Barrie Axford. He thought it "highly unlikely that these flows and signs, whether in mass communications or in tourism, will produce 'a' global culture, given the diversity of their reception and use by local audiences. However, because they permit forms of social interaction not tied to place or limited in time, it is proper to see them as contexts or frames of reference with which new identities may be formed, and new understandings of the world fashioned" (Axford 1995: 157).

The WCCD regional consultations gave abundant proof that one of the outstanding preoccupations referred specifically to the new global reach of American cultural industries. Thus, implicitly, the whole idea of a "world culture" was left outside the margins of debates on culture. Instead, the issue that could already be identified then as a very serious concern—cultural diversity—grew into one of the major cultural challenges in the new millennium. This term was used because it encompassed the many different expressions that had been used in different countries in the recent past: 'tribalism' in Africa, "interethnic relations" in Latin America, 'nationalities' in socialist countries, "regional cultures" and "minority cultures" in the European Union (including the "autonomic movements" in countries such as Spain); "culture wars" in the US; the "politics of difference" in intellectual debates. All of these had different constituencies, cultural motivations, emblematic representations and political aims, yet in all of them the arbitrary sign that characterized the debate was a line dividing one culture, or several, from other cultures. They all had to do with cultural boundaries, which therefore had a bearing on political, economic and social advantages and positionings, as well as on emotional security and spiritual well-being.

While arguments about cultural diversity flourished around the world, what kind of debates were being held at local level? I will close with two examples.

6.4.1 Regional Debates: "Identities Are not Pacts with Gods but Negotiations Between Individuals"

Very distinctive and original thinking about culture policies was surfacing in Latin America and the Caribbean at this time. At the Meeting of Ministers of Culture sponsored by the Inter-American Bank held in Rio de Janeiro and Sao Paulo on 2–5 September 1998, President José Sarney opened the debate by stating that popular cultures, such as beach cultures, carnival cultures and street cultures, must not be destroyed by globalization. The Brazilian anthropologist Roberto DaMatta made an important statement saying that "Today, individuals are not defined by family or nation but as individuals in a civil society organized by associations, work, neighbourhood and so on. There is an indigenization of modernity with experiences of identity which are still ongoing: for example, the cult of the 'Santeria'—an Afro-Cuban-Brazilian tradition that still includes African deities." He argued that "identities are not pacts with gods but negotiations between individuals seeking to fulfil their egotistical interests".

Sérgio Paolo Rouanet, a former Minister of Culture for Brazil, argued that nation states have created a narrow functional modernity or globalization that needs to be opened up into an emancipatory modernity that also comes from the Enlightenment. He saw this as universal, a civilizational project. But he then posed the question:

> Who is the subject of global culture? Different versions state that it is an extension of symbols and values of the North expanding to the whole world. Another version says that global culture is not the sum of cultures, nor is it leading a hegemony but, instead, it offers a new culture, no longer just MacDonald's or fast food but films made in Australia with Japanese financing.

He argued that global capitalism is fundamentally cosmopolitan. The new globality is taking root again in the local. To take refuge behind cultural particularisms leads to a parody, a false consciousness or a cultural nationalism, against the United States and Europe, which is essentialist and protectionist. Universalism leads to hybridization. Rouanet concluded his presentation by saying that we have to move towards a post-identities identity: not to be objects of globalization but subjects of universalization. The Cuban delegate who followed put forward a diametrically opposed view by stating that the "packaged culture" (*la cultura enlatada*) of globalization must be opposed. Instead, "a culture of resistance must emerge to defend the cultural identity of our peoples".

The delegate of the Organization of American States (OAS) then presented the OAS's official policy. He stated that globalization is part of the process of expansion of European culture which has been going on for five centuries. Present-day individuals recognize fewer symbols but these are present in more places, expanding through the channels of entertainment, primarily music and films. Thus, globalization may not be a negative process but positive for many peoples. He argued that the case of Latin America is different: here globalization points towards the conservation of identity, not homogenization, including the restoration of values that had previously been marginalized.

Different views on culture and globalization could be perceived among the representatives of the Caribbean countries. In the opening session, the delegate from Guyana began by saying that, in his country,

> We have disparities in development and in state. We are a new nation. We are from a poor country with needs for schools and housing, so how can we lobby for culture? We still have to fly to Miami to come to Rio de Janeiro: aviation routes, radio, television, the music industry are still colonial. So we must discuss how we are going to sell our products in the world market; even the European Union finds difficulties in selling their cultural products in the United States.

In a later session, he also argued that "We haven't recognized that indigenous peoples are scientists of ethnobiology, biodiversity and especially of democracy. We are still debating who is discovering who, who discovered who." Echoes of this position were heard from other Caribbean delegations. The representative of Barbados stated plainly that "In the Caribbean we have other concerns. We don't want to see culture as an esoteric subject but to look at culture and economics, the expansion of markets for American products, equal opportunity of access to the marketplace." She argued that what was needed at these meetings was "to develop a Latin American and Caribbean position and, if not Latin American, at least a Caribbean position".

6.4.2 Local Debates: Basque Culture in the Twenty-First Century

Bayonne is a medieval city at the mouth of the river which brought so many Basques down from the green mountains of the Pyrenees to the port from which they dispersed throughout the Americas. It is filled with many Basque monuments and is home to the Bayonne festival, which draws thousands of Basques to celebrate their culture in music, dancing and parades of remembrance at the beginning of August each year.

In the French Basque region the landscape looks Basque because houses keep to a traditional Basque architecture. In fact, this architecture was reinvented at the end of the nineteenth century by a Commission for Basque Traditions, which also promoted traditional sports (including jai alai and Basque pelota), arts and crafts, music and dancing. A minority in the French Basque area still speak this mysterious language. Once thought to have been an Altaic or Uralic language, it has since been established that it belongs to none of the Indo-European linguistic families. Instead, it is related to the extinct Etruscan language and to the Navajo language, which traces its ancestry to a northern Siberian region. This distinctiveness of the Basque people is also confirmed by mitochondrial DNA, which links them to northern Asiatic groups.

My own surname, Arizpe, is of Basque origin,[2] related to the Basques who colonized northern Mexico and California, creating the "New Biscayne" ("Nueva Vizcaya") in New Spain in the sixteenth century, with the town of Arizpe as its capital. This was part of the reason why I was invited to participate in the seminar "What Basque Culture for the Twenty-First Century?" in Bayonne, on 5–6 November 1999. It began with a verbal explosion, perhaps reminiscent of the violent explosions that ETA, the Basque terrorist group, had been causing in Spain for many years. The French Representative to the UN General Assembly protested that he had been accused of advocating "ethnic cleansing" and 'balkanization' when he spoke out for Basque culture. His proposal for a Basque university had been labelled "a belated desire for modernity", yet he demonstrated the incongruity of considering archaeological heritage as French, and Basque heritage as ethnology.

Antoine Spire, a French television anchor, replied that if an identity closes itself up against globalization, it is doomed: there are no 'pure' cultures. I then proposed that cultural rights can be defended only on the basis of a universal acceptance of the political principle of human rights and respect for the individual. Therefore, the universalization of such rights is a condition for the demand of cultural rights for a group holding a distinct identity. Eguzki Urteaga, a young French Basque university student, agreed with me. Instead of saying that Basque is a menace to universality, one may say the opposite: that being Basque is the condition of being universal.

Then Erramun Bachoc of the Basque Cultural Institute angrily recalled the fact the Basque people had been excluded from modernity by the policies of the French state, and Spire quickly agreed that it was the failure of those who formulated such

[2]From the Guipúzcoa region in the Basque country in Spain and also, written as Harispe, from the valley of the Aldudes in the French Pyrenees.

policies. However, did not the Basques accept this by turning their backs on cosmopolitanism, against exchange with others, and digging deeper into their isolation and their backwardness?

Questions followed from the floor, one of which, addressed to me and posed by a young man in Basque, had to be translated by one of the members of the panel. I first asked, in French, what language I should answer in. Not in French, came the response, because it is a colonial language. In Spanish then? No, because that is a colonial language for *me*. In Nahuatl (Aztec) then? No, although I do speak it from doing anthropological fieldwork in Mexico. Which of these is *my* language? In my ancestry, I have the Basque language, but also Swiss and Spanish, and I choose to have Nahuatl and English. I asked him how it was that we were going to communicate, and I argued that we need a lingua franca. The languages we have come to us through happenings in history—how can it be otherwise? Are we to choose by world rotation the languages that we will use for international communication?

The Vice-Minister for Language Policy of the Basque Country in Spain responded that, "We don't want Basque to be only a heritage, we want a living language and we want it to be a source of cultural creativity. It is the will of all those wanting to speak Euskara that will bring this about, but we need a linguistic policy. Even if it doesn't exist, it exists." One of the participants in the audience had strongly insisted that the Basque identity depended on the continuity of the Basque language. How true was this? A recent survey had provided evidence to show that differences in perception of identities are closely linked to, but not determined by, linguistic proficiency. In the "Basque country" (encompassing both the Spanish and the French regions), 71% of respondents who were Spanish-Basque or French-Basque bilingual individuals considered themselves exclusively Basque. Among those bilinguals who could understand Basque but did not speak it fluently, only 36% considered themselves Basque, while 25–55% considered that they had dual identity: both Basque and either Spanish or French. In contrast, of those who spoke only French or Spanish, only 11% considered themselves purely Basque, while 22% classed themselves as solely French or Spanish, and 56% said that they were both Basque and either Spanish or French (Gouvernement Basque 1999). Importantly, this showed that language and identity are not tied together and that bilingual people in particular may align themselves with a single cultural identity or with both of those linked to the languages they speak.

We have come full circle back to the importance of language in sustaining culture. The diversity of languages in our world is the backbone of our cultural diversity. And it is to that diversity that we now turn in more depth: as discussed and championed in the UNESCO Universal Declaration on Cultural Diversity and in the World Culture Reports.

References

Alonso Cano, G. (2001). The MAI: on the threshold of a cultural war? In: UNESCO (1998). *World culture report 1998: culture, creativity and markets* (p. 77). Paris: UNESCO.

Appadurai, A. (1996). *Modernity at large: cultural dimensions of globalization*. Minneapolis: University of Minnesota Press.

Arizpe, L. & Alonso, G. (2002). *Cultura, comercio y globalización*. Buenos Aires: Consejo Latinoamericano de Ciencias Sociales. First published in English as *Culture, trade and globalization*. Human Development Report Working Papers. New York: UNDP, 2000.

Axford, B. (1995). *The global system: economics, politics and culture*. Cambridge: Polity Press.

Barber, K. & Waterman, C. (1995). Traversing the global and the local: fújì music and praise poetry in the production of contemporary Yorùbá popular culture. In: Daniel Miller (Ed.), *Worlds apart: modernity through the prism of the local* (pp. 240–62). London: Routledge.

Creative Industries Task Force (1998). Creative industries: mapping document. London: Department for Culture, Media & Sport.

Clinton, Hillary. 1999. Speech.

The Economist (2002). Tongue-tied: the imminent death of a rich African language. *Economist*, 17 October 2002. http://www.economist.com/node/1394769, accessed 3 October 2016.

Escobar, A. (1995). *Encountering development: the making and unmaking of the Third World*. Princeton: Princeton University Press.

Featherstone, M. (1990a). Global culture: an introduction. In: M. Featherstone (Ed.), *Global culture: nationalism, globalization and modernity* (pp. 1–14). London: SAGE Publications.

Featherstone, M. (Ed.) (1990b). *Global culture: nationalism, globalization and modernity*. London: SAGE Publications.

Gouvernement Basque, Gouvernement de Navarre et l'Institut Culturel Basque (1999). *Enquête sociolinguistique au pays basque*. Bilbao: Gouvernement Basque.

Hannerz U. (1992). *Cultural complexity: studies in the social organization of meaning*. New York: Columbia University Press.

Miller, D. (Ed.) (1995). *Worlds apart: modernity through the prism of the local*. London: Routledge.

Sahlins, M. (1985). *Islands of history*. Chicago: University of Chicago Press.

Serageldin, I. (1998). Investing in Culture: A View from the World Bank. Remarks Delivered at UNESCO Symposium on Culture and Development, Stockholm, Sweden, 30 March 1998. Available at: http://www.serageldin.com/Speech/Details.aspx?ID=29.

Statistics Canada (1997). *Canada's culture, heritage and identity: a statistical perspective*. Ottawa: Ministry of Industry.

UNESCO (1997). *Intergovernmental Conference on Language Policies in Africa, Harare, Zimbabwe, 17–21 March 1997: final report*. Paris: UNESCO.

UNESCO (1998). *World culture report 1998: culture, creativity and markets*. Paris: UNESCO.

Wallerstein, I. (1974). *The modern world-system, vol. 1: capitalist agriculture and the origins of the European world-economy in the sixteenth century*. New York: Academic Press.

World Bank (2000). *Culture counts: financing, resources and the economics of culture in sustainable development. Proceedings of the Conference, Florence, Italy*. Washington DC: World Bank.

Joint General Assembly of the International Council for Philosophy and Human Sciences and the International Social Science Council, hosted by the Chinese Academy of Social Science, Beijing, 2004. *Source* Photo was provided with permission by the Chinese Academy of Social Sciences

Chapter 7
Culture and Conviviability

This chapter considers the new phases in international transactions on culture that took place at the turn of the millennium. In the first decade of the twenty-first century, the range of topical discussions on the concept increased, following the publication of the two UNESCO World Culture Reports: one on Culture, Creativity and Markets (UNESCO 1998), the other on Cultural Diversity, Conflicts and Pluralism (UNESCO 2000b), both of which were soon after overtaken as UNESCO shifted attention from culture and development to cultural diversity, the challenges of culture in the marketplace and cultural creativity. Two new international conventions—on intangible cultural heritage and on the diversity of cultural expressions—were adopted by UNESCO member states in 2002 and 2005. By then culture had permeated discussions in other arenas and institutions, touching on the "Dialogue of Civilizations" (an international initiative sponsored by President Khatami of Iran) and playing a central part in the conference on science and culture at the Pontifical Academy of Sciences.

The year 2004 was a marker in the consolidation of international political platforms on culture. This was set in motion by three major international actors: the Global Economic Forum of Davos introduced topics such as religion and culture, creativity, identity, and culture in markets; the United Nations Development Programme took up questions of cultural diversity, cultural liberty and human development; and the Rockefeller Foundation sponsored a project on how to understand and create value in economics, ritual and art.

In parallel with these transactions on culture, another debate emerged in non-economic social science on the effects of neo-liberal economic policy in societies. As Vice-President, and then President, of the International Social Science Council, I was directly involved in this debate and my notes on the relevant meetings will therefore be included in this chapter. As joblessness and inequality began to take their toll, social research showed that young people in developing countries were suddenly opting in unprecedented numbers for drugs and for dealing in the black and grey economies, as well as other criminal activities, a phenomenon whose impact received far more widespread attention when it spread to youth and

L. Arizpe Schlosser, *Culture, International Transactions and the Anthropocene*,
The Anthropocene: Politik—Economics—Society—Science 17,
https://doi.org/10.1007/978-3-642-41602-6_7

large segments of the population in developed countries in the 2010s. While policy statements increasingly referred to "failed states" and "failed governments", sociologists, political scientists and anthropologists highlighted the danger of "failed societies". My own contribution was to insist that 'conviviability'—that is, the capacity to relate to others, a capacity that finds its source in multiple strands in many cultures—should be a key concern in development.

Culture, consciousness and the soul: Pope John Paul II gives an audience to the scholars attending the Seminar on Culture and Science at the Pontifical Academy of Science, Vatican City, 2002. *Source* Photo: Vatican Photographic Services

7.1 The UNESCO Universal Declaration on Cultural Diversity

As part of the preparation and celebration of the United Nations Year of Dialogue among Civilizations in 2001 (see Sect. 7.3), two round tables of Ministers of Culture were organized by UNESCO: the first, "Cultural Creativity in a Globalized World", was held on 2 November 1999, with a second, on "Cultural Diversity 2000–2010: Challenges of the Marketplace", following on 11–12 December 2000. The background document for the latter advocated "a UNESCO Declaration to promote cultural diversity in the era of globalization as a strong political principle in a world in deep transformation, offering guidelines at world level on the interactions between different actors and operators of culture—creators, public institutions, civil society, enterprises of production and distribution, etc." (UNESCO 2000a: 1). For the purposes of this book, it is necessary to clarify the background of the concept of "cultural diversity". The fact that it was suddenly picked up in the early 1990s, after the fall of the socialist regimes, and propelled into the highest international political and development policy transactions is, in itself, an interesting political phenomenon that needs deeper explanation, but that lies beyond the purview of this book. During the nineties, some authors attempted to provide an explanation. For example, Craig Calhoun made the point that the 'old' politics "was the result of hegemonic ideologies differentiating the 'properly political'—and, therefore, most explicitly contestable—dimensions of life from others, and, accordingly, obscuring the workings of power and power struggles in other realms of life … . In short, class politics was partly and also necessarily, identity politics" (Calhoun 1995: 231–232). Calhoun also pointed out that "The politics of identity has often appeared as a new politicization of everyday life, a shift away from some more traditional politics of interests. Not only is the personal increasingly politicized, some analysts note, politics is increasingly aestheticized" (ibid.: 231). This is related to the context-dependent nature of cultural diversity, which gives priority to the way in which cultures are 'framed' and 'displayed' according to contexts and to who the players are in a given setting.

In my view as an anthropologist, "cultural diversity" is a descriptive term that refers to the inherent capacity of all human beings to create meaning and symbols which, patterned according to group formations, may, at a given point in time, produce cultural boundaries. In a pre-twentieth-century world such cultural boundaries tended to coincide with geographical, ecosystemic or historical demarcation lines and were held in place by cultural and religious gatekeepers. These boundaries are arbitrary insofar as they become visible only when an individual or a group displays behaviour that marks that boundary.

To place this in statistical perspective, a matrix that included every cultural behaviour of all individuals in the world would presumably show a continuum of variation, with clusters creating mountains and valleys. A three-dimensional model of such clusters would show how such patterns are generated at different scales. The assumption must be made, as anthropology has shown, that such variation is not infinite but seems to have a finite set of rules and elements. In other words, the contemporary explanation for this finitude is that the genetic make-up of human beings, without taking into account mutations, provides a basic grid of capabilities that no contemporary human groups seem to have overridden. Recent discoveries and hypotheses in genetics, neurology, evolutionary psychology and generative grammar indicate that all human beings have certain capabilities, such as generating language through a commonly shared cognition—a momentous theory that is generating its own complex debate.

My own view is that human creativity, even if channelled through certain parameters, cannot be walled in. In the same way, cultural diversity cannot be reined in, unless this is done by deliberate political, economic or religious design. Where this happens, there is a loss of diversity and, even more importantly, of creativity. And it is creativity that will pull the world out of the dangers of environmental collapse. That is why I oppose a multiculturalism or interculturality that defends previously existing cultural diversity and thus curtails the single most important trait of human survival and progress: creativity.

Returning to the background document to "Cultural Diversity 2000–2010: Challenges of the Marketplace", in the section on "Ideas for a General Perspective", two major new points in the discussion were highlighted. The first was that "cultural diversity must be conserved and promoted on the basis of human rights and the right to self-determination as it is a manifestation of the creativity of the human spirit which is a source of inspiration and of aesthetic satisfaction" (UNESCO 2000b: 3). The second was that: "cultural diversity has an increasingly fundamental role to play in the development of the knowledge economy. When they are truly a reflection of diversity, the major cultural industries that promote linguistic diversity and artistic expression, such as media, have a positive impact on pluralism, competitivity and employment" (ibid.). This was further reinforced by the following point, which began "Culture and, in particular, cultural industries, are an essential source of creation of wealth in the world, as culture has become one of the motors of development."

The major concern of many member states—that cultural products could not be considered in the same way as other merchandise but required dealing with according to the "cultural exemption"—was accommodated in point 2.9:

> Globalization may, in this way, allow for the establishing of closer links between cultures and enrich them but it may also create challenges to cultural diversity: a) the evolution of multilateral policies have an impact on cultural diversity… hence the importance of the current commercial agreements in the World Trade Organization on cultural goods and services. b) development of the new information technologies may also have an impact on cultural diversity. The audio-visual industries and media offer powerful new tools for the expression of cultural identity yet may also represent destructive forces for the most vulnerable cultures from an economic and social viewpoint … c) although all member states must take up this challenge starting out from distinctive cultural perspectives, the common world context of development requires that a series of principles be established providing a coherent framework for action to conserve and favour cultural diversity at all levels, local, regional, national and international (ibid.: 4).

In accordance with these recommendations, the UNESCO General Conference adopted the UNESCO Universal Declaration on Cultural Diversity on 2 November 2001. Its preamble, in accordance with definitions given in several previous intergovernmental meetings stated that the General Conference was "Reaffirming that culture should be regarded as the set of distinctive spiritual, material, intellectual and emotional features of society or a social group, and that it encompasses, in addition to art and literature, lifestyles, ways of living together, value systems, traditions and beliefs" (UNESCO 2001: 1).

The Declaration was carefully crafted on the basis of the round table meetings and work by highly qualified academics; it was coordinated by Katerina Steniou, the UNESCO Director of the Department of Cultural Diversity. It gave a precise, knowledgeable vision of diversity, bringing in, although not developing, the new terms and connections of the international debates on culture. Under the headings of

"Identity, diversity and pluralism" (previously discussed in the second World Culture Report), "Cultural diversity and human rights", "Cultural diversity and creativity" and "Cultural diversity and international solidarity", it addressed the centrality of culture to all the peoples of the world and the role of diverse cultures in enriching our lives, the need to recognize cultural rights, the different ways in which creativity can be inspired by culture, and the necessity for government policies to be put in place to support a wide diversity of cultures in the face of market forces.[1]

The question of defending cultural diversity from globalization led to a very large number of initiatives that cannot be assessed here. One illustrative example was Jeremy Rifkin's proposal to establish a World Cultural Organization to help represent diverse cultures and put cultural protection on an equal footing with the World Trade Organization (Rifkin 2001). The previous year, in his book *The Age of Access*, Rifkin had argued that, given that ideas, inventions and logos had a major influence on all aspects of development, culture was the central process of late capitalism (Rifkin 2000). In fact he labelled the contemporary phase of capitalism as cultural capitalism. Another example was the civil society initiative that became the International Network for Cultural Diversity. This organization proposed that governments must not enter into any agreements that constrain local cultures and policies that support them, and that a new international agreement should be created, to provide a permanent legal foundation for cultural diversity.

7.2 The World Culture Reports

One of the recommendations of the World Commission on Culture and Development, the publication of the UNESCO World Culture Report, quickly went into effect. The 1998 report was subtitled "Culture, Creativity and Markets", the 2000 report "Cultural Diversity and Pluralism".[2] I had strongly advocated such publications, since the Culture Sector of UNESCO was the only one not to publish a regular report. I came to the project with all the enthusiasm of an anthropologist who envisages only the collection of data. I soon found out that many other concerns were at issue, both overt and covert, all of which would surface along the way.

The 1998 Report began by proposing a definition of "living culture": "When we speak about culture, we are looking at ways of living as individuals and ways of living together. A 'living culture' is one which, almost by definition, interacts with others, in that it involves people creating, blending, borrowing and reinventing meanings with which they can identify" (UNESCO 1998: 15). The Report

---

[1]The full text of the Declaration is available at: http://www.unesco.org/education/imld_2002/unversal_decla.shtml.

[2]The report committee was chaired by me and consisted of Louis Emmerij, Keith Griffin, Yoro Fall, Elizabeth Jelin, Mohan Rao, Paul Streeten and David Throsby. The executive director was Ann-Belinda Preis and the staff researchers were Paola Leoncini-Bartoli and Isabelle Vinson.

emphasized "culture's connections" in the same way that Marilyn Strathern refers to "becoming enlightened about relationships", as mentioned in Chap. 1. It explored the themes of culture, creativity and markets, and strategies to create statistical indicators for culture and development, a goal towards which there had been little progress after three years. What would such an indicator measure? The heuristic concept of "cultural development" immediately runs into intractable ontological and political problems about whether cultures 'develop' or 'evolve'. "Cultural production" gets bogged down in capitalist or non-capitalist forms of production of cultural goods. "Cultural freedom" is soon petrified in its definitions by the dictates of specific political ideologies. "Cultural repression" would require micro-observations of forms of psychosocial distress related to culture, and these soon become unmanageably subjective. The plan to develop such indicators was quickly abandoned and efforts were concentrated on collecting statistical data on culture and development. At this point I discovered that, in many (if not most) countries in the world, data on cultures and languages (not to mention religion) are the most politically sensitive statistical data of all. All available data and statistics on culture were aggregated for the world, for regions and for subregions. The work on these indicators of culture and development was continued at a seminar held during the World Bank meeting "Culture Counts", which took place in Florence in October 1999 (see Sect. 6.2.1).

The second World Culture Report dealt with "Cultural Diversity, Conflicts and Pluralism" (UNESCO 2000b). The first section set out the argument that diversity—including not only diversity related to culture but also diversity linked to gender, race and sexual preference—need not threaten stability, as long as citizens are able to adhere sufficiently to national values and cultural practices to secure general compliance and support for the institutions of governance. The economic aspects of the conservation of cultural heritage, and the construction of knowledge through the new information and communications technologies were also explored.

The slow and painstaking statistical work following on from the 1998 Report finally made it possible to include a table specifying the official, national and five most spoken languages in a majority of UNESCO member states. This was a startling eye-opener in many cases and help ground many programmes in a more incisive way. For me, one of the most important aspects of the table was the extent to which it demonstrated that cultures are the infrastructures of political power, each defined by and defining the other (ibid.: 312–15). Through the use of innovative techniques, new data were created: for example, by mining one of the deep archives on ISBN registrations in UNESCO, a table of "Translations and Books in Foreign Languages" was obtained and had quite an impact through showing which countries were open to literature and scientific publications from all other countries (ibid.: 372–5).

A few years later, after strong opposition from several member states to the Reports, for explicit reasons that in some cases did not correspond to the sotto voce conceptual and political ones, the World Culture Reports were suspended indefinitely, on the general basis that UNESCO was publishing too many reports.

7.3 The "Dialogue Among Civilizations"

As mentioned in Chap. 1, with the media impetus given to Samuel Huntington's 1993 thesis on "the clash of civilizations", the term 'civilization' was inserted into high-level international debates. Many of us at UNESCO were surprised by the implicit acceptance of the term and the implicit understanding that both nation states and cultures were subordinate domains to civilizations. Although the term had been discarded among anthropologists (for reasons explained in Sect. 1.7), the Iranian President Mohammad Khatami led an effort to proclaim the year 2001 as the United Nations Year of Dialogue among Civilizations (UN Resolution A/RES/53/22, 16 November 1998). Such is the irony of history that this was the year when the destruction of the World Trade Center in New York also made plain how realist political forces were planning a different pathway for culture in international affairs.

On 16 November 2000, the UN Secretary-General, Kofi Annan, issued a Report on the United Nations Year of Dialogue among Civilizations, in which he expressed his gratitude to a number of eminent persons who had agreed to reflect on the issue of diversity as the "concept underlying a focused reflection about dialogue among civilizations" and the means for managing diversity so that it would no longer be perceived as a threat (United Nations General Assembly 2000: 2). Led by his special representative, Giandomenico Picco, this distinguished group of thinkers and artists (of which I was a member) met for a number of discussions and summarized their findings in a book, aptly entitled *Crossing the Divide: The Dialogue of Civilizations*[3] (Picco et al. 2001).

In the document given to us to initiate our discussion, Mr. Picco had written that, for some, the United Nations Year of Dialogue among Civilizations "implied bridge-building between Islam and the West; for others, a new embrace among the inheritors of the great historic civilizations; and for still others a means of communication among the great religions of the world" (Picco et al. 2001: 14). His argument was that "Wars are initiated by a mindset that perceives diversity as a threat. They are the responsibility of those individuals who wage them: for history does not kill, religion does not maim and institutions do not destroy buildings, only human beings can do that" (ibid.). Again, as in Huntington's formulation, civilization becomes a surrogate for religions. What Mr. Picco failed to observe is that, in some major religions, engaging in violence and "just wars", killing and maiming are fundamental values of their doctrines.

A very interesting aspect of the ensuing discussion, however, was introduced in the first meeting of the group, in Vienna in December 2000. Mr. Picco explained

---

[3]Those involved in the discussions were Prince Hassan Bin Talal, Richard von Wiezacker, Jacques Delors, A. Kamal Aboumagd, Hannan Ashrawi, Ruth Cardoso, Leslie Gelb, Nadine Gordimer, Serguet Kapitza, Hayao Kaway, Tommy Koh, Hans Kung, Amartya Sen, Tu Wei Ming, Javad Zarif and myself.

that the main questions for writing the book dealt with the reassessment of the concept of enemy, the reassessment of individual accountability and the discussion of two attitudes to different civilizations: the perception of difference as a threat and an approach based on a spirit of tolerance. Javad Zarif countered that we must not classify civilizations into two—those that accept diversity and those that do not. We must avoid defining a new enemy. I made the point that, as far as anthropologists were concerned, the term 'civilization' had been thrown out in the 1950s. I argued against emphasizing this particular way of aggregating human beings, who live equally in cultures, in nations and in the world. Sergei Kapitza concurred, declaring that to speak of 'civilizations' was not useful, since there is only a single civilization of human beings.

Later in the meeting, Zarif raised the question of how to derive a web of global values, which appeared to be our task. Kapitza suggested adding new words such as 'altruism'. I interjected that we must start out by first recognizing an important commonality: that of protecting the biosphere so that human life remains viable. Picco again reiterated that the common denominator should be human dignity and the United Nations Charter. Zarif also stressed that we needed to present a positive approach that promoted equity not superiority. Moreover, he saw it as important to recognize that diversity covered both state and non-state actors. I agreed with him and emphasized that we need to develop justice related to diversity so that it is compatible with equality in economic and political opportunities.

The culmination of our project came with its presentation by seven of us to the 56th UN General Assembly on 10 November 2001. Just two months earlier, of course, the al-Qaida attack of 9/11 had further exacerbated both the relevancy and the urgency of the subject we were writing on.

After our presentations, Ambassador Louis Michel, President of the European Council at that time, spoke of the "monstrous attacks of 11 September that had deliberately sought to spread hatred among peoples and to provoke confrontation between religions and cultures". He cautioned that, in the spirit of the dialogue among civilizations,

> we must absolutely prevent two dangers. Everything must be done so that there is no confusion between the struggle against terrorism and that against certain troublesome oppositions. And this danger is real. We must also do everything so that terrorism does not touch, at all, the level of democratic demands and quality of our societies.

Ambassador Michel continued,

> The only real response is precisely a dialogue, and today it is a dialogue of cultures, a dialogue among civilizations, which cannot be reduced to a dialogue among religions … What are the principles on which a true dialogue of cultures may be established? First of all, the equal dignity of all cultures, and their vocation to interpenetrate and to enrich one another in a spirit of tolerance and mutual respect. The other foundational principle is the need for cultural diversity. The right to difference and to identity.

He went on to add a very important consideration:

> This is the perspective of the European Union on the dialogue among civilizations. It must be carried out without taboos. We have the right to ask questions of the civilization that is

ours and ask questions of others. Others also have the right to ask us about why there are certain inequalities.[4]

Ambassador Michel's remarks are important since, in the final pages of *Crossing the Divide* the question is posed "Does reconciliation embody a new global ethic?" (Picco et al. 2001: 210). That question is then developed: "If conflict resolution is a 'legitimate' role for an international institution, is reconciliation a more appropriate role of the new 'power structure' of our time and age, names, for individuals, the non-governmental groups and the very local institutions?" (ibid.). Thus, by lowering the gaze of analysts towards individuals and "very local institutions", the book's conclusion dispenses with the need for higher-level 'civilizations' or 'religions' to examine their own fundamental values.

As transactions on culture and cultural diversity expanded to other spheres, many other countries became involved in this discussion. In Berlin, the Haus der Kulturen der Welt (House of World Cultures) sponsored a conference on "The Global Dimension of Cultural Policy", immediately after the UN General Assembly meeting. Sarat Maharaj began by stating a modernist concept of cultural practice: that the notion of crisis is integral to art, to allow the artist to break out of established practice.[5] He was taking a very different perspective from the conservationist one that prevailed at the time in other meetings and writings. He then highlighted the point that culture is about *heterogenesis*: that is, the creation of difference. This contrasted with the way that, concurrently, in my area of work, culture was being defined as a sameness, a fundamentalism, an essentialism. Knut Nevermann argued that globalization implies secularization as a condition; only then can you start talking of cultural freedom. If that is the case, however, how do you provide representation? Where is the common platform for global understanding? In England the chador becomes a masquerade of religion, while France has dealt with it in more powerful terms. So these are partial models that do not yet fit.

7.4 Two New International Conventions on Culture

As mentioned above, many international meetings took up topics related to culture, stimulated by the Universal Declaration on Cultural Diversity, the UNESCO World Culture Reports and the Group of Eminent Persons for the Dialogue of Civilizations. These generated the momentum to create two new UNESCO international conventions: the Convention for the Safeguarding of the Intangible Cultural Heritage, adopted in 2003, and the Convention on the Protection and Promotion of the Diversity of Cultural Expressions, adopted in 2005.

---

[4]The final four sentences were translated by this author directly from the French version of his speech, since they were omitted in the informal translation into English.

[5]The accounts of this and other meetings mentioned in this chapter draw on my notes of the discussions.

The coalitions that led to the adoption of these two conventions were drawn from a wide range of actors: governments, indigenous and aboriginal associations, free market advocates, civil society organizations, new information and communications enthusiasts, cultural survival activists, artists, and anthropologists and other social scientists, all supporting and intervening in the procedures set in motion at UNESCO. Since a broad range of publications already exist on these conventions, only the key points of studies related to this endeavour will be mentioned here, in particular those related to anthropological discussions.

The first series of studies and debates in UNESCO—on the protection of "oral traditions and languages", proposed by African and Arab countries, and on 'folklore', proposed by Bolivia—began in 1971, as recorded by Serge Gruzinski, and culminated in the 1989 UNESCO Recommendation on the Safeguarding of Traditional Culture and Folklore (Gruzinski 1993: 4). Gruzinski explained how, at a meeting of experts held in 1984, attempts to suggest that the typological approach be taken by some of the participants "ran up against the difficulty of reconciling a folklore approach with one inspired by cultural anthropology ... placing the emphasis on the social and cultural relations which exist within any society" (ibid.: 11). He also pointed to an issue that was only just beginning to be discussed at the time: the suggestion that the Western scientific 'tradition' of referring to universal categories should be relativized (ibid.).

Georges Condominas, a French anthropologist and one of the pioneers in building up the concept of intangible cultural heritage in UNESCO programmes, argued in 1993 that "the products of intangible cultures might be said to constitute living material whose contours fluctuate and are determined by specific social contexts and modified, in the course of history, by the societies concerned which themselves evolve" (Condominas 1993: 10). Importantly, in several instances he referred to "recently created mixed cultures", which could provide an opportunity to observe the formation of new languages born out of the contact between two or more cultures or subcultures. Gruzinski similarly made a strong case for the revitalization of traditional and popular cultures and for taking into account "new cultures" that burgeon in urban areas, assume hybrid and syncretic forms, are rooted in diasporas and are not able to lay down durable roots since they have no area of their own in which to do so (Gruzinski 1993: 20). He rightly emphasized the manipulations and aberrations in defending traditional popular heritage that exalt "ethnic purification", going on to say that "The perversion of respect for ethnic units is all the more likely as the rapid disappearance of traditional models is encouraging a reversion to attitudes which belong to the past: those attitudes exploit the ideological exaltation of old models which are turned into the foundation of an aggressive and destructive policy" (ibid.: 21). His foresight can only be admired as we reflect on the irrationality and brutality that has arisen in this second decade of the millennium as the result of religious terrorism and cultural 'retribalization'. The question that remains is, if insights such as these were already being mentioned in 1993 (as also in the ICAES Congress in that same year, quoted in Chap. 1), why were they not followed up in policies in the new millennium?

With Sakiko Fukuda-Parr and participants at meeting to discuss the 2004 Human Development Report and her staff to discuss the 2004 HDR on *Cultural Liberty in a Diverse World*, United Nations Development Program, New York, 2004

In a 1954 UNESCO publication Claude Lévi-Strauss made the distinction between ethnography (description of cultural data), ethnology (synthesis of data about an ethnic society) and anthropology (comprehensive knowledge that applies to all human development, with conclusions that hold for all human societies) (UNESCO 1954). On comparing the definitions given in the 1989 Recommendation on the Safeguarding of Traditional Culture and Folklore and in the 2005 Convention on the Protection and Promotion of the Diversity of Cultural Expressions, it seems to me that ideas evolved from ethnology to anthropology. The 1989 Recommendation had provided the following definition:

> Folklore (or traditional and popular culture) is the totality of tradition-based creations of a cultural community, expressed by a group or individuals and recognized as reflecting the expectations of a community in so far as they reflect its cultural and social identity; its standards and values are transmitted orally, by imitation or by other means.

In comparison, in the text of the 2005 Convention, diversity was defined as "the manifold ways in which the cultures of groups and of societies find expression", calling attention both to cultural heritage and to the creative arts as conveyors of cultural meaning. In terms of its policy directives, the Convention aims to raise awareness of the need to integrate culture into development as an alternative to the dominance of narrowly defined economic policies.[6]

Importantly, it must be said that the prolonged and complicated political and diplomatic negotiations systematized and concentrated the transactions that had previously been taking place since the publication of the World Commission on Culture and Development Report, the Stockholm Conference Plan of Action, the UNESCO Culture Reports, the Universal Declaration on Cultural Diversity and the numerous international meetings that were convened, especially in the 1990s and also in the first years of the new millennium. The sphere for debate expanded beyond international agencies and governmental institutions to include business, social science, development organizations and religious institutions. An important initiative was taken by the Pontifical Academy of Sciences, who invited me to attend a conference on "The Cultural Values of Science", held in the Vatican City on 8–11 November 2002.

7.5 Consciousness, Culture or Soul? The Pontifical Academy of Sciences

As the debates recorded in previous chapters have shown, culture and religion were variously taken to be adversarial, or complementary, or part of a zero-sum game. As mentioned in previous chapters, I have always argued that Samuel Huntington used "cultural clash" as a surrogate for "religious clash" as he bulldozed his way through historical and anthropological evidence to advance his argument that the world could be carved up into 'civilizations' (in his later writings, 'cultures'); and he referred to all of them in terms of religions. This debate was again taken up by major actors on the international scene in the first decade of the twenty-first century, among them, the Vatican.

The arcane, hushed buildings of the Vatican appeared to me to be the world of the unspoken, of ritualized expressions that created such a dense haze that reality was not present. For an anthropologist, the strangeness of the visible social relationships belied an extremely complex underground of rules and prohibitions. I was struck to find priests in dark robes in twos and threes walking purposefully along corridors and reception areas all the time—because, I discovered, they could not be alone together in their own rooms. My fieldwork questions immediately found the answer: priests were forbidden to be together inside any of the sleeping rooms, with no further explanation given. Meanwhile, I could not help but notice that all meals as well as other services were provided by women, most of them with faces and

[6]The genealogy of the Convention for the Safeguarding of the Intangible Cultural Heritage for the period 1995–2000 is discussed in Arizpe (2015).

skin colour which indicated they came from developing countries. Soft fieldwork inquiries with them failed totally, since they seemed to be very afraid of breaking some rule.

The starkness of furniture and the lack of decoration at St Martha's House, where we were lodged, was highly noticeable, especially when we were invited to view the unbelievably beautiful art in the central buildings, including the Sistine Chapel and the Pope's apartments. Here, the light was extraordinary, with every passage, room and staircase an ascent to beauty and the divine. The awe that invaded one, and the stillness, seemed to call for an elevated spirit.

The conference itself brought together many extraordinary scientists and philosophers, among them Paul Ricoeur and Wolf Singer. I could not help but think of the myriad transactions of beliefs and ideas that, for more than a millennium, had been negotiated and arbitrated within these walls—transactions that had changed Western history and its influence on the place of humans on the planet. It seemed to me that this was the cauldron of beliefs that now had to change.

The discussions were uneven and confined along parallel lines of doctrine and science. In the online summary, HE Monsignor Marcelo Sánchez Sorondo notes that "all anthropologists agree that culture should be seen as a set of learned ways of behaving and adapting as opposed to inherited patterns of behaviour or instincts … Culture is a typical characteristic of man who is not rigidly guided by determining laws which establish him within a given horizon. On the contrary, he is a self-interpreting animal, a self-made man. He never ceases to express himself and to give himself a name, and this development, at the centre of which is to be found man's freedom, is called 'culture', which is different from nature" (Pontifical Academy of Sciences 2002).

One exchange stands out in my memory for providing a succinct metaphor of present-day positions on culture. Wolf Singer, the very distinguished neuroscientist, had just finished explaining the most recent discoveries about the structure of the mind from a neuroscientific point of view. He described a three-tiered structure of the human mind, unique among animals, whereby a person will perceive stimuli at a first level of consciousness, then respond to that stimuli at a second level, and then be conscious that she or he is responding to the stimuli and be able to assess the response. He concluded: "And this is consciousness." I immediately called for the floor and said: "Ah, that is culture." Then a Catholic scientist jumped up and said: "No, that is the soul."

Our three reactions summarize the traditional ways in which culture has been understood in the Western world: the view of the exact sciences, that of the social sciences and that of Catholic doctrine. It is indeed interesting to compare the three terms in their general use in the written literature through the use of n-grams, as explained in Chap. 1. As noted there, the frequency of use of the word 'culture' overtook that of 'consciousness' in 1925, well before the extraordinary new discoveries of neuroscience; it also overtook that of 'soul' in 1939, 'religion' in 1966 and 'spirit' in 1971, although the resurgence of religion in many places in recent years will probably have altered this statistic.

Samuel Huntington's use of the term 'civilization' as a euphemism for religions, since he then referred to world regions mainly in terms of religious areas, opened the door to introducing the political agendas of their contemporary leaders and institutions into many international meetings on culture.

Since the eighties, religious institutions had begun to mobilize through the ecumenical movement, initially among the Christian churches, then encompassing other religions, notably Islamic leaders in Muslim countries and Hindu leaders. For instance, in the Vatican, a few years earlier, Cardinal Joseph Ratzinger, Prefect of the Congregation for the Doctrine of Faith (who attended the Pontifical Academy's conference and subsequently became Pope Benedict XVI), had organized a meeting of historians and theologians which confirmed the will of Pope John II to reflect on the primacy of the bishop of Rome as a visible sign of unity among Christians.

In effect, as discussed by religious leaders both at the Vatican and in Davos (see Sect. 7.6), the question of the authority and the primacy of the Pope was still the major roadblock to the reunification of the Christian churches. The Syrian, Coptic and Armenian churches, which did not accept the conclusions of the Council of Chalcedon on the double nature (human and divine) of Jesus Christ, had separated from the Church of Rome in the fifth century. For Orthodox Christians, the major objection was the central role of the Pope in the governance of the Catholic Church. This was also the objection of the Church of England, although another major divergence has been the ordination of women. The fundamental issue in all these disagreements was the difference between the jurisdiction of the Pope pertaining to theology and the power of the Roman Catholic Apostolic Church of Rome as a man-made political institution.

Nevertheless, the ecumenical movement of Christian churches was an invitation for other major religions to develop politically strategic cooperation and interventions at the highest international level. Agreement between the Vatican and leaders of traditional Islamic doctrinal groups has become apparent in the last two decades, in their coalition against reproductive rights and women's advancement, but also, as could be seen at UNESCO, in undermining concepts of universality, equality and human rights. So too has the agreement of the Vatican with the development policies of the Bretton Woods institutions, although certain sectors of the Roman Catholic Apostolic Church actively criticize neo-liberal economic policies, especially in Latin America.

7.6 Culture at the Global Economic Forum, Davos

The surge in interest in culture led to many meetings held during the Global Economic Forum, where the focus, generally, was on practicality and applicability, the sessions therefore sometimes opening up very interesting and unusual discussions related to culture. As all the sessions were open to all participants, the fact that culture became a recurrent theme at the annual meetings indicated that there was high demand for such discussions.

It is not possible to describe the transactions in the debates on culture at these meetings because of the very varied format of the sessions. Some were debates by distinguished speakers, others workshops, others discussions during meals, others conducted at separate tables. None required keynote papers or systematic responses to centrally defined issues but rather involved open discussions, bringing in specific business or government experiences offered by participants and only minimally conducted by scholars. Moreover, since sessions were open to all participants, the mix of people present at any particular session was very varied. The themes related to culture were expressed in non-academic terms, closely related to the concerns of industry, trade and economic growth. Short reports of sessions were available for participants summarizing the discussions, or streamlining them.

I first received an invitation to the Global Economic Forum in 1996, while I was Assistant Director-General for Culture at UNESCO, and became a member of the Academic Advisory Group from 2003 to 2006. At the meeting, I was asked to introduce the session on "Artists and Change", and so I tried to bring together the threads of the references to culture that I had heard in recent international meetings and at the Davos session. I began by asking whether there is a global culture. I noted that the vice-president of one of the major multinational corporations[7] had said the previous day that corporate culture may be the bridge between cultures, and that cultures divide us while economic interdependence unites us. I disagreed with the idea that culture divides us because there is and always has been a universal culture: that created by great art. I said that Ravi Shankar speaks to all of us through the form of a millennium-old music tradition of a given culture. Nadine Gordimer has made us live through South Africa's poignant history as if it were that of our own country. I argued that this was possible because all human beings the world over have a heart, both literally and metaphorically. It is one of the wonders that transcends cultures. How could we not be burnt by the sun of tragedy for the father and daughter in Nikita Mikhalkov's last film or the pain of the loss of love that made the painted face of the actor transparent in Chen Kaige's film?[8]

I went on to ask, if we all have such universal hearts, how are we doing in cultural globalization? My answer: not well, not well at all. The previous day the CEO of a major multinational corporation had said exactly that in discussing how parent companies were still determined by being Italian or American and so on. I argued that we cannot untangle our need to belong, to have a cultural heritage, and the imperatives of living in a world which is creating new codes daily on the Internet. However, a common complaint that had arisen in several sessions was that so much trash was being generated through telecommunications and the Internet. The reason why was very interesting: the explosive expansion of global telecommunications and informatics had multiplied the number of possible transmissions by thousands, but the number of works of art to be transmitted had not similarly

[7]As I do not have permission to cite the name of the corporation in question, I will only refer to it generally, but this does not alter the thread of ideas it represents.

[8]The artists mentioned were all present in Davos.

multiplied. I said that the solution was to stimulate creativity. Vast amounts had been invested in communications, but correspondingly little in helping human beings be more creative. While there was a dynamic market for works of art already created, there needed to be greater investment in cultivating creativity in all areas.

As the Assistant Director-General for Culture of UNESCO, I argued that the problem is that sometimes, even when there is investment, creativity is hindered because of bureaucratic intermediaries, or because some artists create only for themselves and this is interpreted as irresponsibility on their part. However, the other side of the coin is that artists need freedom to be able to create. The dilemma is how to foster an atmosphere of freedom for creativity while ensuring that the resulting creations are meaningful to society. As economic globalization advances, people and cultures redefine their place, their identity and their new borders. In art, this transition pre-dated globalization: Ravi Shankar was changing the borders of Indian music in the nineteen-sixties. All the artists present at the Forum not only dealt with such change in their work but actively contributed to a transition which involved taking cultural traditions and reinterpreting them.

The choice of themes at the Global Forum evolved over the years during which I was involved. While in 1996, when I was first invited to attend, culture was barely mentioned in the programme, by 2003 it accounted for at least two or three sessions a day. In the years between, specific topics had ranged from global trade in cultural goods,[9] via the protection of cultural heritage, to culture and economics.

The Report on the 2003 session, on "The Values of Globalization", illustrates the concern expressed at the Global Forum regarding the violent protests against globalization in Seattle, which led to the use of the term 'values' to try to understand them. Given the centrality of this topic for the book, it is worth quoting several paragraphs of this report (Global Economic Forum 2003).

> When people protest globalization, they are usually objecting to allegedly negative social, political and environmental effects. But globalization has more subtle effects as well. As globalization spreads, and as people come into contact with other nations and cultures, the values embedded in this contact begins to cross borders. As this occurs, other more historically embedded values are being endangered.

One value imparted by globalization, implied but no less significant for that, is homogenization. Global logos for multinationals and their products look the same no matter where they appear. So do the values that are implicit in the entertainment programming that is broadcast around the world. In the face of the globalization juggernaut, local values and identity can be swept away. It has been estimated, for example, that over half the world's 7,000 languages will be extinct by the end of this century.

Another value attributed to globalization is materialism. Today's globalization is rooted in free trade, and that depends on people buying things. In the marketing of consumer goods, for example, global corporations attempt to stimulate demand in

[9]Global trade in cultural goods increased from US$95 billion in 1980 to US$388 billion in 1998 (exports plus imports), according to UNESCO (2000a).

new markets that may not have been exposed to them in the past. Some feel this "manufacture of desire", as it has been called, is not a positive impact. Globalization is also deeply secular—not anti-religious, but not spiritual either.

If globalization were to be more inclusive, what other values might it embrace? First, it might champion local values, and second, it might encourage the seeking of meaning through experiences other than the consumption of material goods. And of course a deeper investigation would uncover other values as well" (Global Economic Forum 2003: 1–3).

By 2003, with religious leaders replacing labour leaders, NGO activists and intellectuals among the participants at the Forum the topics to be discussed at sessions had also changed. The following example of the session topics for the 2004 programme illustrates these changes:

Religion, Culture and the World Stage (plenary): "The World Economic Forum is organizing the Council of 100 Leaders, a multi-stakeholder community dedicated to strengthening understanding between the West and the Islamic World. The agenda points are: How do we differ in responding to, and understanding, the challenges of globalization and modernity? (2) What is preventing and impeding dialogue? (3) How can we best build bridges able to bear the weight of our differences?"

Breaking the Taboos (session): "Current events have forced religious leaders across the religions to confront topics that have been previously considered 'taboo'. The agenda points are: (1) How are religious leaders responding to the taboo issues of homosexuality and female participation in organized religion? (2) Is there a 'correct' way for religions to address the challenges that are posed by modernity?"

Protecting the Past: Cultural Heritage in Zones of Conflict (session): "In Afghanistan, the Balkans and the Middle East, cultural heritage has been one of the many victims of conflict. The consequences of theft and destruction of cultural and religious objects, and museums, galleries and churches, are far greater than the simple financial costs. The agenda points are: (1) What role does cultural heritage play for the historical memory of a country? (2) How should the global community, especially government and business, respond to the destruction of cultural heritage? (3) How should we deal with regimes that have tried to destroy cultural heritage? (4) How should the international community address the illicit traffic in items of cultural heritage?"[10]

Green with Jealousy, Red with Anger: On Emotion and Colour (meal): "What role does colour play in understanding and depicting emotions?"

Can Art and Culture Be Owned? (session): "Many indigenous, ethnic and religious groups are reclaiming their cultures and laying claim to art objects on display that, in their opinion, do not belong in institutions or corporations. The agenda points are: (1) Can culture be owned by a particular group or does it belong to humanity? (2) How can museums and galleries find a balance between cultural sensitivity on

---

[10] I joined three high-level male speakers on the stage for this session. I was the only one to refer to the recent theft of cuneiform and other valuable items from the Museum of Baghdad.

the one hand, and providing a vibrant locus for education on the other? (3) Should businesses be able to own works of art?"

Understanding and Building Creativity (meal): "Creativity is an oft-admired and sought-after ability in all areas, from business to the arts. But despite creativity's high value, there is no easy method for achieving it. The agenda points are: (1) How is the media landscape evolving, and how will it look in 2020? (2) What does this revolution mean for traditional media suppliers, producers, and viewers? (3) How should the mainstream media respond if it is to remain relevant in the future?"

A whole section of the programme was dedicated to 'Values', which, among many other sessions, had the following ones related to culture:

Choose Your Own Identity (meal): "It is broadly agreed that geography and citizenship are no longer sufficient to determine identity—culture, religion and ethnicity also play important roles. The agenda points are: (1) Is there a "gold standard" for determining identity? (2) Can we choose our own identities, or is identity something that is chosen for us? Can identities be selected and discarded depending on needs? (3) How do political and economic externalities impact our choices of identity?"

Is the Religious/Secular Divide More Dangerous than the Inter-Religious Divide? (session): "Much emphasis has been placed on the differences between religions. Less attention has been paid to the differences between the religious and secular worlds, a schism which is becoming ever more apparent. The agenda points are: (1) What are the main differences between secular and religious values? (2) Do liberals of different religions have more in common with each other than liberals and conservatives of the same religion? (3) Has the focus on inter-religious differences led to a disregard of the religious/secular divide, to its detriment? (4) How are secular states responding to the challenge of integrating people whose primary identity is religious in nature?"

The Culture of Conflict (meal): "Conflict, from the interpersonal to the geopolitical, is omnipresent. The reasons for it are sometimes clear and, more often, uncertain. The agenda points are: (1) What conditions encourage hostility and conflict? (2) Are certain cultures more prone to conflict than others? (3) Is there a particular culture of conflict?"

The Vanishing Intellectual (meal): "The 'intellectual' seems to be a vanishing figure in public life. The agenda points are: (1) Why has the intellectual figure begun to disappear and what can societies do to encourage the growth of intellectualism? (2) What role does government play in inhibiting or encouraging intellectual growth? (3) Must a public intellectual necessarily oppose government? (4) How does the role of the public intellectual correlate to the corporate strategist?"

Three trends stand out in this programme. The first was that business leaders were keen on bringing culture into the discussion of governance and the resolution of conflicts that were interpreted as stemming from individuals' behaviour, emotions and perceptions, with little connection to the expansion of markets, trade and cross-border mobility. The second trend made visible the emerging question of cultural diversity in everyday settings: economic investment, the workplace, consumption patterns in the market, housing and urban services, and so on. The third was that, having chosen as the framework for discussion the concept of values, the unease with which cultural and religious values were perceived to be in tension (and, in some cases, their underlying incompatibility) overshadowed all discussions. The missing elements in this discussion were the values pertaining to democracy, citizenship and nationality, all of which would become the unspoken yet key accelerators of heightened conflicts in the next decade.

The programme of the Global Economic Forum shows that the real clash was actually between the need of business for borderless and unfettered world markets and the pressures on governments to reconfigure citizenship to take into account 'ethnic' values and claims, which were leading to national "culture wars". At the same time, the unspecified and confusing "cultural values" of multiculturalism on the one hand and the secular traditions of nation states on the other were clashing with the doctrinally bound values of religions. As the Forum grappled with these issues in Davos, another clash came to the fore: the conflict between development theories applicable to all peoples of all countries and the sudden claim that they should take into account multiculturalist claims. This complex debate was taken up in numerous international meetings and programmes. The most important ones will be considered in the following pages. They demonstrate the next phase of the debate on cultural diversity during the first decade of the new millennium.

7.7 Cultural Liberty in a Diverse World: The 2004 Human Development Report

In the 1970s, the influential International Society for Development established a forum that allowed political leaders, international functionaries and researchers to debate the emerging issues in international development; their conclusions would then influence the policies of international UN agencies and international NGOs. A group of participants then set out to create an alternative methodology measuring the impact of development policies in different countries, one which would provide a more comprehensive view of the educational, gender, health and other social dimensions of development beyond the statistics of GDP as the sole indicator of economic growth.[11] Richard Jolly called this new model "development with a human face". Accordingly, and with the support of researchers from many countries, Amartya Sen, Mahbub ul Haq, Richard Jolly, Sakiko Fukuda-Parr and Paul

[11]The history of development theories and policies that led international efforts in this area may be consulted in the publications of the United Nations Intellectual History Project, led by Richard Jolly, Louis Emmerij and Tom Weiss (see: http://www.unhistory.org).

Streeten, among many other distinguished authors, created one of the most important initiatives in development policies: the Human Development Report (HDR).

As I had been invited since the beginning of the 1990s to participate in several of the annual meetings that were held as brainstorming sessions on the topics taken up by the Human Development Report, I began to insist that the theme of culture be considered, and it was finally chosen for the 2004 Report. At the end of a year of work on the data and analysis for this issue of the Report, Sakiko Fukuda-Parr, then director of the HDR, told me that that it had perhaps been the most complex analysis the Report had ever taken up. Conceptual complexities, divergent theories, the dearth of statistics and the reticence of some governments to provide basic data on languages and cultures in their countries because of political sensitivities made the task of producing a coherent analysis, based on reliable statistics and survey data, very challenging. Nevertheless, in July 2004, the Human Development Report on *Cultural Liberty in Today's Diverse World* (UNDP 2004) was launched in Brussels, Belgium.

In the first meeting with the authors of the global background papers for the issue of culture, organized in New York on 24–25 September 2003, Amartya Sen gave the introductory paper. In previous writings, Professor Sen had already made reference to culture and its associated concepts of human rights, identity and diversity. In his 1999 book, *Development as Freedom*, he argues that "processes such as participation in political decisions and social choice cannot be seen as being—at best—among the means of development (through, say, their contribution to economic growth), but have to be understood as constitutive parts of the *ends* of development in themselves" (Sen 1999: 291). "In pursuing the view of development as freedom, we have to examine … the extent to which people have the opportunity to achieve outcomes that they value and have reason to value" (ibid.). Providing such opportunities means enhancing human capabilities through education, learning, and skill formation and, foundationally, providing them with substantive freedoms to lead "more worthwhile and more free lives" (Sen 1999: 296). Professor Sen also highlights a key consideration by noting that "responsibility requires freedom" (Sen 1999: 294), a major point which will be brought to bear on the discussions of the Anthropocene in the conclusion of this book.

Previously, Professor Sen had also made the case against cultural insularity, insisting that cross-cultural communication and appreciation are more prevalent than is typically acknowledged. "The culturally fearful often take a very fragile view of each culture and tend to underestimate our ability to learn from elsewhere without being overwhelmed by that experience" (Sen 1999: 243). He also notes that recognizing diversity *within* different cultures is extremely important, pointing out that "the fact is that in any culture, people seem to like to argue with one another, and frequently do exactly that—given a chance. The presence of disidents makes it problematic to take an unambiguous view of the 'true nature' of local values" (Sen 1999: 247).

At the meeting of the Human Development Report in 2003, Professor Sen opened the discussion by highlighting two distinct problems. He asked whether individuals should have the opportunity to be included in a group and whether groups have the right to preserve their culture. He offered the example of French Algerians in Paris: their desire to live an Algerian life is discouraged; at the same time they want to be like the French. Both issues were present and therefore, should be discussed together. In the background paper he wrote for the meeting, Professor Sen emphasized that, in taking up the concerns about culture, "a new understanding can fruitfully supplement old wisdom, but it must not supplant what was genuine in earlier discernments. In fact, considering cultural deprivation *along with* well-established examples of economic, social and political deprivations (rather than seeing them as contrary considerations) may be useful in understanding how interlinked the different types of human deprivations are" (Sen 2004: 2). Further, in his presentation in New York, he stated that "redistribution has to go together with recognition, but you cannot redistribute life things". He summarized this as a question of equity and cultural liberty: equal participation in culture is part of equity. He concluded by saying that it is important to emphasize that there is heterogeneity in all cultures.

I was asked to comment on Sen's paper, so I said that culture is constitutive to society yet it has specific instrumental uses, which may be analysed following his theoretical propositions on human capabilities. I also made the point that cosmopolitanism, in terms of peoples having different languages and cultures living together, was not invented by the West but goes back to other histories such that of the as Sanskrit era. My response highlighted the importance of Sen's insight that reason is always present before identity, in the sense that there is always a choice in accepting whether or not to belong to a group having a certain identity.

Arjun Appadurai was also asked to comment and he stressed that we must not only make explicit that is there an "arrow of time" but also ask Amartya Sen to develop the idea of an "arrow of life". He agreed with Sen that there was a "global flooding of local cultures", but felt that when resistance to this happens, the self-appointed guardians of those cultures usually have national interests in mind. He spoke of "predatory cultures" (what Kwame Anthony Appiah had termed the emerging "cultural tyrannies"), such as the Serbian and Hutu cultures, whose main rhetorical case was that they could not subsist with another culture but must make it disappear.

In the ensuing discussion, Neila Kabeer pointed out that the context must be taken into account. For example, in Bengal there is no gender consciousness because there is no reason to highlight it but when women come to the UK they are victimized by the tyranny of Muslim leaders. Alfred Stefan argued that we need to promote multiple and complementary identities, such as the Belgian and Catalan identities, since, if people are given the opportunity, they choose dual identities. He also quoted research that in Yugoslavia the social distance between Croats and Serbians had narrowed but from the moment that Milosevic was elected the gap began to increase again. So nationalists and secessionists were the first people to be killed because they had dual loyalties. He concluded by saying that "democracy is not possible without a federalism of multiple complementary cultural and ethnic groups".

In my own presentation, I said that we must be careful that the debates do not lead to making culture another conditionality of the West towards developing countries, as several ambassadors had repeatedly told me. The 2004 HDR needed to look at the coincidence of inequality and poverty with the boundaries constraining cultural minorities. I also strongly criticized the possibility of sliding into the "flat culture syndrome" whereby reified cultures have no layers (time), no internal dynamics of change (agency) and no selection of cultural elements for purposes of representation, display and framing. Finally, I raised a question that I was asking in many forums: why is the 'ethnicism' (the political use of ethnicity, as opposed to 'ethnicity') of minorities good and the nationalism of groups bad, especially since many nations are based on ethnic heritage? I concluded that there is a problem of scale in cultural diversity, which tends to feed extremists on the right. Arjun Appadurai agreed that there is a correlation between the geography of anger and the effects of globalization. His counter-hypothesis is that there is another complex dynamic involving the way in which states, even democratic ones, actually mediate globalization effects.

From the point that 'culture' was adopted as a concept for policy formulation, anthropologists have constantly opposed its reification: that is, turning it into a concrete, motionless object. Such a notion underlies the metaphor of a "mosaic of cultures" as fixed, impermeable entities. In the second World Culture Report, we proposed instead the metaphor of a "rainbow river", following Archbishop Desmond Tutu's image[12] of South Africa as a "Rainbow Nation", of cultures flowing, exchanging and mixing. I also suggested that "the liberty of citizens" be mentioned together with "the dignity of persons" as the aim of the defence of cultural diversity. The debate on these terms was developed at greater length in the 2004 HDR section on "Cultural Liberty and Human Development".

At the final background meeting for the 2004 HDR, held at the Rockefeller Centre in Bellagio, Switzerland, I gave a paper entitled "Notes on the Framing and Display of Cultural Diversity in Globalization". The paper began by cautioning that "Very different processes and movements are being pulled together under the heading of 'cultural diversity'." I saw the most important of these as: mobilizations around cultural demands and cultural minorities, indigenous peoples and migrants to keep and develop their cultural legacies; market strategies by governments and local cultural industries to gain a comparative advantage in world markets; trade negotiations around investment and marketing of cultural content commodities; differences in the proportion, form and content of the representation of different cultures in television, film and multimedia production; patents of intellectual property registered by native artists, entrepreneurs and indigenous peoples; and the information gap in relation to the participation of people from different cultures in the Internet, although the the number web pages in other languages has grown exponentially.

---

[12]Later elaborated upon by Nelson Mandela.

As a summary of the substance of the 2004 Human Development Report, as well as of the salient political, social and cultural debates in the international sphere at the time, in the following pages I quote excerpts from my address at the Report's launch. This took place on 7 July 2004 in Brussels, in the presence of Crown Prince Philippe and Princess Mathilde of Belgium,[13] and of members of the European Commission and of the ministries in the Belgian government. At that time, as President of the International Social Science Council (ISSC), I was on the receiving end of numerous research reports, academic debates, political insights and institutional shifts in the world of social science.[14] This gave me a vantage point from which to point to important trends in scientific, public and academic debates and to what new research was needed in the social sciences in conjunction with the natural sciences.

I began my address by explaining that the 2004 HDR proposed a new perspective based on the concept of cultural liberty as part of human development (Arizpe 2004: 2). I noted that Amartya Sen explicitly states that the denial of cultural liberty, exclusion from social interactions and rejection of one's sense of identity figure prominently in the deprivations that human beings have reason to resist and want to remedy. And I stressed that, through placing cultural issues within the broad framework of human freedoms and values, we can see the promise of a well-founded, rather than ad hoc, appreciation of the cultural dimension of human lives (ibid.).

After mentioning the billions of television sets around the world, the billions of webpages already channelling information and communications at that time, and the approximately 150 million cross-border migrants and 600,000 tourists travelling around the world, I concluded that, paradoxically, a world that was opening up so many channels of communication was at the same time also fostering the loss of both languages and cultures. At UNESCO, the question of how to manage relationships between cultures led to many different kinds of recommendations and programmes where we were working with many people who wanted to conserve their traditions but also with others who were more interested in working towards a "transformative engagement" with modernity and globalization (ibid.: 3–5).

I said that "The choice must be theirs, as the 2004 Human Development Report clearly states. What we must guard against is what I call the 'flat culture syndrome'. Neither historically nor at present have people belonged to single, flattened cultures. They have created and chosen to belong to cultures which interact, combine and exist at different scales of belonging. Enabling conditions must allow for these layered multiple practices—for example, as the Report states, public recognition of the use of three languages: one international language, one national or regional lingua franca and a mother tongue. In some cases they will be the same, in others they may be different. Such a model recognizes the realities of cultural layering,

---

[13]Now King Philippe and Queen Mathilde.

[14]I held the office of Vice-President of the ISSC in 1993–1994 and 1998–2002 and that of President in 2002–2008.

instead of imposing an idealized, abstract view of cultures as homogeneous units somehow suspended outside real political and social structures. Monolingualism and religious orthodoxy are exceptions rather than the norms and have occurred mainly among geographically or politically isolated people" (ibid.: 5).

I continued, "As Amartya Sen has put it beautifully, does the championing of cultural diversity demand support for cultural conservatism, asking that people should stick to their own cultural background and not try to move to other life-styles?" (ibid.: 6). Sen's answer was that such a position would "immediately deliver us to an anti-freedom position … diversity will then be achieved at the costs of cultural liberty". I therefore argued that "the possibility of choice is important in preventing what Anthony Appiah has called 'new tyrannies' in the form of newly asserted identities which can tyrannize by eliminating the claims of other identities which one may also have reason to accept and respect" (ibid.: 5). To stop the new tyrannies from developing, I reiterated that we need "freedom to create", as I had argued since the 1998 Stockholm Meeting on Cultural Policies for Development.

I explained that "The 2004 Human Development Report brings [these issues] to bear in today's world in which religious fundamentalist, violent and extreme groups are on the rise in many regions … . When the state fails, coercive movements may step into offer education, insurance or law and order. It may be necessary for the state to go beyond restrictive measures to contain intolerant ideologies and coercive movements. States must have the political will to take a stand against intolerance. This may be particularly necessary when women's rights are not ensured. It becomes even more manifest in cases in which women ask to be given even the most basic liberties, as in the case of Afghanistan" (ibid.: 8).

On the question of war, I noted that "the UNESCO preamble to the Constitution reads 'wars begin in the minds of men'—not 'in the minds of women', please note—and they begin, in the first instance, with the categories that are used to classify others. Societies bring up many, many kinds of categories, but only those to which a political purpose is attached become reasons to go to war. Ethnicity and religion have been managed under a rational political philosophy that allowed negotiations to be conducted in terms of state citizenship. Pushing forward these categories as relevant political categories is creating consequences that are unforeseen" (ibid.). I gave the example of how Huntington's equation of civilizations with religions in his writing on the "clash of civilizations" in fact weakened the West: "He erased from view the achievements of the West that have made it the pervasive, structuring cultural force in the world. This did not come from religion but from science and political philosophy. It was Western philosophy, science and social science that enthralled the world by proposing a political framework to manage the historically endless strife between cultures and religions" (ibid.: 9).

In selecting passages of my address while writing this book, I focused on those that reflected the transactions which were being carried out at the time. I had therefore planned to leave out the following section, which refers specifically to Mexicans in the United States. However, the new political relevance of this topic now 2017 seems to me makes it important to quote the corresponding reference to it in my 2004 address in New York. In my critique of Huntington's arguments, I mentioned that he had also published an article entitled “The Hispanic Challenge”. In this article, “he asserts not only that the ever-growing presence of Hispanics in the United States poses a challenge for American society but goes as far as to say that Mexicans, in particular, may actually lead to the segmentation of American society. He is right in terms of the growth of migration and demographics of the Latino population.[15] And he is right that Mexicans carry with them the force and imagination of a millennia-old culture that has kept enriching itself by mixing with many other European, African and Asian cultures. However, scientific studies on the issues he raises have shown that he isn’t right on several issues. By the third generation, descendants of Mexicans do assimilate to the American way of life. He is wrong also about bilingualism. A study by Alejandro Portes showed that in south Florida and southern California there is a positive association between bilingualism and better academic performance (Portes/Hao 1998). Also, without exception, fluent bilinguals outperformed limited bilinguals as well as English-only students in standardized tests and grade point averages, even after statistically controlling for parental status and other variables, as Rumbaut and Cornelius found in their study in San Diego, California in the 1980s (Rumbaut/Cornelius 1995)” (Arizpe 2004: 9).

As I had written a chapter on the rural exodus in Mexico and Mexican immigration to the United States for *The Border that Joins* (Brown/Shue 1982), I could not let Professor Huntington’s arguments pass without additional comment. So I continued, “One can use the same argument that has been used by free market [advocates] by saying that nobody forces Americans to eat Mexican food. They prefer it. So how free is the market? His book is, in fact, very timely. At the time when joblessness has increased in the United States his book points towards Latinos and particularly Mexicans. Is this another blueprint for racial and cultural wars? What seems hard to understand it that nobody will benefit (from such a blueprint) because such wars will disrupt markets, distort voting patterns, tear the social fabric of communities even more. It will, in fact, dampen all the self-correcting mechanisms that capitalism has evolved” (Arizpe 2004: 9–10). Only a decade later, very sadly, the consequences of this opinion, highlighted politically, has only deepened the consequences of these processes which have been opposed by democratic political forces but are at present being given free rein by neo-conservative and especially in the US, by the Republicans and the Alt-Right groups.

---

[15]The United States administration formalized the term ‘Hispanics’ to refer to Spanish-speaking individuals in their territory. However, another term, ‘Latinos’, is widely used by Latin American and Caribbean residents and migrants in the United States on the basis that it does not leave out the Portuguese-, French- and Creole-speakers in the country. This transaction has, of course, important political connotations.

7.8 The Rockefeller Seminar on "The Value of Culture and Art"

At the turn of the century, studies on the economics of different aspects of culture and heritage were given priority under the sponsorship in many public and private institutions. Among them, the Getty Conservation Institute held a meeting on Economics and Heritage Conservation in Los Angeles in December 1998, organized by Marta de la Torre and Randall Mason, to identify concepts and approaches for evaluating both the economic and the cultural values of heritage, with the aim of investigating links between them. The report of the meeting refers to the quandary that

> Methods of economic valuation increasingly dominate society's handling of the value of heritage, while the same methods are unable to account for some of the most salient values and virtues of heritage—namely, historical meaning, symbolic and spiritual values, political functions, aesthetic qualities, and the capacity of heritage to help communities negotiate and form their identity. In short, heritage cannot be valued simply in terms of price (Getty Conservation Institute 1999: 2).

The report produced two conclusions: first, that the variety of values ascribed to any particular heritage object is matched by the variety of stakeholders participating in the heritage conservation process; and second, that 'heritage' is an essentially collective and public notion (ibid.: 2–3). While economic thinking and concepts make indispensable contributions to the understanding of the role of conservation in society, not all heritage values can be put into a single, traditional economic framework, nor can they all be measured in monetary terms. As a participant in the meeting, it seemed to me very important that the consensus was reached that "aspects of conservation are best modeled as a continuing and contingent process, … [observing that] a process-centred model contrasts with the conservation field's traditional focus on products and outcomes" (ibid.: 4).

These concepts were further discussed in the much broader and more complex field of how to understand the value of culture and art, at the seminar organized by the Culture and Creativity Program of the Rockefeller Foundation at its Bellagio Centre in Switzerland on 12–18 February 2004. As it is not possible to summarize in a few paragraphs the rich debates that took place between economists, anthropologists, artists, museum curators and experts in high art, only the discussions concerning the main themes of this book will be included here, especially those related to economics and culture. In contrast to the meetings noted earlier in this chapter, which were sponsored by government, business and religious institutions, the transactions at this meeting were free from immediate policy considerations and envisaged long-term trends, as demonstrated in the book that was published, with deliberations and a considerable number of iterations in writing and rewriting its text.

David Throsby, a distinguished economist who has worked continuously with UNESCO in its culture programmes, opened the meeting, noting that parallel systems of assigning value exist in economics and anthropology and among artists.[16] The other coordinator of the meeting, Michael Hutter, emphasized that value is not only related to individual preferences but also shaped collectively. Value judgements make a difference which goes beyond use and exchange value, which leads us to explore processes of valuation: not only how values are defined but how they are observed. He argued that people can change their vantage point if they do not only look through economics. There are different value areas, with a battle going on between economy and art. The new "media economics" expands this discipline to the new information and communication technologies and to intangibles. He mentioned three further points: visual culture is becoming a new discipline, as the 'eye-economy' now connects to certain logics and structures; we have moved from postmodernity to 'contemporaneities'; and markets are "fields of contestation" about value.

The question was also brought up of whether culture wars were only in art or also in ethnic relations, since people value art differently. In some cases people may disagree on the public benefit of art. The case was made that, for example, Charles Saatchi, the owner of the *Sensation* exhibit, increased the pecuniary value of his collection at the expense of its non-pecuniary effect on the public. In the discussion, the controversy of the *Sensation* exhibit was referred to as the "consumption of an event", as one move in the cultural wars, of the society of the spectacle (to quote Guy Deborde). *As cultures become identifiers, people's opinions of a given event will follow real or imagined instructions from gatekeepers since it is so difficult to develop one's own criteria about art as it becomes more and more inscrutable.*

Several participants remarked that one can find excellence in traditions from different communities, so the challenge for observers is to penetrate multiple art forms and aesthetics. This was countered by another speaker, who said the claim that, if the distinction between folk art and high art becomes a barrier, then you get a segmented market; there is a difference between saying this is an artist doing Chicano art and saying this is a Chicano doing art. Another delegate made the important caveat that, if everything is acceptable, then "you leave us with no discriminating device".

One important discussion started with a delegate's suggestion that the economist's concept of utility maximization can be used as an explanatory model. If I want a kiss, I can accept that I am trying to maximize my level of happiness, but I can't explain why a kiss makes me happy. We shouldn't confuse the map with the road itself. An explanation is not the same thing as an underlying reason. Someone else commented that, since utility maximization can be applied to anything, then "economics can integrate everything", including suicide. Another participant noted that even very good studies of demand function are never able to explain all variability, particularly with regard to art or love, and that economics can explain what we do

[16]Michael Hutter and David Throsby co-edited the book that was published in the wake of the Bellagio meeting: see Hutter/Throsby (2008).

with objects provided that it takes the self as a given, but runs into problems when an exchange involves the self. People look at art because they want to change their self.

In explaining exchange as the fundamental principle of societies, and the gift as the mechanism that binds people together—widespread than systems where value is signified with a particular symbol, whether cocoa beans, gold or paper money—one of the anthropologists stressed that there is a value in simply belonging to a society. If you belong to it, you believe its conventions, in the same way as believing that paper money is backed by gold. An objection was made, pointing out that connecting power in culture may become a constraint for people who want to be free. Another speaker mentioned an unease with culture as a site of negotiation because it presupposes separate agents that can arbitrate those interests, whereas reciprocal relationships have to do with bodily and affective relationships. A closing point in the discussion was that opting for tradition or not depends on the available alternatives.

In the context of the economics of contemporary art, the discussion whirled around the central question was how value is formed between cultures and hybrid forms. One response was that it is the Western market that defines value: low when it is ethnographic, collective and high when it is 'authentic', following abstract Western individual value. In postmodern terms, it was noted, Aboriginal artists are doing something entirely different from their traditional rituals for a market which happens to appreciate abstract art. The question then shifted to whether the communities would be destroyed by the high prices paid for their art: a community that usually received $200 for a painting now receives $200,000 and this is hard to come to terms with. Two further speakers mentioned that the market in the West will always look for art that is beyond it, and that the value is to be found in the fact that these artists can communicate. A strong objection was then made contesting both assumptions: "I disagree with authenticity as a criterion of value. It is a concern only for anthropologists and cultural merchants. It is never an aesthetic concern." A closing remark was that in economics value is explanatory and in anthropology it is descriptive.

As the discussion continued, in summing up, one participant said that economics has not dealt with Man but with institutions and exchange; the specific economic approach is not about money but about taking a specific point of view from individual incentives to action: motivation, as opposed to a holistic approach. From the standpoint of the sovereignty of the individual, cultural stories can be perfectly economic: they transmit information but also affection, enjoyment and so on.

In concluding the discussion in this session, one participant insisted that "the notion of cultural value must be reduced to information, even if some of it is nonsense. My point is that it is reasonable to think of economic value as that assigned by an economic model and we have a measure which is money. But that which does not fit into this model is not deemed to be important." Another participant summed up by saying "so maybe economics should not try to look at conflict resolution on *what* your values are, but rather on *how* you resolve the conflicts".

In the last session, when asked what we had learned in the seminar, several answers were notable. "What did I learn? How people think, and an intuition about incommensurability, do you mean this scale? Some start with art on the outside and others work from the outside towards art." "Economists came saying 'we're systematic' but we achieved a balance in the middle with the ethnographers." Another answered "I was struck by three points: that culture is a process, not an outcome; that an original is the manifestation of process; and the idea that cultural goods continue to generate value." One participant summed up his experience in the seminar by saying "I was shattered then re-formed thanks to the good will of the people here … I am against definition because it ossifies, but we must be careful of contextualizing each time we use the terms 'culture' and 'art'." Another agreed: "I am against definitions. I learned that cultural value is a matter of procedure."

7.9 "Failed Societies", 2002–2006

In the first decade of the millennium, a parallel international debate to that on culture and cultural diversity began to take shape as social science studies started to document the emerging phenomenon, arising in many countries, of increasing youth involvement in drug trafficking, rioting and petty criminality. At the ISSC the member organizations and associations began to sound the alarm. My mantra as President of the Council at that time, which I repeated in every congress or forum to which I was invited to give a presentation, was: "Economies are growing but societies are collapsing". This made for difficult discussions since there was such a barrage of opposition in the media and in academic policy in some countries to the terms 'social' and 'society' that it was difficult to talk about issues related to the psychosocial functioning of societies as a cohesive whole. Research in Europe was still framed as social science but in the United States the "human sciences" moved towards behavioural, perceptual and health factors. Margaret Thatcher's quote that "there is no society, only individuals", taken out of context, found echoes in many political and academic spheres. At the same time, the 'social' became invisible under the umbrella of cultural diversity, which set up a classificatory scheme based on difference. The point I kept making was that this scheme could not explain the similarities in the processes that increasingly destabilized civil societies.

Institutions began to react to such a state of affairs. The World Commission on the Social Dimensions of Globalization was set up by the International Labour

Organization (ILO) and published its report "A Fair globalization: Creating Opportunities for All" in February 2004. A joint meeting sponsored by the ILO, UNESCO and the ISSC was held in Paris on 7 July 2004. The phrase that became the mantra of this meeting was "we must be positively critical, not only politically correct".

Gonzalo Abad, a UNESCO programme officer from Ecuador, summarized the main innovative points of the Report. He pointed out that it focused on the regional dimension of social integration, advocated parliamentary control of globalization in individual countries, recommended coherence in policies, and suggested both that a forum be established for debates and transactions in the social sphere and that a report be published periodically on democratic social globalization. Abad also mentioned that the UNESCO Medium-Term Plan had as one of its major objectives the 'humanization' of globalization and the pursuit of "education for all".

Ruth Cardoso, a distinguished Brazilian social anthropologist and a prominent member of the Commission, remarked that it was a decisive moment for the world because contradictions were becoming so visible. There was a global social movement which was not anti-globalization but anti-inequality and anti-poverty. She argued that there was a need for another theory of global change, alongside a need for greater democracy at all levels, particularly with regard to women.

Aminata Traore, a Senegalese social leader, strongly denounced the replacement of the human with the market. She saw the world as falling into a practice of "pensée unique", one-dimensional thinking. The cultural challenge (*enjeu culturel*) was that the world was now made up of symbolic violence. Globalization was not responsible for all that was wrong but it had made promises that it could not keep.

My responsibility as President of the ISSC led me to promote such debates. At the "Fair Globalization" Conference in Berlin on 22–23 November 2006, I began by highlighting the fact that "there is increasing international concern about 'failed states', and in some cases 'failed economies' but there is never mention of 'failed societies'". I then stated that "it is of greatest concern, then, that by now the liabilities of an unjust globalization are becoming most visible in the social arena and in cultural and religious conflicts" (Arizpe 2006). I also came back to the question of the social and the cultural. I explained that the concept of "social economy" meant different things in Germany, France, Sweden and the United Kingdom, just as that of "social development" meant different things in Europe, in Latin America and among *altermondialistes*. "For policy purposes", I continued, "a 'social model' may be defined as the set of principles, rules and institutional arrangements that organize social relations (Lefebvre/Méda 2006: 9). But let us be aware that such a definition also applies exactly to the term 'culture'."

The second aim of my presentation was to show how, in some cases, the retreat of governments from social protection was caused by extremism in cultural and religious communities:

> There are certain demands for safety nets, the building of 'social capital' through social activities, as well as elements of adaptive roles and identities which, if not provided by the state or by the political process, may be recast in cultural and religious idioms. One of the questions is, why this shift in language? The answer has to do with the weakening of the capacity of liberal democracies to produce political ideas and platforms. Additionally, the marginalization of the non-economic social sciences from the real discussion of development policy or the creation of a knowledge base for societies which must operate in a new globalized framework has left the individual with no intellectual or ideological basis with which to construct a view of the future, let alone of their social roles, adaptive identity and prospects of attainment. Cultural and religious discourses have been brought into fill this ideological and social vacuum. However, when such entities enter into the political process, they become one more political actor whose beliefs and actions begin to be judged as those of other political actors. When their beliefs are, instead, posited as "non-negotiable", then an era of "hyper conflict" will become inevitable (Arizpe 2006: 2–3).

I ended my paper with a call for the better use and management of resources for sustainability:

> Every prospective study mentions the need for greater social organization in managing more knowledge about ecological functionalities. This will require rebuilding social organization along different lines than many traditional ones. Warring tribes of different political, cultural or religious persuasion will hardly be able to negotiate such cooperation for sustainability. We urgently need a new, coherent proposal for a development which is social, global and sustainable (ibid.: 5).

7.10 Conviviability

My involvement since the 1980s with the Society for International Development (SID) had made me aware of the increasing importance of such a social, global development and so, in my search for a new vocabulary to build a sustainable future, in 1997 I presented a brief paper entitled "Conviviability: the Role of Civil Society in Development" (Arizpe 1998b). Fully aware that I was inventing a new term, non-existent in English, but taking advantage of the openness of the English language to neologisms, I explained that a similar term exists in Spanish (my native language) and in Nahuatl, the Aztec language spoken in Meso-American civilizations. The motive of my proposal was to call attention to the fact—which had been made blatantly evident in my research on social perceptions of deforestation in the Lacandon rainforest in Mexico (Arizpe 1996), and in the revelation that I had experienced while carrying out long-term fieldwork in monolingual Nahua-speaking villages high up in the Sierra in Central Mexico (Arizpe 1972, 2006)—that sustainability will never be achieved if 'conviviability' among peoples concerned with the environment is not attained first. I will outline that research briefly here, to show how such terms from indigenous languages are directly relevant to our present cosmopolitan quandary.

A meeting of farmers who were granted land to settle in the Lacandon rainforest, Marqués de Comillas, Chiapas, Mexico, 1992. Photo was taken by Lourdes Arizpe

7.10.1 What Is Conviviability?

As mentioned in Chap. 5 (footnote 3), I use the term ‘conviviability’ as a translation for the Spanish notion of *convivencia*, which is conveyed neither by the English ‘conviviality’ nor by the simple notion of “living together”. Its roots are in the verb *vivir* (“to live”) and the noun *vivencia* (“lived experience”), which means a specific event of living. *Convivencia* is used frequently in daily life in Mexico and, while it is recognized mostly as a social occasion, one in which friends, food and feasting come together (in much the same manner as ‘conviviality’), it goes deeper, as the *sharing* of life experiences, acknowledging and accepting the value of other people’s culture.

My 1998 paper on conviviability summarizes many of the main issues that were discussed at the time and I proposed a new analytical concept, so parts of it will be cited here. In it I argued that other factors that are crucial for human development must now be targeted in development policies, including democratic governance, the organization of civil society, the eradication of poverty and culture in development. Working with civil society touches on all these aspects, yet we lack an analytical concept that would allow us to understand all these factors together. “This concept should also provide the goal for working together towards sustainable development in the 21st century. While better terms could perhaps be suggested, I would like to put forward that of ‘conviviability’ for such a purpose” (Arizpe 1998b).

Conviviability also refers to the need to redefine the codes and attitudes through which human beings with different ideals of the good life are able to live together compatibly so that sustainable development is ensured. In this setting, the discussion of the role of civil society becomes central: how can societal processes be channelled as the nation state transforms itself? The legitimacy of the state comes from its duty to keep civil peace: that is, to prevent opposing communitarian groups, be they political parties, capital and labour, religious sects or local communities, from coming into grievous conflict. The next move must therefore be to discuss how civil society and cultural enclaves can relate to each other while living together in the framework of a single state.

Given what I was hearing at many international meetings at this time, I thought it important to point out that

> Great confusion has been generated in the last few years by the implicit assumption taken by many writers that the revitalization of cultural identities is a reawakening of pristine cultural forms which had remained dormant in the 20th-century nation-states and which take up their old autarkical form. This is not the case. Most often, it is the building anew, with the remembered elements of traditional cultures, of an identity recomposed to mark out their stakes in the social and political territories that the transformation of the nation-states and the context of globality are opening up. In opening up the state from within, a most pertinent question arises: how far do we go? And the answer must be: only as far as a stable, consensual state can continue to function (ibid.).

It is striking that ten years later we were all talking about "failed states".

7.10.2 *A Different Concept of Human Sociality*

The notion of conviviability first came to my mind when reading a song (really, more of a rhetorical exhortation) of the Mexica (Aztec) Lords recorded in 1580, which reflects how the Nahua of central Meso-America had conceived of people as *coming into being through social interactions*. The song contains the words "I come, I appear! Friends, I come as created by a song, I come as shaped by a song … I have come to create a company of friends."[17] In the Nahua languages generally (there are many variants), the words associated with the English 'being' have strong connotations of connection to others. '*Nonteicniuhtlaco*' (here translated as "I have come to create a company of friends") is a polysemic term which, additionally, as in

[17] I have given a slightly different translation of this quotation from that supplied by Bierhorst (2009: 83). He translates it as: "I come, I appear! Friends, I come created as a song, I come fashioned as a song … I've come to assemble a company of friends", and this seems to apply more to comradeship in arms than simply to friendship. My own interpretation is based on contemporary Nahuatl spoken in communities in the state of Morelos.

agglutinative languages such as Finnish, it attaches prefixes and suffixes to extend its meaning. This term is made up of *nonte-* ("I come") and *icniuhtla*, formed from the root word *icniuh* ('brother'), and means "to cause people to be friends" or, more fully, "to cause others to become acquainted so they will love one another".[18] If ever a definition has been given to the process of conviviability, this is it.

This concept (to "cause people to be friends") chimes strongly in my mind with the fundamental norm that I found in all my fieldwork experiences in Mexico, whereby people are extremely generous, both to their community and to outsiders. This norm, just like many Aztec words that are Nahua neologisms in Mexican Spanish, seeped into the *mestizo* (mixed) Mexican society. It is intricately interwoven with values of gift-giving, sharing, supporting, nurturing, helping or simply expressing empathy towards others. These attitudes may be conceptualized as generosity, solidarity, cohesion, loyalty and affability. In the cultural narrative of Mexicans, a traditional saying is "one gets rich by giving".

Yet this concept goes deeper into the definition of oneself. In the very traditional Nahua villages I lived in, there is no notion of 'self' in the Western sense. Instead, one is everyone else. One 'becomes' with and through others. This fascinating notion reminds us anthropologists of Philippe Descola's categorizations of cultural norms of internal and external identifications with sentient beings and the environment. It also ties into Tim Ingold's idea of life as 'wayfaring' in all senses of the term.

Another question is whether there may be degrees of obligation in relation to conviviability. In Mexico, obligations to collateral kin are absolute; towards ritual kin (such as godmothers and godfathers) they are very strong; even towards close friends they are much stronger than in other countries. In my view, however, there is a norm predisposing Mexicans, more than other nationalities, to, at the very least, affability towards all other people.

In the cultural narrative of Mexicans, it may be said that "one gets rich by giving". This holds for giving things as gifts, food as nurture, and time as an offering of friendship. For every gift, another, even in a different form, is expected but not demanded. For every favour, a gentle look and a broad smile will suffice. In Mexican society, the dense, intricate web of giving and being in debt prolongs the real and the ritual kinship and friendship towards all horizons. Such an ingrained, confident predisposition towards conviviability will hold in Mexican migrants in the first generation and will only become more targeted towards certain groups in the second generation. By the third generation, young Mexican descendants have the same predispositions as American youth.

On 7 September 2017, an earthquake with an intensity of 8.4 on the Richter scale hit south-east Mexico, heavily damaging the indigenous regions of Chiapas and Oaxaca and the central valleys, including Mexico City. Twelve days later, on 19

[18]Translations by Bierhorst of the definitions in the first Nahuatl dictionary, written by Alonso de Molina.

September, a drill was conducted in schools and all public buildings at 11 a.m. to commemorate the 1985 earthquake in Mexico City. Two and a half hours later, another earthquake hit the central valleys. It measured 7.1 on the Richter scale and had even more devastating consequences in Mexico City and nearby regions, since its epicentre hit Axochiapan, a town only two hours away from the city. The outpouring of solidarity, both within Mexico and from other countries, was extraordinary and the mobilization of people, especially young Mexicans, was exemplary. Within half an hour of the collapse of the Enrique Rebsamen school, several hundred volunteers had formed human chains, trying to find trapped survivors, taking away the rubble, and bringing blankets and food. The Mexican army was ready with rescue programmes and specialized teams. People broke spontaneously into singing the national anthem and popular songs in buildings where it was presumed people were still alive under the collapsed cement floors.

Our team of anthropologists at the National University, since we had been working in some of the hardest-hit villages high up on the Popocatepetl volcano, received calls from our friends there asking for blankets and food. So I knocked on all the doors along my street in Mexico City and asked for donations of these goods. The following day my house was inundated with piles of blankets and all sorts of things, including children's toys, which took us a week to take up to the mountain.

When we arrived at the village, Hueyapan, volunteers were already asking people to park outside the village because there were such traffic jams of trucks, cars and pickup trucks filled with goods and young people, old people, children, all wanting to help. The same was happening all over the region. This went on for weeks, when, of course, dissent ensued between volunteers and government institutions, insiders and outsiders, charity organizations and reconstruction specialists and so on.

In any case, it seems to me that this experience supports my contention that Mexican culture gives pre-eminence to conviviability. Yet the intensity of such desire to bond leads, with the same intensity, to the furies unchained by what is perceived as betrayal. As a result, Mexico has one of the highest rates of homicides between friends and family members. It also has the most shameful phenomenon of all, one of the highest rates of 'feminicides', that is, the assassination of women at the hands of husbands and other family members. This phenomenon, as well as the recent upsurge in criminality, brutality and violence, is, however, entangled with other illicit trends which cannot be taken up here. I did present a paper on "The Roots of Violence in Mexico" at the British Academy, London, in November 2016. Statistics included in the paper showed that Mexico has one of the lowest salary levels in Latin America, government social security and extension services have been slashed, and the education system has been downgraded. So, the question now is: how long can historical trends of conviviability survive amid such devastating social fraying of communities?

7.10.3 Culture, the Cosmopolitan Outlook and Conviviability

The English 'conviviality' means a hospitable exchanging of good food, good wine and conversation—sharing good times. In a way, culture encompasses this conviviality, but in a much broader sense. Culture is made up of norms that regulate sharing language, symbols and so much else, as well as constraints to deter the destruction of conviviality. Thus, while conviviality is usually codified in the core of cultures, the norms as to which specific codes of behaviour make living together convivial vary in different cultures—etiquette is an instrument of conviviality—and the quality of conviviality depends on people knowing and accepting those norms. Such norms will only be excused when there is a language barrier involving strangers or foreigners.

As Ulrich Beck has noted, the mixture of native and foreigners in different countries is not new but "the awareness of it, its self-conscious political affirmation ... is the new characteristic of the cosmopolitan vision" (Beck 2006: 21). Beck argues that cosmopolitanism has become the definition of a new era, the era of reflexive modernity as capital tears down all national boundaries (ibid.: 2). In a sense, however, both I and Anthony Appiah consider cosmopolitanism as the name not of the solution but of the challenge (Appiah 2006: xv). One of the challenges, in my view, is to clarify how individuals and other institutional entities (nations, communities, groups) would be organized in such a cosmopolitan vision. As Beck asked: "How can the horizontal cosmopolitanization of places, biographies, families, parents, education, economics, work, leisure, consumption, politics, and so forth, be observed, reinforced, and raised to public awareness?" (ibid.: 177). If given cultures already have a historical norm of conviviability, of living-with-a-difference, the creation of this type of cosmopolitanism would be facilitated.

Several different modes of organization have been suggested. Seyla Benhabib, in analysing the "claims of culture", proposes a process of democratic iteration that would result in the continued broadening of cosmopolitan norms between democratically liberal countries and peoples beyond their borders (Benhabib 2002). Toni Erskine uses the concept of an "embedded cosmopolitanism" and proposes "a most promising combination of a situated account of moral agency and an inclusive sphere of equal moral standing. Nevertheless, ... it cannot promise a *universally inclusive* sphere of equal moral standing" (Erskine 2008: 253, emphasis added). She suggests "a commitment to a fellow-feeling with all that is human" arising from a particularist "ethic of care and responsibility" (Erskine 2008: 248). Could conviviability be one of the routes to achieving this commitment and this ethic?

As I repeated constantly in the 1990s and 2000s, economies were growing but societies were collapsing. Now, however, economies are no longer growing and societies continue to collapse. Reversing this collapse is vital and can only be understood and systematically pursued by creating the notion that economic and environmental sustainability is impossible without social sustainability. We must find the capacity, both as individuals and as groups, to "live together with a difference". The first step in this direction is surely to create a global system of power and counter-powers, as suggested by Beck (2005). At the national or local level, we need systems of reiterative democracy (Benhabib 2002) which will gradually, at every turn of decision-making, invent new concepts and practices of conviviability in an ongoing process leading towards the achievement of a cosmopolitan vision.

7.11 Chengdu, 2013: "Historicized Knowledge, Not Empty Forms Floating About"

At the international conference in celebration of the tenth anniversary of UNESCO's Convention for the Safeguarding of the Intangible Cultural Heritage, which took place in Chengdu, China, on 14–16 June 2013, Koichiro Maatsura confirmed that 153 countries had ratified this Convention, making it one of the most rapidly ratified international conventions by UN member states. As with other meetings described in this book, only the highlights in terms of concepts and ideas discussed will be included here and not the very rich debates about aspects of the implementation of the 2003 Convention in different countries. As might be expected, in revitalizing living cultural heritage, new symbolic outlooks were created and different cultural figures took the lead, while, at the same time, claims to prominence and economic gains among government, local and intragroup practitioners also fostered conflicts.

In the ten years during which the Convention had been applied in many different countries, theoretical concepts as well as practices had changed. Wang Wenzhang, the Director of the Centre for Intangible Cultural Heritage of China, opened the debates by stating that sustainable safeguarding was now the key process in implementing full-scale cultural ecologies. China's intangible cultural heritage protection had been boosted by the change from dealing with specific objects to an approach focused on a holistic and systemic understanding of the country's heritage. From this standpoint, he stated that the evolution of world cultures went hand in hand with the all-round development of humanity. Mr. Wang also argued that there must be a balance between inheritance and innovation. To achieve it he brought up a new concept, that of 'constancy', referring to the collective rules and group practices which makes it possible to promote each intangible cultural heritage practice in terms of generational constancy and timely preservation, salvage and proper utilization, and to employ intangible cultural heritage in the market without damaging authenticity or generating excessive abuse, which means achieving sustainable safeguarding.

Intangible Heritage and Dream: Festival of Intangible Cultural Heritage, as a celebration of the tenth anniversary of UNESCOs Convention for the Safeguarding of Intangible Cultural Heritage, Chengdu, China, 2013. *Source* Photo by Lourdes Arizpe

In the first round table discussion, on "Achievements of the Convention: Changing the Discourse of Intangible Cultural Heritage and Implanting New Concepts", after Koichiro Matsuura had announced that 153 countries had adopted the Convention (meaning that the concept of intangible cultural heritage had taken root), Janet Blake asked about the key terms as they related to safeguarding: what is 'intangibility'; who are the stakeholders; and what are the values being safeguarded? She also stated that there is a paradox in universality and particularity. Augusto Arantes pointed out that, in pursuing the safeguarding of intangible cultural heritage, deeper levels of social embeddedness are involved. Thinking of the intangible in universal terms does not mean that particularity is abandoned. For policy-making the similarities at a deep level are important, in spite of superficial impressions. He saw no contradiction between the two. Mr. Matsuura agreed, remarking that, in tangible cultural heritage, authenticity and integrity meant not changing anything, but that the Convention, in contrast, was much more flexible.

Kristin Kuutma then argued that one of the conundrums is that we try to define something that is changing all the time, that is changed by innovation. The

International Convention is the "Midas touch": what it touches changes into gold. Things look good on paper but dealing with people is sometimes far more difficult. Arantes countered that, if it is meaningful for those who participate in the practice, it will be meaningful for those at the end of the line of stakeholders, so it is a dialogue. Intangible cultural heritage is not about nostalgia but about a richness which is needed to create a new future. There is a receptivity to that approach. The Convention should be related to historicized knowledge, not to empty forms floating about.

In the third round-table debate, "Parallel Universes: Intellectual Property, World Heritage and Cultural Goods and Services", Chérif Khaznadar led the discussion, observing that the anthropological approach present in the Convention makes it more open to participation. Rieks Smeets commented that the implementation of the Convention led to many new cultural spaces being created. When the cultural space of Jemaa el-Fna (the central plaza in Marrakech) was placed under the purview of the Convention, it became clear what intangible cultural heritage was. Now countries were implementing more concrete elements, such as craft, within the Convention. Intangible cultural heritage is so open a concept that it can encompass anything. *Savoir faire* is very important for the transmission of objects but there are also spaces or buildings that are associated with those elements. As with clocks, one must rewind the mechanisms so that they can give the correct time.

Alissandra Cummins, President of the UNESCO Executive Board, remarked that the Board in fact believed that the 1972 World Heritage Convention was insufficient, so it had proposed that UNESCO develop a new convention on "Museum Heritage". This recognizes the fact that the work carried out by museums creates cultural heritage. But the purpose of the convention has to be clarified: is it to protect cultural property, the collections of museums? Collections are subject to different conventions today and so there is enough scope for an overarching convention, which the Board will begin developing by bringing out a declaration, not a convention. Ms. Cummins stressed the importance of documentation, which is dealt with in the UNESCO programme "Memory of the World", and cited as an example the Maroon culture in Jamaica, a secret intangible cultural heritage that we know about only because there is a documentation heritage aspect to it.

My own paper formed part of the fifth round table, "Open Questions and Future Directions".[19] I began by noting that all the belief systems related to nature and the cosmos should be taken into account as the background of all cultural knowledge. It was important to bring to the fore the fact that many intangible cultural heritage

[19]In describing my paper, I will refer exclusively to what I said at the Chengdu meeting. A reworked version was published as "How to Reconceptualize Intangible Cultural Heritage" in Arizpe (2015): 95–99.

practices give meaning to both human animals and non-human animals in ecosystems.

The key question in my presentation, however, focused on asking who was thinking for the world. This is a question that is not answered by multiculturalism nor by the instrumentalization of culture, a process which many people had brought up, but is still unresolved, sliding towards what has been called the creeping conservatism in intangible cultural heritage programmes, as gatekeepers of cultures and religions that are being left behind try to instrumentalize intangible cultural heritage to stop change.

I argued that, to enable a future of open participation and advancement, three assumptions about culture have to change. Firstly, understanding intangible cultural heritage requires a continuous rethinking of its core concepts and applications. Secondly, although there is a belief that a convention without rules will operate in favour of vulnerable people, the opposite is in fact true: the fewer rules, the more abuse by the powerful occurs. Cultures exist in contexts of power relations, and rules then help to keep the balance between those with power and those who are most vulnerable, through concepts and operations. If the thinking does not incorporate this levelling, when no continuous philosophical and conceptual discussion is going on, then debates are reduced to organizational matters; but reorganizing does not solve inconsistencies. Thirdly, intangible cultural heritage is not the end point of a process. The past is not going to teach us how to solve the problems of the present: it provides the resources, but is not the blueprint for the future.

I then brought up another key question: are we really safeguarding intangible cultural heritage? It is good to stand by and guard it. It is better still to reconfigure it—not necessarily to transform it or to transfigure it, because we want to keep the past, but to reconfigure our thoughts and practices about it. So intangible cultural heritage is the starting point for recreating the future. It is an event as understood in quantum physics: a point in a continuum of gravitational and electromagnetic forces. In my own definition, intangible cultural heritage is "a moment in time".

Many countries have been successfully developing this cultural policy in the field of intangible cultural heritage. As an example, the 19–25 August 2017 issue of *The Economist* considered this trend important enough to included a report on how the Chinese government has reversed the policies of the Cultural Revolution in the 1960s and 1970s in order to intensify efforts "to build what [President Xi Jinping] refers to as 'cultural confidence'" (*The Economist* 2017a: 36). The report lists an agenda that touches every aspect of life, from local and national festivals to boosting the use of Chinese medicine, practising t'ai chi, and television shows of higher 'moral' content (ibid.). There is an economic logic to such policies, the report continues, since they protect some Chinese firms from foreign competition and promote new sources of consumption. Following a major UNESCO programme, in April 2017 the Chinese government was interested in developing cultural industries into a 'pillar' of the economy. "China's ancient heritage stands at the centre of its sales pitch to the world, too: becoming 'a socialist cultural superpower' is now an official national goal" (*The Economist* 2017a: 37).

It is important for the conclusions of my argument to note that in some areas cultural policies are foundering while in others, such as in the case of intangible cultural heritage and cultural expressions in cultural industries, they have now risen to a higher level of economic and political relevance in the neo-liberal world context.

It is also worth noting that in the same year that the Chengdu meeting took place, another initiative was held, also in China, in Huangzou, with the view of placing culture at the centre of sustainable development policies. The Huangzou Declaration states that "there is no single development model for all and that different cultural perspectives will lead to different pathways to development" (UNESCO 2013: 2). Two paragraphs later it states, in point 1, "Integrating culture in all policies and programmes for development", that "development is configured by culture and local context, which ultimately also determine their results" (ibid.). Defined in this way, what does 'development' mean if it does not represent a unified model of policies for world economic growth and sustainable human development? What does "sustainable development" mean if it does not refer to an internationally agreed environmental model of sustainability for coordinated action by all countries of the world? Not only does such a position contradict most United Nations and UNESCO normativities and conventions but it creates equivocal expectations in culturally affiliated groups that their own ethnic, local or religious rights will prevail over nation-state laws and international laws and conventions.

While the defence of local environmental territories and resources is important in stemming the depletion of resources by corrupt government officials or abusive business corporations, the most important successful programmes to protect the environment are coalitions, most of them having participants from many different cultural, indigenous, scientific and national backgrounds.

7.12 The UNESCO International Decade for the Rapprochement of Cultures (2013–22)

By the end of the first decade of the millennium, the divisive effects of policies focused exclusively on the diversity of cultures without integrating policies for diversity with national legislative and juridical institutions had been recognized and acted upon by member states of the United Nations. On the one hand, cultural rights were discussed within the framework of the political and legislative agendas in many countries. On the other, programmes intended to prevent the divisive effects of separating populations according to cultures and religions were steered towards interculturality and interreligious dialogue. In the following pages one of the most important initiatives will be described, drawing on the report published by the United Nations.

In 2010, the Year of the Rapprochement of Cultures was adopted by the UN General Assembly. In December 2012, Resolution 67/104 set out the period 2013–22 as the International Decade for the Rapprochement of Cultures (IDRC) and reinforced member states' commitment to furthering interreligious and intercultural dialogue and the promotion of mutual understanding and cooperation for peace. The programme is co-sponsored by Kazakhstan and UNESCO and, in terms of partnerships and resources mobilization, bilateral cooperation is currently being sought from the Abdullah bin Abdulaziz International Programme for a Culture of Peace and Dialogue supported by the Kingdom of Saudi Arabia, the "Baku Process" driven forward by the government of the Republic of Azerbaijan, and the Congress of Leaders of World and Traditional Religions promoted by the Republic of Kazakhstan (UNESCO 2016b: 23). It also links up to major regional African, Islamic, Asian, and Latin American organizations (ibid.).

The introduction to the programme begins with a very different kind of appraisal of situations related to culture in societies in the world today. It speaks of the

> widening moral gap in our societies and the extent to which our societies are ill-equipped to effectively manage and overcome the challenges that continue to arise. Often in parallel with developmental progress, retrogressive patterns have emerged that present persistent threats to peace-building and social cohesion (ibid.: 5).

It goes on to mention "recurrent cycles of deadlock" and that "the litany of polarization within our societies often defined along the lines of 'identity'—religious, cultural, ethnic and otherwise—are [*sic*] increasingly manifested in prejudice, intolerance, racism, xenophobia, discrimination, radicalization and extreme violence" (ibid.).

No one would dispute that this is a realistic appraisal of the current state of social and cultural relations in many countries. In terms of the transactional analysis in this book, which takes a long-term view, the question that springs from such an assessment is: what has gone wrong since the beginning of the decade?

In the UNESCO Roadmap quoted above, mention is made of the "new levels of complexity in addressing challenges" and the reiterated call since the World Commission on Culture and Development made its recommendation "for a more inclusive and participatory notion of development" (ibid.). This is an era "marked by increased divisiveness, and learning to live together in the 21st century presents a new frontier for managing cultural diversity" (ibid.). The key aspect then, is to go back to development, but mainly in order to find ways of managing diversity. The introduction continues by stressing that "The implementation of the Decade will be guided by a move beyond conceptual deliberations towards concrete and practical strategies" (ibid.: 8).

In spite of the above, in terms of 'Vision', the document states that "The International Decade for the Rapprochement of Cultures (2013–2022) embodies a global vision and commitment to unification in diversity" (ibid.: 10). Yet there is no specification of the model of unification in diversity that is envisaged, nor which international or national institutions and actors should be involved. There is even less information on *how* they should be involved, given the diagnostic of how 'ill-equipped' societies are at present to counter the challenges set up by cultural diversity. Nor is any mention made of how the interreligious dialogues would be conducted.

In a separate document on the Plan of Action for the IDRC, emphasis is placed on "the pressing urgent need to take into account and clearly demonstrate new articulations between cultural diversity and universal values" (UNESCO 2016a: 9). However, this articulation would require an intellectual research effort that is not spelled out in the Plan's strategies. Instead, reference is again made to values, in terms of "a commitment to such principles as compassion, conviviality, hospitality, solidarity and brotherhood which are the cornerstones of human coexistence, inherent in all cultures, faiths and secular ideologies" (ibid.). In this framework of co-existence, the Plan proposes three steps for effective engagement: assessment of current national policies to monitor inclusiveness and respect for diversity; setting annual priorities; and establishing meaningful partnerships at national, regional and global levels (ibid.: 10–11). The four "priority areas of action" are set in terms of: (1) promoting mutual understanding and reciprocal knowledge of cultural, ethnic, linguistic and religious diversity; (2) building a framework for commonly shared values which fosters social cohesion; (3) disseminating the principles and tools of intercultural dialogue through quality education and the media (on this point, however, the Roadmap states that "to date rigorous data on intercultural dialogue has remained obscure" (UNESCO 2016b: 13)); and (4) fostering dialogue for sustainable development and its ethical, social and cultural dimensions (UNESCO 2016a: 11–14).

Two major points are worth highlighting here. First, the third priority area for action states that "Rather than emphasizing the differences in culture, heritage, religion or belief among groups, focusing on common elements leads to understanding that there are also similarities worth celebrating" (ibid.: 12). Yet in the rest of the document this is not spelled out: who will identify such common elements—communities, government officials or leaders of movements? If the search for the "new articulations between universal values and cultural diversity" and for such 'similarities' is to be given this priority, the term 'rapprochement' according to its definition in French and English, can lead only to a *shortening* of the distance between separate objects. This still leaves the question of how these reified objects that are being defined and categorized as separate entities can interact.

In the Plan's section on "Modalities for Effective Implementation", the first "main activity" proposed aims to "improve access formal and non-formal quality education for all, free from cultural, religious and gender prejudices and stereotypes" (ibid.: 15). Formulated in this way it raises the question of how—given that many, if not most, prejudices and stereotypes (especially in religious and gender terms) originate in traditional cultures and religions—the two aims of defending diversity, on the one hand, and preventing prejudice and discrimination, on the other, are to be made compatible.

The second point is that the IDRC programme is centred on traditional cultures and religions, leaving out philosophy, science, political science and the other intellectual creations that, since the time of the Enlightenment, have been so vital in fostering ever more inclusive and democratic political ideas and development policies for all through citizen's rights and human rights. Focusing exclusively on 'traditional' values leads precisely to the divisiveness and 'retrogressions' that the programme notes as a matter of concern.

7.13 Protecting Cultural Heritage in Zones of Armed Conflict: Is a New Framework for Action Needed?

As the concept of security broadened from a narrow definition of national security to that of human security more generally, defined by the Human Security Network as "freedom from fear", a new framework emerged for understanding not only threats against human rights but also human well-being in more comprehensive terms (Brauch et al. 2009: 59). Environmental concerns, gender equity, identity and anthropogenic climate change are now included in scenarios for securitization—which does not necessarily mean militarization. "Freedom to live in dignity" is considered an important dimension of human security, defined in many ways, as part of human sustainable development (according to the United Nations Development Programme) and of basic human functionalities and capabilities, as developed by many scholars, among them Amartya Sen and Martha Nussbaum.

The question of whether there is a case for the securitization of culture has always created great controversy and in recent years has been taken up in concepts and legislation related to cultural rights. The very precise question of the protection of cultural heritage has exploded in the media for many reasons, but in particular it represents a shock to the framework of the extraordinary world consensus which surrounds the international conventions established to protect it.

The constant and ever-escalating number of events involving the destruction of cultural heritage has created an opportunity for discussing the international responsibility to protect such heritage. After the first such instance—the bombardment of Dubrovnik during the Yugoslavian civil war—such occurrences have multiplied, including the obliteration of the Buddhas of Bamiyan in Afghanistan by the Taliban, the destruction of Muslim buildings in Timbuktu in Mali at the hands of Islamic militants, and the deliberate and systematic demolition of antiquities in Iraq and Syria by Isis.

Lourdes Arizpe with students at the Bibliotheca Alexandrina, Alexandria, Egypt. April 2010. *Source* From the author's personal photo collection

Leadership in discussing responsibility for protecting cultural heritage has now been taken up by the academies of the US and the UK. A meeting on "The Responsibility to Protect Cultural Heritage in Events of Armed Conflict" was held at the British Academy in Carlton House Terrace, London, on 29 November–1 December 2016. The seminar focused on the antecedents of the creation of such international norms as the Responsibility to Protect doctrine (R2P), the national, regional and global responses to the destruction of cultural heritage, and the role of the military in protecting cultural heritage, including the question of whether United Nations blue helmet units are needed to protect people's heritage from illegal trafficking. Prominent members of the National Academy of Science of the United States and the British Academy, directors of major cultural foundations such as the Getty Foundation, the Andrew W. Mellon Foundation and the Heritage Museum in St Petersburg, the director of the Bibliotheca Alexandrina, and military officers were active participants in the meeting, as well as social scientist members of the American Academy of Sciences, the British Academy and major universities around the world.

Lloyd Axworthy, a former Minister of Foreign Affairs for Canada, strongly advised that "It is time for liberals and cosmopolitans to fight back to stop the turmoil that creates exclusiveness." Referring to Syria, he argued that "it is not enough just to bomb and get out because the public doesn't want it. In fact, we should stick it out until we make sure that the populist extremists are routed."

The search for solid foundations for this recasting of culture in the context of armed conflict follows a razor's edge, which Mariet Westermann, Executive Vice-President of the Andrew W. Mellon Foundation for Programs and Research made evident when she remarked that the title of session 6, "Political Costs and Moral Imperatives: How to Navigate the Dilemma?", would have better reflected the dilemma as "Moral Costs and Political Imperatives".

Meeting of the board Bibliotheca Alexandrina with Ismail Serageldin, Alexandria, Egypt, April 2010. *Source* Photo was made by the Bibliotheca Alexandrina that granted permission

After citing statistics showing that, by 2016, the 1972 Convention for the Protection of the World Cultural and Natural Heritage was the second most ratified international convention (ratified by 165 countries, with 1,052 protected sites and landscapes, of which 841 are cultural) and that the 2003 International Convention on the Intangible Cultural Heritage had been similarly successful, I further noted that the most fruitful recommendation of *Our Creative Diversity* was the international programme of cultural heritage volunteers. Given such a high level of world consensus on the protection of cultural heritage, I called attention to the paradoxes of culture: (1) the more globalization tends to homogenize cultures, the more cultural diversity rises; (2) the more the protection of cultural heritage is publicized, the more it becomes vulnerable to political use; (3) the more cultural heritage is referred to as a capital asset, a property, the more business interests invest in it strictly for economic gain; and (4) the more cultural heritage is proclaimed as important for social cohesion and well-being, the more it can be manipulated for diverse ends in the media and on the Internet. My question then was: in such a turbulent streaming of culture, how do we operate? I suggested that a distinction be made between conflicts that are a positive readjustment, where local peoples are re-engaging with economic, political and social structures that were previously closed to them, and struggles that will amplify and deepen conflicts and wars. The two scenarios require different strategies and policies to preserve cultural heritage. I said that the crucial point was made by Mary Kaldor: whose side are we on? Multiculturalism has led to such cultural retrenchment that both Angela Merkel and David Cameron have rejected it. So why is nobody referring to religious cleansing? I argued that the question we must pose is: what are the mechanisms of legitimacy in caring for and protecting cultural heritage?

A striking new theoretical framework for protecting heritage was brought to the table by Mary Kaldor, Professor of Global Governance and Director of the Civil Society and Human Security Research Unit at the London School of Economics and Political Science. In our contemporary world, she stated, “war is not two sides: it is an attack on civility”. She went on to explain that before wars, people had said “we all have mixed identities”, so war is between the populists who believe in a single, exclusive identity and those with mixed identities: “Through violence, you do create exclusive identities, you gain from violence itself.” Most wars are now directed against civilians, not against the military. The first to be killed are intellectuals, academics and doctors because they are the first to protest. All revolutionary conflicts begin with movements for democracy and against violence. But then young people, especially from rural areas, are recruited for violence, while those in cities look for ways of solving the violence.

How do we oppose such wars? Kaldor argued that an essential element is to defend civility. We must give up the assumption that wars are military opposition. In celebrating diversity, contemporary words are very preachy. People have resisted war; there are many examples but they do not get any attention. So it is very important to protect islands of civility. When research is focused on extremist identities, this reinforces them. Kaldor also gave a deep insight into the fact that such a focus "breaks up communities". She pointed out that the reconstructed bridge at Mostar "did not reconstruct the cosmopolitan, mixed identity of people who lived in Bosnia". As a strategy against this break-up, she emphasized using cultural heritage to focus groups, to "bring people in".

One of the problems that Kaldor mentioned was that security can create a black economy in selling antiquities. However, this could be attacked through the R2P, which need not be linked to military protection. The military are trained to defeat an enemy, not to protect cultural heritage, and so she argued that an emergency force would have to be different: they would need to protect intangible cultural heritage. Returning to the subject of identities, she pointed out that "Brexit is conjuring exclusive identity to gain power". Once people are made different, they are put into a box. This constitutes a danger at the moment, so protecting cultural heritage is central to protecting civil society. The cosmopolitan view is against war on terror and security.

Francesca Giovannini, Senior Program Officer for Global Security and International Affairs of the American Academy of Arts and Sciences, brought many of the issues mentioned together by saying that there is an intellectual opening for creating a framework for protection of cultural heritage, so it is important to push the academic and political discussion forward. She mentioned two essential conditions for the construction of this framework: firstly, champions are needed, and an entrepreneurial state to take the lead in continuing this endeavour; secondly, an intellectual debate must endure, even if policies change.

After the meeting reported above several important initiatives have been undertaken at the international level and in some countries. An interesting counterpoint to the way in which cultural policies on intangible cultural heritage in China have expanded: in that same issue of *The Economist* of August 2017, a report describes what happened in the Middle East. It begins by stating that "The Middle East is used to ruins" (*The Economist* 2017b: 39), citing episodes of destruction from the ruin of Nineveh at the hands of the Babylonians to that of Jerusalem by Hakim and that of Baghdad by the Mongols. 'Still,' the report goes on, "the desolation of the past three years is probably the worst on record. According to the United Nations, half of the old city of Mosul, in Iraq, and a third of the old city of Aleppo, in Syria, are rubble. Hundreds of minarets, monasteries and monuments have been toppled. Of the world's 38 endangered cultural-heritage sites, 22 are in the Middle East, says UNESCO…" (ibid.). The report continues "The jihadists of Islamic State (IS) like to boast of their role in the wreckage. They have filmed themselves razing ancient temples, churches and mosques" (ibid.). It does point out that "… the American and Russian armies, along with their local allies, have inflicted at least as much damage in their war on the jihadists" (ibid.). A new recent event is also the damage to the national museum in Dhamar, to the Great Dam of Marib—an ancient engineering wonder—and to the al-Qasimi complex of mud-brick towers in the old city of Sana'a, all of them perpetrated through air strikes by the Saudi-led coalition in Yemen.

Given all these events, the report then explains that pressure for international action is rising: "In March (2017), the UN security council reaffirmed that attacks on cultural sites are a war crime. Last September (2016) Ahmad al-Faqi al Mahdi pleaded guilty in the International Criminal Court (ICC) to crimes of cultural destruction in Mali. The jidadist leader admitted to ordering attacks on Muslim shrines in Timbuktu" (ibid.). Judges from the ICC then awarded reparations, setting a precedent for claims in the destruction of other heritage sites.

The Economist also looks at the aftermath of such destructive events, pointing out that "Poverty, despair and a collapse of civic pride hasten the vandalism" (ibid.). It mentions that local truces and national and international police are not always able to stop smugglers of valuable treasures. Another consequence of such destruction, rarely mentioned, is that reconstruction may actually become "the final act of destruction". It mentions the case of land prospecting in Aleppo, where the building of tower blocks may soon destroy layers of an 8,000-year-old archaeological site, and it quotes an archaeologist who said that "Bulldozers … can be even more damaging than tanks" (ibid.).

7.14 Summary

This chapter has shown how cultural policy and cultural rights became a mainstream international concern as governments, business organizations, scientific organizations and civil society groups took up the challenges of culture, each group having different powers, interests and outlooks. There was a clear shift in the assumptions, aims and expected outcomes, as transactions on culture led to different agendas, while seeming to retain one and the same agenda.

Cultural Changes in Women's Life Stages. Report of Seminar held by UNESCO at the United Nations World Conference on Women, Beijing, 1995

References

Appiah, A. (2006). *Cosmopolitanism: ethics in a world of strangers*. New York: Norton & Co.

Arizpe, L. (1972). Parentesco y economía en una sociedad nahua. MA dissertation, Instituto Nacional Indigenista, Mexico City.

Arizpe, L. (1981). The rural exodus in Mexico and Mexican migration to the United States. *International Migration Review*, 15, 4, 626–650. Reprinted in: P. Brown & H. Shue (Eds.) (1983), *The border that joins: Mexican migrants and U.S. responsibility* (pp. 162–183). Totowa, NJ: Roman and Littlefield.

Arizpe, L. (1996). *Social Perceptions of Deforestation in the Lacandona Rainforest*. Ann Arbor: University of Michigan.

Arizpe, L. (1998a). "Convivencia": the goal of conviviability. In: UNESCO, *World culture report 1998: culture, creativity and markets* (p. 71). Paris: UNESCO.

Arizpe, L. (1998b). Conviviability: the role of civil society in development. In: A. Bernard, H. Helmich & Percy B. Lehning (Eds.), *Civil society and international development* (pp. 21–24). Paris: OECD Publishing.

Arizpe, L. (2004). Cultural liberty in today's diverse world: the 2004 Human Development Report. Address delivered at the official launch of the 2004 Human Development Report, Brussels, Belgium, 7 July 2004.

Arizpe, L. (2006). An international perspective of social development models. Paper presented at the Fair Globalisation: A Coherent Policy for More Employment and Decent Work Conference, Berlin. 22–23 November 2006.

Arizpe, L. (2008). The ritual and the promise: why people value social ritual. In: M. Hutter & D. Throsby (Eds.), *Beyond price: value in culture, economics, and the arts* (pp. 141–158). Cambridge: Cambridge University Press.

Arizpe, L. (2014). *Lourdes Arizpe: a Mexican pioneer in anthropology*. Heidelberg: Springer.

Arizpe, L. (2015). The genealogy of cultural heritage. In: L. Arizpe, *Culture, diversity and heritage: major studies* (pp. 101–117). Heidelberg: Springer.

Arizpe, L. & Alonso, G. (2002). *Cultura, comercio y globalización*. Buenos Aires: Consejo Latinoamericano de Ciencias Sociales. First published in English as *Culture, trade and globalization*. Human Development Report Working Papers. New York: UNDP, 2000.

Beck, U. 2005. *Power in the global age: the new global political economy* (trans: Cross, K.). Cambridge: Polity Press.

Beck, U. 2006. *The cosmopolitan vision* (trans: Cronin, C.). Cambridge: Polity Press.

Benhabib, S. (2002). *The claims of culture: equality and diversity in the global era*. Princeton: Princeton University Press.

Bierhorst, J. (2009). *Ballads of the lords of New Spain*. Austin: University of Texas Press.

Brauch, H. G., Oswald Spring, Ú., Grin, J. et al. (Eds.) (2009) *Facing global environmental change*. Heidelberg: Springer.

Brown, P. G. & Shue, H. (Eds.) (1982). The border that joins: Mexican migrants & U.S. responsibility (Maryland Studies in Public Philosophy). Lanham: Rowman & Littlefield.

Calhoun, C. (1995). *Critical social theory: culture, history, and the challenge of difference*. Oxford: Blackwell.

Condominas, G. (1993). The evolution of anthropology today and the appropriate role for UNESCO in relation to research establishments and universities in the area of the intangible cultural heritage. Paper presented at the International Consultation on New Perspectives for UNESCO's Programme: The Intangible Cultural Heritage. Paris: UNESCO.

Erskine, T. (2008). *Embedded cosmopolitanism: duties to strangers and enemies in a world of "dislocated communities"*. Oxford: Oxford University Press.

Getty Conservation Institute (1999). *Economics and heritage conservation*. Los Angeles: Getty Center.

Global Economic Forum (2003). "Report on the session on The Values of Globalization". Document made available to participants at the 2003 Global Economic Forum, Davos, Switzerland.

Gruzinski, Serge. (1993). Protection of the intangible cultural heritage: survey and new prospects. Paper presented at the International Consultation on New Perspectives for UNESCO'S Programme: The Intangible Cultural Heritage. Paris: UNESCO.

Lefebvre, A. & Méda, D. (2006). *Faut-il brûler le modèle social français?* Paris: Seuil.

Picco, Giandomenico et al. (2001). *Crossing the divide: dialogue among civilizations.* New York: Seton Hall University.

Pontifical Academy of Sciences (2002). The cultural values of science. http://www.pas.va/content/accademia/en/events/2002/culturalvalues.html, accessed 3 October 2016.

Portes, A. & Hao, L. (1998). E Pluribus Unum: Bilingualism and Language Loss in the Second Generation. Levy Economics Institute Working Paper No. 229. Available at: https://ssrn.com/abstract=121374.

Rifkin, J. (2000). *The age of access: the new culture of hypercapitalism.* New York: J. P. Tarcher.

Rifkin, J. (2001). World culture resists bowing to commerce. *Los Angeles Times*, 2 July 2001.

Rumbaut, R. G. & Cornelius, W. A. (1995). *California's immigrant children: theory, research, and implications for educational policy.* San Diego: University of California Center for US–Mexican Studies.

Sen, A. (1999). *Development as freedom.* New York: Oxford University Press.

Sen, A. (2004). Cultural freedom and human development. Background paper for the Human Development Report 2004, presented at the New York meeting of background paper authors, 24–25 September 2003.

The Economist (2017a). "The Socialist Party is redining what it means to be Chinese." Issue of 19–25 August 2017, page 36.

The Economist (2017b). "War in the Arab world has devastated the region's heritage". Issue of 19–25 August 2017, page 39.

UNESCO (1998). *World culture report 1998: culture, creativity and markets.* Paris: UNESCO.

UNESCO (2000a). *Study on international flows of cultural goods, 1980–1998.* Paris: UNESCO.

UNESCO (2000b). *World culture report 2000: cultural diversity, conflict and pluralism.* Paris: UNESCO.

UNESCO (2001). UNESCO Universal Declaration on Cultural Diversity. http://www.unesco.org/education/imld_2002/unversal_decla.shtml, accessed 3 October 2016.

UNESCO (2013). *The Hangzhou Declaration: Placing Culture at the Heart of Sustainable Development Po*licies. Available at: <http://www.unesco.org/new/fileadmin/MULTIMEDIA/HQ/CLT/images/FinalHangzhouDeclaration20130517.pdf.

UNESCO (2016a). *Action plan: 2013–2022: International Decade for the Rapprochement of Cultures.* Paris: UNESCO.

UNESCO (2016b). *Roadmap: the rapprochement of cultures.* Paris: UNESCO.

United Nations General Assembly (2000). United Nations Year of Dialogue among Civilizations: report of the Secretary-General. UN A/55/492 World Commission on the Social Dimensions of Globalization (2004). A Fair Globalization: Creating Opportunities for All. Geneva: International Labour Organization.

Chapter 8
Recasting Culture

> *(Culture) may be used as an element of explanations of why people think or behave the way they do, but it cannot be a substitution for analyses that take into account economic, social and biological factors.* Adam Kuper, *Culture: the Anthropologists' Account (2003).*

The media commotion about the turn of the century and the start of the new millennium—full of awe and panic over "The end of the world is here!", "Apocalypse now, really"—and the more sober 'Y2K' scare of having all computer systems crash on 1 January 2000 were not simply entertaining distractions. They brought to the surface millenarian fears that were thought to have been left behind by 'civilization'. Most importantly, they made visible the spread of a collective awareness that the human world was now being represented in a novel way. Previous chapters have described the threads of transactions that emerged in meetings and programmes, developing new fields, vocabularies, procedures and positionings in this cultural process of 'worlding', a word which is the active component in the remaking of a cosmopolitics, in the sense described by Isabelle Stengers and her collaborators (Stengers 2005).

The evolution of the positionings of major global institutions and organizations in the first decade of the millennium has been described in Chap. 7. The United Nations and UNESCO, the business sector at Davos, religious institutions such as the Vatican, the scientific development community, broad and diverse global social movements, and countries interested in opening up places for themselves in international transactions on culture all generated specific currents that developed in response to particular situations and the positionings of others. It became apparent that societies in the twenty-first century would have to reconfigure their own cultural imaginaries and their sense of historical time, while artists and cultural practitioners needed to reimagine and represent afresh their place on the new world political and diplomatic stage. As cultural constituencies reassembled, drew in new members and

L. Arizpe Schlosser, *Culture, International Transactions and the Anthropocene*, The Anthropocene: Politik—Economics—Society—Science 17,
https://doi.org/10.1007/978-3-642-41602-6_8

clarified their positions, a deep intellectual recasting of the term 'culture'—as well as of other terms involved in globalization, such as 'development', 'equity' and 'sustainability'—began to gather momentum in creating the philosophical cartography of this new cosmopolitics.

Again, culture has become a core instrument in this process—to such an extent that David Steigerwald stated in his 2004 book *Cultural Vanities: The Paradox of Cultural Diversity in a Globalized World* that "it is not too much to say that the foremost dynamic in the so-called postmodern mind over the last generation has been the conviction, now hardened into dogma, that culture determines history" (Steigerwald 2004: xi). For this reason alone it has been vital for different constituencies to rethink their place in contemporary intellectual spaces. It would be impossible to summarize the profusion of debates and publications related to culture in this first decade and a half of the new millennium. Instead, if we follow the method of trying to provide an ethnography of the evolution of thinking about culture in the international sphere, the materials themselves now make it necessary to take up the intellectual threads in published writings and to document the threads of cultural transactions in this period (many of which were first generated in the live discussions described in previous chapters) that have recast culture in the political realm in recent years.

Two specific dates marked this change in the perspective on culture. After 2001, as Steigerwald harshly puts it, "a period that began with people singing the Disney anthem 'It's a Small World After All' ended, shockingly, with religious fundamentalists flying airplanes into the tallest symbols of that cheery little world" (ibid.: xix). The shock of 9/11 slowly seeped in during the years following and it changed the course of transactions on culture.

Yet the second threshold, 2004, as mentioned in the previous chapter, provided a new, optimistic momentum in the field of culture. The United Nations Development Programme set out a new policy definition, that of "cultural liberty", in their report *Cultural Liberty in Today's Diverse World*, the World Commission on the Social Dimensions of Globalization published its report *A Fair Globalization: Creating Opportunities for All*, and the Rockefeller Foundation *Seminar on the Value of Culture and Art* concluded that value is this domain is more than price.

As countries moved to reposition themselves in the shifting framework of globalization or to create their own position—for example, the Baltic States and the Central Asian countries—culture became a centralizing policy instrument for recreating a historical and nationalist sense of community.

As previous political cartographies of cultures in the world slipped their moorings, the decades-old intellectual process of recasting this concept came to the fore in the first years of the new century, with strong, critical views from many anthropologists and other social scientists. As explained further on, Adam Kuper argues against the idealism and relativism that has influenced these debates, especially in American cultural anthropology: "culture could be a euphemism for race, fostering a discourse on racial identities while apparently abjuring racism" (Kuper 2003: 22). The rise of white supremacy and neo-Nazi movements in several developed countries in 2017 has brought this highly sensitive issue to the surface once again.

In addition, unprecedented scientific and technological breakthroughs have turned cultures into experimental practices in co-productionist modes of knowledge where digital culture is introducing a novel layer of communication whose interaction with on-the-ground cultural practices we are only beginning to analyse. This is taken up in a subsection of this chapter on digital culture.

When I began to write this chapter on "Recasting Culture" my intention was to show how the scientific debates on culture had been running in parallel to the public rendition of the discourses on culture as evidenced in the meetings sponsored by governments, United Nations agencies, the private sector, social science organizations, religious institutions and civil society associations described in the previous chapter. In so doing, I had hoped that the contrasting enunciations, invocations, references and spins on ideas about culture would highlight the counterpoint between the public and the scientific narratives and discourses. I found such a wealth of material in the anthropological and social science publications dealing with culture between 2000 and 2017 that I preferred to include references to a very broad range of new ideas about culture rather than to discuss each one in depth. For comprehensive assessments of this literature I refer readers to Kuper (2003), *Culture: the anthropologists' account*; Kahn (2001), *Culture, multiculture, postculture*; and Eagleton (2000), *The idea of culture*; and for a new vision of the future of culture, Fischer (2009), *Anthropological futures.*

Science, it is often said, focuses on the infinitely big and the infinitely small. Given the trends in the last twenty years, it also focuses on the infinitely narrow. By this I mean that once the intellectual, political and policy space of culture was defined in the nineties, an invisible barrier surrounding it seems to have led to an ever greater endogamy of ideas within that space. This pattern is certainly one of the discoveries of this wide-ranging review of international transactions on culture. This chapter shows how scholars and governments are trying to break out of this barrier of *idées reçues* to discuss other aspects of culture. In particular, a subsection is included briefly describing the discoveries of cognitive science and the experiments in digital culture that will no doubt move thinking on culture in directions we cannot foresee.

UNITED NATIONS RESEARCH INSTITUTE FOR SOCIAL DEVELOPMENT
(UNRISD)

ETHNIC VIOLENCE,
CONFLICT RESOLUTION
AND CULTURAL PLURALISM

Report of the UNRISD/UNDP
international seminar on
Ethnic Diversity and Public Policies
(New York, 17-19 August 1994)

UNRISD

Geneva, February 1995

"Ethnic Violence, Conflict Resolution and Cultural Pluralism", research report by United Nations Research Institute for Social Development, Geneva, Switzerland, 1995

8.1 Culture in the Scientific Sphere

8.1.1 Rethinking the Cultural

One of the discussions which pioneered debates among anthropologists on culture in a global setting was a session entitled "Rethinking the Cultural: Beyond Intellectual Imperialisms and Parochialisms of the Past", held at the American Anthropological Association meeting in Washington in December 1994. In his introductory remarks, Robert Borofsky, the organizer of the session, began by asking whether we can we make the concept more interactive and processual so as to take account of the diverse connections and power relations between what are often called the 'local' and the 'global' social, as well as the subtle dynamics in which change overlaps with continuity and individual diversity overlaps with collective sharing.

Borofsky argued that culture is not a reality—something that exists in the real world—but a concept we use to describe a range of phenomena. Following Roger Keesing, he preferred the adjectival phrase "the cultural" rather than culture, to emphasize this point. He said that the concept of culture is a historically situated idea that draws on both a tradition of social criticism concerning the effects of Western industrialization and also on particular forms of nineteenth-century European nationalism. Applications of the concept tend to oscillate between emulative models (focusing on ideals for inspiring people) and communal models (beliefs and behaviours that people reputedly come to share as a result of collective experiences). In both cases, he stated, "the cultural" deals with points of reference that frame interactions relating to real or imaginary communities.

Pierre Bourdieu contributed the reflection that information tends to flow according to, and thus to reinforce, inherited imperial and postcolonial hierarchies. A methodological analysis of the national, imperial and linguistic effects of domination in the emerging global social scientific field is a prerequisite for checking such effects and for fostering a genuine, scientific universalism. Veena Das pointed out that culture has entered the discourse of nation states and of market research, as well as the formulation of international covenants on human rights and the rights of peoples. Examining how the pressure to formulate the notion of culture at a sufficiently abstract and general level allows international instruments to talk of the "right of a group to preserve its culture", on the one hand, and the local examples that inform concrete lives, on the other, provides an opportunity for us to understand the relationship between levels of social life and new positions.

Fredrik Barth took up knowledge as an entry point, arguing that culture reflects "other ways of knowing". Firstly, he said, we use culture to indicate what is salient to outsiders, marking 'differences', so that it leads to exoticizing, representing others' points of view partially and unreasonably. Secondly, such 'differences' are situated outside interaction as a collective, stereotyped behaviour of groups. Thirdly, in public debate 'culture' is used to define discourses on identity, allowing leaders to claim that they are speaking in the name of others, who are thereby given access to the media, but in fact they disempower followers. Our strategy must be to transcend this form of discourse but, even as we oppose reification, it comes back because of the prototype we use of culture as a shared, localized collective. He noted that he had previously argued for knowledge as a modality of culture because it makes culture transitive in the interaction between people. He contended that what is needed is to accept much more native concepts. We should not seek a fictitious cultural authority in others but engage with them, not to remove their knowledge but to look at it using political and moral criteria. He argued that what mattered was to ascertain the criteria of validity in different cultures and the kinds of knowledge they create.

In the ensuing discussion, Gustavo Ribeiro emphasized the need to work on the notion of transnational community as the new information and communication technologies expanded world connections. Yet Heloísa Buarque de Hollanda interjected that the way people perceive 'differences' varies from country to country, pointing out the preference of Brazilians to swallow difference rather than confront it. Nancy Scheper-Hugues asked whether we are suspending the ethical in favour of a diversity of others. She argued that there must be an ethical standard, because culture can be used to classify for purposes of discrimination and exploitation. In the primacy of the ethical, she posed a series of transcendent values, even pre-cultural values, among which were responsibility, accountability and answerability, and she concluded that morality precedes culture. A further comment that was shared and agreed upon was that anthropology had gone from being a species-centred discourse to being an ethno-centered one.

Marshall Sahlins argued that the panic about the concept of culture was due to a lack of analysis of historical anthropology. The great paradox of our age is the confrontation between local forces and global frameworks. Culture is not reducible to power, or to differences. Sahlins accepted that there is a development of a world culture based on Western culture, mainly as marketization of culture, with a decentring, new flow of cultural form from South to North, and new attention to both human rights and the earth's ecosystems. But this development has gone hand in hand with a re-pluralization of cultures, with local cultures (seen as having a value in themselves) finding ways not only of adopting the world culture but also of resisting it. He acknowledged that there is an improvidence of the global system which escapes human control, so that it is experienced as uncertain and uncontrolled. This leads to locals turning to their own cultures, using those cultures to avoid being 'just like white men". In this way, indigenous cultures relive the experience of European local cultures in the nineteenth century. He argued that culture is real in the global world and therefore it is necessary to qualify it, not flee from it. Ten years on, this debate has become much more attuned to the runaway political debates on how to manage cultures in a global world.

During this time, as will become evident in this chapter, many anthropologists have been greatly concerned about the risks of 'essentializing' culture, a form of instrumentalization that it seems to me has led to many of the extremist events linked to culture, from jihadism to white supremacist ideologies. One example of such opposition is that of Ulf Hannerz, who echoed critiques made by many other social scientists in advocating a processual view of culture, while "developing our critique of old and new varieties of essentialism" (Hannerz 2006: 279).

8.1.2 A Hyper-Referential Term that Becomes an Explanation

In 2003, Adam Kuper published his influential book, *Culture: the Anthropologists' Account,* which summarized the most important critiques of social scientists concerning this term, especially when it becomes entangled with politics. He begins by stating that this hyper-referential term, which he suggests should be substituted for more precise terms such as knowledge, belief or art, becomes "most tricky when used not as something to be examined or analysed but as an explanation in itself" (Kuper 2003: 25). Indeed, in this second decade of the millennium, both in cultural policy and in the news media and the social media, the term is used as if it explains, by itself, how people are behaving, without reference to the economic, political or social conditions that drive the context in which they live.

His knowledge of the use of culture in the implementation of apartheid in South Africa leads him to state that "culture could be a euphemism for race, fostering a discourse on racial identities while apparently abjuring racism" (Kuper 2003: 18). In his very detailed historical account of the rise of the genealogy of culture and civilization, he explain, very precisely, the transactions that occurred when the school of thought of the Enlightment was countered by the Romantic movement and provided the background for the contrasting of French and German intellectual positions in this field. He also showed that the challenge of the biological theory of human progress and human difference led to culture's being conceived of as in opposition to biology.

Moreover, he says, the attitude of these peoples to the West, among other things, is not homogeneous, "nor is the emphasis on cultural differences necessarily welcome in post-colonial societies" (Kuper 2003: 18). In other parts of this book, I myself have explained how ambassadors from countries having many languages, religions and ethnic groups, as expressed directly to me as Assistant Director-General of UNESCO, strongly opposed UNESCO programmes on cultural diversity, not to mention cultural rights. Why such opposition did not deter or establish gradations in policies applied to cultural diversity is a matter that was not negotiated through transactions of ideas but certainly by pressures that involved power and coercion.

Kuper makes this points as well with reference to non-native residents in developed countries. "Immigrants to the West might also be troubled by the exhortation to cherish and build upon their differences, when they would, perhaps, enjoy the opportunity to become unhyphenated citizens" (Kuper 2003: 15). It is worth mentioning that this consideration is always omitted from accounts of the expressed wishes of women escaping the restrictions of ethnic and religious communities in their countries of origin who want to blend into their countries of immigration where women are given greater respect, freedom and other opportunities.

The appearance of a field entirely devoted to the study of culture—cultural studies—has made the debate about culture become political again, Kuper argues. It seems to me, however, that the rise in the use of the term 'culture' has almost always been influenced by political ideologies or theories of economic development, especially during the twentieth century. His point is well taken, though, when he considers it even more problematic that in this field "every cultural product is judged, not on aesthetic grounds but by applying the simple test of the radical. It is either oppressive or liberating."

His analyses are particularly helpful in tracing the contradictory views on multiculturalism which he thinks were adopted as an agenda in cultural studies in the United States. "The underlying assumptions of multiculturalism are distinctively American", Kuper states, explaining, among other points, that "the protagonist in the multicultural struggle is not the worker, or the citizen, but the cultural actor. Politics are dictated by cultural identity, and they are about the control of culture" (Kuper 2003: 15). In the end he contends that the multicultural project is ripe with contradictions "because contemporary American anthropologists repudiate the popular ideas that differences are natural, and that cultural identity must be grounded in a primordial, biological identity, but a rhetoric that places great emphasis on difference and identity is not best placed to counter these views" (Kuper 2003: 31).

In the end, he again argues that unless beliefs, rituals, knowledge, moral values, the arts and rhetorical genres are separated from the single bundle labelled culture, "we will not get far in understanding any of it. For the same sort of reason, cultural identity can never provide an adequate guide for living." Finally, he concludes, "there is a moral objection to culture theory. It tends to draw attention away from what we have in common instead of encouraging us to communicate across national, ethnic, and religious boundaries, and to venture beyond them" (Kuper 2003: 33).

8.1.3 Culture in the Context of International Relations

In international relations studies, culture only began to be recognized as a separate domain in the 1990s. The preface to Bøås and McNeill's *Global Institutions and Developments* cites social capital, sustainable development, good governance and poverty as concepts central to the discourse surrounding multilateral institutions (Bøås/McNeill 2004: xv). This helps explain why culture has become such a prominent issue and attracts attention in so many different areas of social, political and economic life.

Barry Buzan argues that "ideas about culture are a means for going beyond the debate on an international system to that of an international society" (Buzan 1993:

327). He claims that society already has this structure and that the logic of uneven development suggests that this pattern will be followed in the future. There has to be some sense of community before a norm of reciprocity can emerge, which is the political foundation for a regime to come into play (ibid.: 350). However, he makes an important point by emphasizing that *a community-based world society cannot become global* so there must be an international society to provide the political framework necessary for world development. Thus the two become symbiotic. If society is to be understood in terms of concentric circles, it might be seen as a regime of regimes, adding a useful element of holism to the excessively atomized world of regime theory. The question then, according to Buzan, would be one of what is legitimate in terms of interventions across concentric rings: for example, deterring nuclear weapons production, or punishing outlaw regimes. Or, I would add, the responsibility to protect cultural heritage in the event of wars, as was discussed in November 2016 at a joint meeting of the American Academy of Sciences and the British Academy and which is described ethnographically further on in this chapter.

The same question needs to be applied to debates on culture as an international concern. How do the ideas about culture discussed in the 1990s trigger new trends in the twenty-first century? As an important consideration, James Hentz argues that, through the persuasive power of an idea, through "knowledge politics", "ideas are part of a larger competition pitting the relative power of different 'civilizations' against one another" (Hentz 2004: 194). Indeed, both Arturo Escobar and James Ferguson argue that ideas are marketed as technical solutions to developmental problems but they come packaged as political solutions favoured by powerful domestic actors (Escobar 1995; Ferguson 1994).

Structure by itself can only explain the possibilities of action. Hentz states in his chapter on "The Power of Ideas", "It is well known that neorealist and neoliberal approaches in international relations are unable to explain how interests are formed: neoliberals argue that interests are endogenous, part of the political process, while neorealism treats them as exogenous" (Hentz 2004: 195). He also notes that Ikenberry and Doyle have argued that neo-realism is preoccupied with the consequence rather than the cause of structure, while rational choice focuses on the reasoning behind actions (cited in ibid.). Bøås defines constructivism as "the ability of multilateral institutions to establish a global consensus around certain 'ideas' [which] is conditioned by an intersubjective process of learning. But this process can neither be understood outside its structural context, itself contested, nor abstracted from domestic political contexts" (cited in ibid.: 194). For this reason, I have tried to clarify the process whereby ideas were first brought to the table of representation of governments, intergovernmental organizations and civil society organizations, and then negotiated to create a core international discourse and slightly different political positions, which are clearer in the formulations of national policies related to culture from the beginning of the new millennium. The meetings and programmes described in Chap. 7 demonstrate these different variants in the understanding of culture and cultural diversity for policy purposes.

A further crucial distinction that constructivism makes is that not only is it important to explain the intersubjective nature of interests but that structures and agents reconstitute each other in a process of iteration, as Audie Klotz has explained (Klotz 1995: 451). Importantly, she emphasizes that the interests of states do not exist to be discovered by disinterested rational actors. Rather, they are constructed through a process of social interaction, a statement that pithily summarizes the proceedings described in this book. As the international work on culture evolved, spurred on by the publication of the WCCD Report, different governments, UNESCO staff groups, intellectuals, and art and culture practitioners adapted, reformulated or changed their positions.

In the UNESCO 2009 report *Investing in Cultural Diversity and Intercultural Dialogue*, a much more explicit processual definition of culture is given. It portrays culture as a complex array of overlapping forms, markers and meanings in constant flux and goes on to state that

> it is increasingly clear that lifestyles, social representations, value systems, codes of conduct, social relations (inter-generational, between men and women, etc.), the linguistic forms and registers within a particular language, cognitive processes, artistic expressions, notions of public and private space (with particular reference to urban planning and the living environment), forms of learning and expression, modes of communication and even systems of thought, can no longer be reduced to a single model or conceived in terms of fixed representations. … [W]hereas the Organization's longstanding concern has been with the conservation and safeguarding of endangered cultural sites, practices and expressions, it must now also learn to sustain cultural change (UNESCO 2009b: 4–5).

It is evident that this definition of culture finally brings together the many different threads developed through transactions in international meetings since the early 1990s. In particular, affirming that UNESCO must now learn to sustain cultural change provides a theoretical boundary in recasting culture. Several other innovations, especially in academic research, are making such a change even more relevant.

8.1.4 From Culture to Identity

According to Terry Eagleton, culture is threatened equally by postmodern and by cosmopolitan culture (Eagleton 2000: 71). High culture, he argues, is itself increasingly being invaded by cultures of identity, to produce a crisis within the academic humanities. If, however, high culture means not so much minority art as certain spiritual values, then postmodernism has been busy zapping away at the moral and metaphysical foundations of the Western world at precisely the moment when these foundations need to be at their firmest. The enormous irony which he sees in this process is worth quoting in full:

> The very free-market operations by which the West imposes its authority on the rest of the world help to breed, back home, an increasingly sceptical, relativist culture; and this then helps to erode the spiritual authority ('Culture') which is needed to lend those global

> operations some veil of legitimacy. … In a three-way interaction, culture as spirituality is eroded by culture as the commodity, to give birth to culture as identity. (ibid.).

On a global scale, Eagleton believes that the relevant conflict is between culture as commodity and culture as identity. His main attack is on postmodernism: he declares that "Postmodernism with its scorn for tradition, stable selfhood and group solidarities, is bracingly sceptical of such politics, even if it is mistaken to see nothing in tradition but the dead hand of history and nothing in solidarity but coercive consensus" (ibid.: 72).

8.1.5 *Culture as an Experimental System*

A more dynamic definition of culture leads to considering it as an experimental system. Michael Fischer, defining it as a methodological tool for inquiry, believes that it might best be understood "in terms of its historically layered growth of specifications and differentiations refined into a series of 'experimental systems' that … allow new realities to be seen and engaged as its own parameters are changed" (Fischer 2007: 3). Such a tool, at the same time experimental and systematic, leads towards both intersubjective and institutional accounts which were not adequately accounted for in previous conceptualizations.

Fischer goes on to say that this definition of culture as a unit of research is "designed to give answers to questions that we are not yet able to ask clearly" (ibid.: 32). This is precisely the path we must take in thinking about culture in the new framework of the Anthropocene. We must begin to give form to questions whose answers will feed back into new questions with greater precision: "cultural analysts need to continue the development of the rich tradition of dealing with causality begun by Marx (the feedback loops of commodity fetishism), Weber (multicausal factors at historic conjunctures), and Freud (layered and interacting narratives with different temporal connectivities)" (ibid.: 38–9). As Fischer sums up, "Cultural analysis of the sort that is alive to the multiple discourses that compose cultural fabrics will find alternative possibilities for alliances and coalition where it might have been thought there was only dichotomy and opposition" (ibid.: 41).

8.1.6 *Culture in the Scientific Co-production of Knowledge*

Sheila Jasanoff has argued that science and technology are central to the culture and politics of modernity, and thus science must be a core part of our discussion of culture. Yet, "Scientific knowledge, in and of itself, is not a transcendent mirror of reality. It both embeds and is embedded in social practices, identities, norms, conventions, discourses, instruments and institutions—in short, in all the building blocks of what we term the *social*. The same can be said even more forcefully of technology" (Jasanoff 2004: 2).

In the co-productionist framework, science is understood as neither a simple reflection of the truth about nature nor an epiphenomenon of social and political interests. According to Jasanoff, "Co-production can therefore be seen as a critique of the realist ideology that persistently separates the domains of nature, facts, objectivity, reason, and policy from those of culture, values, subjectivity, emotion, and politics" (ibid.: 3). She notes that work that examines the cultural practices of science and technology "has asked how the supposed universality of facts and artifacts fares in disparate political and cultural settings, as well as how different domains of research and development acquire and retain particular cultural characteristics" (ibid.: 6).

8.1.7 Culture in the "Web of Life"

Today we know that most organisms are not only members of ecological communities but also complex ecosystems themselves, containing a host of smaller organisms that have considerable autonomy and yet are integrated harmoniously into the functioning of the whole. So there are three kinds of living systems—organisms, parts of organisms and communities of organisms—all of which are integrated wholes whose essential properties arise from the interactions and interdependence of their parts (Capra 1996: 33).

Similarly, cultures develop within communities which usually have smaller communities, all of them complex systems of relationships that are constantly being reconfigured. This explains the co-evolution of meanings and practices as people having different ways of life interact with one another. Acting together gives collectivities a much greater capability for adaptation to different environments—both natural and social—and to achieve the aims they set for their individual and collective survival.

These communities make up a "web of life"—a term coined by Fritjof Capra but "an ancient idea, used by poets, philosophers, and mystics through the ages to convey their sense of the interwovenness and interdependence of all phenomena" (ibid.: 34). A large part of this web of life has another web superimposed and interwoven with it: the "web of meaning", formed from the language capabilities and extraordinary properties of human minds.

The human mind has been studied extensively by Humberto Maturana, who pioneered the Santiago theory of cognition. He argues that self-awareness is tied closely to language, therefore making a study of communication essential to our understanding of consciousness. According to Maturana, communication is a coordination of behaviour among living organisms through mutual structural coupling. It can be witnessed, for example, in birdsong, where a single mating song may in fact be formed from a duet between the two partners (Maturana/Varela

1987). Where humans differ, however, is in the level of their ability to communicate and to learn through communication.

Culture is inseparable from understanding the nature of the mind as a systems phenomenon—that is, an integrated whole that has properties that none of the individual parts have (Capra 1996). In more recent years, culture has been understood as a concept that bridges mind and body. People, in their bodies and minds, tend towards homeostasis, a dynamic balance that is kept stable by a self-balancing and self-reinforcing feedback. The multilinear flow of cultures must therefore take into account the circular causality in the way each individual and every group creates such feedback loops.

8.2 Cultural Achievements and Emerging Paradigms

As cultural conflicts and outright wars increasingly dominate the news, it is vital to counter this bias by referring to the great achievements in the protection of cultural heritage, cultural history and the knowledge of cultures. A shining example is the reconstruction of the Bibliotheca Alexandrina, thereby recapturing Alexander the Great's dream. Through the boundless determination of its Director, Ismail Serageldin, previously (as mentioned in Chap. 6) a Vice-President of the World Bank, the Bibliotheca has constituted itself as one of the major regional hubs for literary works and electronic communications, and is now one of the intellectual jewels of the Middle East. It is also an irreplaceable training centre for young people from Egypt and the Middle East, many of whom locked hands and surrounded the Bibliotheca to protect it from potential attack during the turmoil that enveloped Egypt in 2010.

On 7 November 2006, as a member of the Board of Trustees of the Bibliotheca, and as President of the International Social Science Council, I co-coordinated a joint symposium with the International Council for Philosophy and Human Sciences (ICPHS) entitled "Emerging Paradigms in the Social Sciences". It is worth briefly describing its most relevant points, as they show the close parallel between issues of social organization and issues of culture.[1] Michel Wierworka, a French sociologist, outlined the major changes in social science research: (1) a new concern with particularisms and identity which could be traced to the new relevance of cultural diversity; (2) the continuing examination of globalization and its impacts; (3) the "development of individualism" (to use his term), focused on economics and subjectivity; (4) the decline in conflicts between the working class and proprietors.

He also warned us that the "red suburbs" of Paris—where riots with car-burnings had recently erupted (importantly, not yet linked to ethnic or religious conflicts)—were not full of people who felt excluded. Subjects want recognition and respect and this has to do with subjectivity. But who is the subject? It is vital to build up the

[1]Once again, quotations are drawn from my notes taken at the meeting.

creative dimension of human beings. This is connected to cultural diversity because people now choose their identity. So there is a relationship between subjective individuality and collective identities. The question for social sciences is "What do we do with a global space with more and more actors?" He concluded by stressing that the important issues include international finance and international justice.

In the ensuing discussion, emphasis was placed on how conflicts are always multidimensional. In earlier times there were no electronic or media devices to capture this multidimensionality and so these clashes were simply seen as macro-conflicts such as the Cold War. Another topic discussed at length was that some people oppose classical universalism, the Enlightenment, and particularism, with the salient point being made that these concepts now had to be seen not as opposed but as complementary.

The American economist Robert Solow brought up another issue: that there is now a change in terms of policy of choice. Among the different demands made of governments, a choice must be made between equity, utility and entitlement. Building and maintaining social capital requires societal agreement on choice. Those of us interested in the theme of culture continued discussing this after the session, commenting on how close the perspective of culture is to that of "social capital" as developed by Putnam (1995), which had been taken up by the World Bank as a term—in contrast to that of 'culture'—more susceptible to being dealt with through economics.

8.3 Culture, Globalization and '*Planetarisation*'

A major undertaking that has become a central reference in this field is the "Cultures and Globalization Series", edited by Helmut Anheier and Yudishtir Raj Isar: *Conflicts and Tensions* (2007), *The Cultural Economy* (2008), *Cultural Expression, Creativity and Innovation* (2010), *Heritage, Memory and Identity* (2011) and *Cities, Cultural Policy and Governance* (2012). The volumes track trends in all regions of the world and cover the broad range of issues identified in their titles. A major contribution to international debates on culture, the series brings in thinkers and specialists from many countries, thus expanding the lists of subjects and experiences that are to be taken into account in international and national debates and policies. It takes up the role of cultural difference in politics and governance, touching on all the major current political issues on this topic around the world.

It is not possible to summarize here the points made by the authors in all these volumes, though their impact will be evident for many years to come. However, as an illustration of the topics covered in the series, in the foreword to *Heritage, Memory and Identity*, the French historian Pierre Nora points out that "what is particularly striking is the way in which [history, memory and identity] have shifted [in the last thirty years] and in the same manner, from the individual to the collective" (Nora 2011: ix). Nora explains this shift in an interesting way, using Walter

Benjamin's distinction to affirm that, as result of changes during the last third of the twentieth century, there is now "less transmitted memory and more and more acquired memory" (ibid.). He argues that this has happened as a result of the misuse of the term 'memory', which "was spontaneously adopted to describe the emergence of self-awareness on the part of … minorities, but is today widely misused. In relation to history it has even taken on an aggressive, polemical, protest-driven dimension" (ibid.: x).

In their introduction to the same volume, Anheier and Isar present, on the basis of "the multiplexity of 'culture'" (Anheier/Isar 2011: 1), a general conceptual framework in which, of the three terms forming the focus of the book—heritage, memory and identity—"the one that is recognized across the world as a domain of public *cultural policy* (in the sense of the remit assigned to ministries or departments of culture) is *heritage* or, more precisely, *heritage preservation*" (ibid.: 3). They explain that "this consists of the valorization and preservation by individuals and groups of traces of the past that are thought to embody their cultural identities" (ibid.). However, this brings up "the fraught issue of conceptualizing 'culture' as the fixed property of particular groups" to justify, as in the Balkans, present violence and war, an exercise which extends to the selective rewriting of histories and their reshaping as an image for tourists (ibid.).

8.3.1 Globalization, Time and Space

In anthropology, debates on culture and globalization began as early as the 1970s, when the first expansion of capitalism to developing countries started to have an impact on the livelihoods and habitats of autochthonous and indigenous peoples. By the turn of the millennium, globalization constituted a major challenge to anthropological research (Lewellen 1999; Inda/Rosaldo 2002; Pieterse 2004). According to Lewellen (1999: 11), globalization is still seen from three different perspectives: the sceptics, who say that globalization does not exist or has been vastly oversold; the evolutionary view, that it represents only a change in degree, not in kind, in terms of the working out of the logic of capitalist expansion; and the hyper globalization thesis, which states that the world is experiencing something entirely new and fundamentally different.[2] In the emerging anthropological model of globalization, there is little that is not contested.[3] The importance of anthropology in defining the issues of globalization was emphasized by Katy Garner and David Lewis, especially in showing the social consequences of capitalist development (Gardner/Lewis 1996).

---

[2]For other definitions of globalization, see Robertson (1992); Streeten (1994).

[3]Lewellen prefers the term 'model' rather than "meta-narrative" as used in Englund/Leach (2000: 227).

Inda/Rosaldo (2002) suggest that globalization implies something more radical about the world than the mere fact of increasing global interconnectedness—it points to a basic reorganization of time and space. I believe that globalization goes beyond reorganization: it challenges us to reconceptualize the nature of human bonding in terms of the creation of ethnic, cultural and religious groups in a political context. And we know that this implies a reconceptualization of human nature itself. Early anthropology posited a human nature constructed by cultural practices in relatively isolated social environments.[4] Now, however, on the one hand, the socialization of children occurs, with high frequency, through a pluricultural ecology of communications; on the other, individuals have to contend with school, work, governance and leisure settings in which they must interact with people who possess varying degrees of cultural difference. Hence, the classic anthropological concepts of human nature and social interaction must change. This book is about that change, as reflected in the discussion which took place in many countries and which came to a head in international debates in the 1990s. What was being constructed in such debates was certainly a new cultural world order, but it was also a new, cosmopolitan concept of human nature, human agency and human relationships.

However, the interactions of different kinds of cultures with globalization vary enormously and sometimes even display contradictory trends. Where national cultures have been created purely as civic cultures attached to states that are not fully stabilized, as in some African countries, they are waning. In contrast, national cultures in countries based on historically important major cultural units will not be strengthened by the weakening of the state. Small-scale cultures have been able to organize movements and place their demands on the national stage, thanks to globalization, while others, no longer having the safety net and political backing of a national state, are finding themselves more vulnerable in a global economy that is leaving many regions and social groups out of development.

8.3.2 *The Multidirectionality of Cultural Flows*

Rather than considering culture as a different process from that of globalization, many authors point to the increasing multidirectionality of cultural flows. Countries in the North are sites of extraordinary cultural heterogeneity as migrants, crafts and images from the South establish themselves there (Inda/Rosaldo 2002: 20). Meanwhile, permanent immigrant communities create networks of cultural flows that entangle all cultural groups in the world (Arizpe 2004; Hannerz 2002). Thomas Eriksen and other Nordic anthropologists prefer to talk of their empirical material in terms of "transnational flows" which have origins and destinations, having been

[4]To a certain extent this held even for individuals in urban neighbourhoods, since travel and communications were fairly restricted.

instigated by people (Eriksen 2000). While I agree with them that "the global only exists to the extent that it is being created through ongoing social life" (ibid.: 5), my argument is that the outcome of such transnational flows is creating a scale of cultural thinking and cultural politics which is more than its constituent parts.

Marilyn Strathern states that we cannot expect to find wholeness or integration of cultures (Strathern 1991; see also Amit/Rapaport 2002: 13). This must now be taken as the primary assumption in studying cultures, as they are quickly becoming a site not only of contention but of the re-joining and re-assembling of cultural practices brought or taken by practitioners criss-crossing the world.

Traditionally, anthropologists worked with an understanding of culture that "presumed static, homogenizing, territorialized entities and artifacts in need of 'preservation'" (Vertovec 2011: 250–1). It remains the case that in many popular formulations, especially those of international cultural policy, 'culture' is conceptualized as a bounded unit involved in 'clashes' of one culture with another, or in 'dialogues' of the simple exchange of information, driven by the desire to understand each other better. A new attempt to break down these confrontations led to the formulation of the idea of 'intercultural' practice, whereby there is some form, rarely explained, of interpenetration of information from one culture to another. From here, there is a relatively small jump to the other concept frequently mentioned in cultural analysis. This is 'hybridity', which is not new in its substantive definition since it has been historically and controversially discussed in many countries and regions in past times.

8.3.3 Cultural Diversity in Globalization

In the 1990s, as expressed in the many transactions described in previous chapters, the threat of cultural homogenization as a consequence of globalization together with information and communications technologies had fuelled research, policies and mobilizations that, by the beginning of the first decade of the new millennium, had differentiated into various kinds of responses to this purported homogenization. By then, many authors were noting that, in fact, far from imposing homogenization, globalization was bringing about greater cultural retrogression, fragmentation, or forms of neo-indigeneity.

David Steigerwald tackles this question head on in *Culture's Vanities*. In no uncertain terms he declares that "Sometime in the last quarter of the twentieth century, the developed world appeared to slip from the Age of Materialism into the Age of Culture" (Steigerwald 2004: xi), an age in which "measuring everything from the accumulation of wealth to the exertion of military power against some cultural background has become more than just a way of raising interesting questions; it is now the litmus test of 'cutting-edge' thought, against which most writing is measured and judged" (ibid.). This, indeed, is a thesis to explain the narrative arc laid out here, on why culture has become such a prominent concept in recent times.

Steigerwald argues that Raymond Williams elevated the concept of culture in *Culture and Society*, believing that, in an increasingly democratic world, the old notion of culture ought to give way to the more generous conception of it as "a way of life" (Steigerwald 2004: 237). Steigerwald believes that Williams's hopes were dashed because "the only cultural standard of our day is whatever sells … It is equally clear that culture as a way of life has lost its democratic character" (ibid.). It is important to bear in mind that this thread in transactions on culture was already being made explicit in 2004.

Steigerwald's paradox of cultural diversity in a globalized world consists in communities and individuals reacting to globalization by reasserting their cultural heritage and thus creating more cultural diversity. He even goes beyond this to call attention to a reverse of the diversity paradox: "The less anti-globalization activists emphasize culture, the more likely they are to act in ways that support cultural diversity" (ibid.: 240). His hero, in this sense, is José Bové, the French sheep farmer who vandalized a McDonald's and created a movement against Monsanto in defence of small farmers and a culture of good food based on enhancing excellence in the skill of taste.

We would do much better, Steigerwald asserts, "to champion a vigorous cosmopolitanism that draws its inspiration from distinct but bygone ways of life while committing itself to new forms of cultural expression judged by strict standards of well-done work" (Steigerwald 2004: 238). How could this be done? His answer: "Such an effort at cultural reclamation inherently returns to the virtues of our two traditions of cultural thought, to a clear-headed anthropological understanding of what it takes to maintain cultural integrity and to a sensible aesthetic that, once again, takes the cultivation of skill seriously" (ibid.). He continues by giving a straightforward political perspective on the matter: "Doing so will not reverse the counterrevolution that the Right has sprung in the last generation, but it might remind progressives that culture is not the same as economic and political power and see them agree to reengage those realms in ways that might return the Left to relevance" (ibid.).

8.4 Gender and Global Cultural Restructuring

The intrincacies of culture and power are also central to transactions on the changing roles of women in contemporary societies. As argued in this book, when ethnicity is used to build fences, restrict cultural exchanges, it becomes ethnicism, a form of political ideology. Similarly, the sexual dimorphism that characterizes the human species and is not present in other non-human species, that is, the observable genetic and phenotypical differences between women and men, as well as other variants in both women and men, is reflected universally in divergent ways of building social models for categories of persons. At present, after the Second Wave of feminism and as a new Third Wave is opening up, a much more radical redefinition of gender has become a core issue in recasting the role played by culture in this new movement. In this section, the rise of the women's liberation movement since the nineteen-sixties to the present will be briefly summarized, and the open doors that questions about gender have generated will be situated in transactions on culture.

While right-wing and religious movements have insisted that the recent women's movement grew out of a change in values and hs brought about a breakdown in society, evidence shows that it followed closely along the lines of other important shifts. Two simultaneous discoveries were made: "the pill", that allowed control over the time and spacing of births, and the demographic realization that unchecked population growth was becoming a threat to human survival, given the limits to growth and to the carrying capacity of ecosystems. For the first time in human history, the major purpose of living socially swerved from ensuring physical survival to ensuring survival through purposeful living.

It is interesting that this coincides in time with the rise in the use of culture in the way expressed superbly by Marshall Sahlins: "Before, culture was just lived. Now it becomes a self-conscious project" (Sahlins 1994: 2). In this sense, the role of women became very clearly a new, self-conscious project. Because population now became a matter to be managed through personal decision, women's lives also became open to the possibility for women to become full participants, political and economic actors, in their societies.

Two other processes contributed to this liberation of women from their traditional sociobiological roles. The first was the interest of the business sector in employing more women workers because of the advantages they gave firms: lower pay than men, weak or absent unionization making them easy to fire and greater dexterity in manual tasks. This was especially evident in situations where young rural women, with little education and yet willing to participate in paid labour in the manufacturing and servicing sectors, arrived in droves at factories which were sometimes makeshift buildings that shifted to new areas "when we run out of girls", as one employer in the maquila—mainly assembly industries along the Mexican border—candidly admitted (see Arizpe 1984). In spite of the many negative aspects of this phenomenon, as my own fieldwork research showed, young women relished the opportunity to get away from their parents' homes, and it gave them educational, work, consumer and travel opportunities that they had never had before (ibid.).

The negative aspects, especially when both business enterprises and governments were unconcerned with the outcome of such a major shift in development, led in many cases to city, community and family collapse and unprecedented rates of criminal activity. In the case of Mexico, it also fed the desperation of migrants to travel to the United States illegally, as well as to a rapid rise in criminal activity and violence connected with drugs and weapons all along the border.

There was another negative aspect of such a transition. Cultures and religions that were based on repressing women's capabilities used the rise of feminism as a justification to retrench into conservative traditions, proclaiming that women were the primary holders of precisely those traditions. This backlash was also fuelled by the unevenness that was created in societies, especially in those where the women's movement became mainstream, when the role of men did not change in unison with the new participation of women. While women became more proactive, the backlash against this change has fostered the exacerbated virility that has been on the rise in some countries in the last few years, leading to higher rates of feminicides –as, notably, in Mexico–, racism among white supremacists and the Religious Right in the U.S. and a culture of guns and brutal violence in films, television series and videogames. Of course, this is also a consequence of recent realpolitik movements for world hegemony that seek power through might.

Generally, because of the uneveness of progress in balancing the gender roles of women women are saddled with a triple labour burden: unpaid domestic chores, care of children, the ill and the elderly, and paid labour. Women's roles such as political leadership, social organizing, leading and participating in sustainability projects are activities also add to the weight of women's responsibilities in the contemporary world. It must be taken into account that women belong to different social classes, ethnic, racial and other cultural groups, so special attention has to be paid to their specific challenges. In particular, aboriginal and indigenous women face the additional challenges of discrimination.

In the Introduction to the aptly-titled book *Is Multiculturalism Bad for Women?*, Joshua Cohen, Matthew Howard and Martha C. Nussbaum argue that the central question in the debate on cultural aspects of women's emancipation is how to "understand a commitment to equality in a world of multiple differences" (Cohen et al. 1999: 5). The answer suggested by Susan Okin is to strive for "a multiculturalism that effectively treats all persons as each other's moral equals" (Cohen et al. 1999: 5). In her chapter in this book, entitled "Culture beyond Gender", Saskia Sassen takes a different position. She says that Susan Okin's framing of the argument for feminist rights as contraposed to group rights because "the norm in most cultures is an inequality between men and women" is won by the feminists (Sassen 1999: 76). However, Sassen asks, "what if 'culture' cannot be made to pivot so exclusively on the oppression of women?" (Sassen 1999: 77). She goes on "If we overlook the joint presence of (and relations between) a dominant (or 'host') and minority culture, we may be overlooking the many sources of pain and rage produced by intercultural engagements. Engaging, whether by necessity or by choice with a dominant culture may lead in turn to the 'need' (in both men and women) for the refuge of one's culture also in areas other than gender" (Sassen 1999: 78). Therefore, "rather than rejecting group rights as such, the analytic and political focus may well have to negociate intracultural gender inequalities at the center of Okin's concerns and intercultural oppression that frequently lies at the origins of the experienced need for group rights" (ibid.).

The distinction highlighted by Sassen is very important in general terms for understanding the cultural dynamics which affects cultures internally, *intraculturally*, and also in their connections with others, *interculturally.*

Other issues discussed recently in this area are brought together in the book by Marianne Marchand and Anne Sisson Runyan, *Gender and Global Restructuring: Sightings, Sites and Resistances*. They argue that gender is central to the restructuring of a globalizing world, both in the way neo-liberal political and economic powers take advantage of gender oppression and also in the resistance to these policies. Specifically they criticize the view of globalization as a unidirectional, masculine (active) process where women are seen only as its victims. At the end of their book, they conclude that "the restructuring of 'the intimate', that is, where gender, sexuality and racial identity formations and relations are most visibly organized and emotionally felt, is fundamental to the processes of global economic and political restructuring" (Marchand/Runyan 2010: 239). Events as recently as 2016 have indeed confirmed how insightful their interpretation is, among them the vote on Brexit in the UK and the 2016 election in the United States. Their interpretation is leading to new analyses of the relationship between culture, emotions and political choice, such as that of the Israeli sociologist Eva Illouz, a specialist in the analysis of emotions in capitalism. In an interview, she declared that "emotional populism is a threat to democracy" (Illouz 2017: 24). In their book, Marchand and Sisson were optimistic that a 'post-liberalism' would see growing resistance to and disruption of neo-liberal capitalism and a return to state intervention in the market economy.

A very important movement that took on many of these challenges is the DAWN, Development Alternatives with Women for a New Era, a network of feminist scholars, researchers and activists from the economic South working for economic and gender justice and sustainable and democratic development. Since the founding of DAWN in 1984, in which I participated, the network has been recognized as a significant agent in the development of south feminist analysis and a key participant in the global/feminist and women's movements. "A political and ideological 'south' location remains vitally relevant in the era of globalization, albeit a radically different South standpoint from that taken in relation to the 'older' global political economy".[5]

Social vulnerability derived from discrimination has also been a research focus in issues related to gender and culture. In the specific case of 'forced' migrations, Úrsula Oswald Spring describes how women are left behind, with no rights or government support, to deal alone with agricultural work and to care for children and their extended families, in the context of insecure land and water rights and insecurity in the face of criminal activity (Oswald Spring 2009b: 8). In cases of natural disasters, Oswald Spring cites studies illustrating the social vulnerability of women: when the tsunami hit Aceh (Indonesia), men survived in a ratio of 3:1; in Sri Lanka sixty-three per cent of the victims of weather events were women; the proportion was the same in the Kashmir earthquake (Oswald Spring 2017). In order to take such processes into account, Úrsula Oswald Spring and Hans Günter Brauch have proposed a broader framework for analysing gender in contemporary processes. Entitled 'HUGE', this wider understanding of security encompasses human, gender and environmental security concerns (Oswald Spring 2009b).

In the field of gender, identity politics has opened up deep debates on the genetic make-up of sexuality and its transduction to social and political self-ascription and

[5]Dawnnet.org/about. Accessed December 15, 2017.

representation. In the Chicano movement in the United States, authors such as Gloria Anzaldua began transacting the concept of 'borders': borders between cultures, in this case American and Mexican traditions, and the borders between women and men, as well as between heterosexual and lesbian women. After the Christopher Street events which led to the emergence of a 'gay' movement among men, and of transgender and other self-ascriptions, the concept of sexual orientation shifted towards sexual preference. In a few years, this led to a worldwide cultural movement, Pride, representing a wide-spectrum coalition, whose acronym keeps shifting in terms of Lesbian, Gay, Bisexual, Transgender and other specificities. More radically, 'queer' theory rejects all binary ascriptions related to sexual preferences.

These cultural movements are important as representations of diversity understood as going beyond cultural diversity to emphasize choice in all spheres of personal life and style of living, including sexuality, physical and mental capacities, family composition and reproductive rights. They are aspects of the recasting of all cultural backgrounds to the institutional and organizational structures that were broadly defined in the context of industrial capitalism. They are also connected to policies of human development that situate personal choice as the central instrument for freedom and for development in world capitalism. At a more fundamental level they are part of the undercurrent whereby "human nature" and "becoming human" are deep challenges that need to be addressed in terms of sustainability, equality and equity.

8.5 Diversity in International Policy Frameworks

The previous chapter described the adoption of the 2001 Universal Declaration on Cultural Diversity, the 2003 Convention for the Safeguarding of the Intangible Cultural Heritage and the 2005 Convention on the Protection and Promotion of the Diversity of Cultural Expressions. For the purposes of an ethnography of transactions on culture, it is important to record the debates that followed the application of the Conventions.

Major actors in international transactions on cultural policy continued to demand new normativity and policy instruments, and an important meeting was held in 2008. Mme Francoise Rivière, then Assistant Director-General for Culture, convened a meeting of experts to brainstorm on "The Interactions between the Convention for the Safeguarding of the Intangible Cultural Heritage (2003) and the Convention on the Protection and Promotion of the Diversity of Cultural Expressions (2005)" at UNESCO on 31 March and 1 April 2008. A quick and simple assessment of the two normative conventions yielded the following comparative features: first, the 2003 Convention focused on processes whereby persons safeguarded their traditions, which were then transmitted through traditional means; second, the 2005 Convention concentrated on products of creativity in a framework of modernity, which were then transmitted through modern devices.

Cherif Khaznadar, a former Director of the Maison des Cultures du Monde (House of World Cultures) in Paris, and former President of the Assembly of State Parties to the 2003 Convention, began the discussion, saying that confusion arises because there is no choice between difference and complementarity. One must be different to have a dialogue; tradition and creation are not opposed, since tradition is created every day. Let us not forget, he said, that these Conventions are there to detach ('dégager' in French) culture from markets.

Antonio Arantes, a distinguished Brazilian anthropologist, agreed that the subject matter of the two Conventions was intertwined. He asserted that emphasis should be placed on questions of the social base of the agents in their application. The agents of the Convention on Intangible Cultural Heritage work in small face-to-face groups that are collective and communal. The Convention on Cultural Diversity, by contrast, is implemented by individuals who are artists, cultural producers and people from industry, who act in complex networks through contractual relationships and institutions. The procedures are also of a different nature. Intangible Cultural Heritage safeguards products between generations and gives value to distinctive aspects of identity, enhancement and the preservation of objects. Cultural Diversity allows a flow of market commodities and exchange relations; it deals with markets rather than places; there are international flows of capital, and modes of organization for production based on knowledge of sophisticated technologies. Furthermore, there are differences in scale, social relationships and volume of production. Finally, the notions of professional culture expressed in them are very different: intangible cultural heritage is dealt with by anthropologists, cultural diversity by social scientists and economists.

In her turn, Chiara Bortolotto, a young Italian anthropologist, highly skilled in juridical questions on intangible cultural heritage, pointed to the sequences of political and cultural relations between the institutions managing the two Conventions and the persons and agents involved in the protection and promotion of heritage and diversity. The political discourse that people activate in the promotion of cultural diversity is implicit in intangible cultural heritage yet it must be presented as "artistic craftsmanship" to be valorized. But this may create barriers against crafts from other countries. Commoditization is intrinsic to the production of crafts. This may apply to the same object but may emphasize its diverse facets. For example, in gastronomy, the "Mediterranean Diet" is a way of life but it has also always been a commercial activity.

Paul Nchoji Nkwi, a professor of African anthropology the University of Yaounde, Cameroon, in turn argued that there is a distance between customs and tradition. Customs are the most changeable, so individuals can change them—for example, dance. In contrast, traditions are immutable, so hardly change over a very long time. The Conventions must therefore emphasize the strategic value of heritage. Culture will not survive without the gatekeepers.

In the ensuing discussion, Khaznadar asserted that all cultural heritage is created by individuals, to which Arantes countered that we do not see individual expressions but those developed in groups. Yet he added that he did not like the word 'collective'. Khaznadar replied that, even in the collective formula, there is always a person who guides it, as in the Khatayatam dance. I interjected that a distinction must be made between corporate and collective. Corporate traditions do not allow for innovation and so those cultures are left behind, while others create responses to unprecedented conditions. Toshiyuki Kono, a professor of law from Kyushu University, explained that in Japan until the 1950s support was for individual performers. Kabuki theatre, for example, needs musicians and performers; some are boring, while others are great performers. Only the masters were considered very highly, and received many honours and prizes. The discussion continued, with Bortolotto clarifying that in intangible cultural heritage the 'Masterpieces' programme was important but then, importantly, a new view of representivity was added.[6] Art is the "savoir-faire" technique, which is the basis for the "Living Treasures" programme in Japan.[7]

As I was Chair in this session, I then suggested that we were dealing, not with objects or performances, but with "moments in cultural time". Peoples, cultural practitioners and governments all decide at some point, for political or social reasons, to give a specific identity to a cultural practice. With the passage of time, these become heritage. In the discussion, Leoncio Orellana, a Chilean architect, pointed out that in Latin America there is now an inflation of world heritage because recognition from the outside creates prestige. When it is a question of memory, everyone wants it, but there may also be a conflict between, for example, a social project and a cultural project. Culture in this sense is always economic. Related to this, Vladimir Cosma, a Romanian composer, made the explicit point that culture has always been a commodity. "When I was young," he said, "I wrote music that was called 'The little woman worker on her tractor' and now I continue to compose and it's another alienation, but it has always been like that."

Going back to the session of the meeting on the interactions between UNESCO Conventions, Azzedine Beschaouch, a UNESCO programme officer, pondered how a Muslim mosque filled with Italian and Maltese symbols is inscribed on the World Heritage List as a Muslim monument. Cultural diversity is often present in both physical and intangible cultural heritage. He mentioned the famous puppets of Sicilian origin who perform identical plays, with the single difference that in Sicily the Christian king wins, and in Tunis, the Great Turk. It has to do with having a recurring template but never the same performance ("avoir un cadre répétitif et jamais une répétition"). (I might mention a similar case where, in ethnographic fieldwork in southern Mexico, we found that in the representation of the so-called 'Conquest' of Mexico by an indigenous group, Hernan Cortés, the military commander of the Spanish forces, is poisoned and so the indigenous Aztec leader, Cuauhtemoc, becomes the ruler of Mexico.)

[6]The Masterpieces in Intangible Cultural Heritage Programme was initially developed at the end of the 1990s as a first step towards establishing an International Convention. It was inserted into the Representative List of the final 2003 Convention.

[7]"Living Treasures" was the first programme, set up in South Korea and Japan, to honour the great masters in all activities in the arts and crafts; it inspired many countries to set up similar programmes.

Rex Nettleford, former Minister of Culture of Jamaica, called attention to the possibility that places that once thought of themselves as homogeneous are now realizing their diversity, so the question arose of whether Romania would contaminate Europe or enrich it. How will 'homogeneous' Europe deal with Caribbean heterogeneity? In the Caribbean, diversity and oppression (particularly of Africans) has always led to creativity. Rome and Greece were crossroads of civilization. In the same way so many cultures have crossed in the Caribbean. In 1492, with the expulsion of Moors and Jews, Spain lost its intelligentsia. But life goes on because of the meeting of the wisdoms of different cultures.

Another important intellectual contribution was made at the meeting a year later on "Conceptual and Operational Guidelines to Integrate Cultural Diversity and Intercultural Dialogue Principles" held at UNESCO from 6 to 8 July 2009. Organized by Katerina Steniou, then Director of UNESCO's Division of Cultural Policies and Intercultural Dialogue, and David Throsby of Macquarie University, Australia, with the collaboration of Paul Nchoji Nkwi, of Cameroon, the meeting established a new outline of perspectives on cultural diversity. In the preliminary programme the following topics were noted for discussion: "no one may invoke cultural diversity to infringe on human rights guaranteed by international law, nor limit their scope" (from the UN Declaration on Cultural Diversity); pluralism is the political response to cultural diversity; the recognition of multiple belongings and identities; culturalization of social issues and the risk of segregation in the name of particularism. Global ethics were also brought back to debates, as was cultural nationalism. A clear statement was made, referring to "cultural diversity as a connecting or divisive force within and between societies".

In the *Global Synthesis of Papers Commissioned by UNESCO*, the "Rationale for a New Cultural Policy Profile" includes "a recognition that cultural identities, however they are defined and interpreted, are resilient and resistant to standardisation—indeed, their very diversity is one of their most important characteristics"; agreement that cultural interrelationships are far more subtle and complex than implied by simplistic stereotyping models such as the Huntington thesis of the "clash of civilizations"; an appreciation of the values attaching to cultural diversity; a recognition of "the increasing importance of intercultural dialogue as a means towards the reduction of cultural tensions and conflicts and towards a greater sense of understanding and respect both within and between nations"; an awareness of the relationships between culture and development; and an assertion of the "fundamental role of human rights as a cornerstone of any cultural policy" (UNESCO 2009a: 2).

Several contributors refer to what had by then become common knowledge: as Koichi Iwabuchi argued,

> globalization is, in fact, constantly giving rise to cultural diversity. Globally disseminated cultural products and images are consumed and received differently with the specific cultural framework formed by the political economic and social contexts of each locality and by people's differing statuses depending on their gender, ethnicity, class, age and other factors. At the same time, in each locality these products and images are reconfigured through a process of hybridization … Meanings are negotiated locally, resulting in the creation of new products that are more than mere copies (Iwabuchi 2008, cited in ibid.: 3).

Thus the *Global Synthesis* states that "identity politics may not have been caused by globalization but it seems to be one of its unintended consequences" (UNESCO 2009a: 3).

The document affirms that a cultural policy needs to nurture the positive contributions that global cultural diversity can make to societies. In his paper, Paul Nchoji Nkwi writes that

> While the diversity of culture simply implies the obvious multiplicity of socio-cultural patterns or identities, cultural diversity takes the whole notion of culture beyond the bounded limits or confines of a given society within a broader national context typically purveyed by classical sociological and anthropological thought. Therefore, cultural diversity looks at the simultaneous disintegration and integration processes in the modern state which transcend the orthodox notion of a state/society boundary but equally exists on the larger scale of trans-national or trans-societal formations (cited in ibid.: 6).

Intercultural dialogue was introduced as an explicit policy objective at this time, defined as "a process involving open and respectful exchange or interaction between individuals, groups or organisations with the aim of developing a deeper understanding of diverse perspectives" (ibid.: 8). It can play a central role in "understanding, appreciating and assessing different forms of knowledge, wisdom and spirituality. In so doing, it enriches the intellectual, emotional, artistic and religious landscapes of the world" (ibid.). Reference to religions here stands out as part of the discourse on interculturality. The authors of the commissioned papers cautioned that intercultural dialogue "for all its good intentions, is not necessarily easy or straightforward. It requires an abandonment of stereotyping and of reducing a cultural group to a single characteristic such as religion or language" (ibid.: 10). Again, the reference to religion and not to culture is unusual for UNESCO. In the previous chapter, we saw that the programme that emanated from this new political priority gave rise first to the Year and then to the International Decade of the Rapprochement of Cultures, which "reinforced Member States' commitment to furthering interreligious and intercultural dialogue and the promotion of mutual understanding and cooperation for peace" (UNESCO 2016: 4).

In recent years diversity has become a mainstream concept to refer to many emerging phenomena, for example, in gender identities (LBGTQ—lesbian, bisexual, gay, transgender and queer or questioning), new national and cultural groups, new tribalisms and new lifestyles. It has also been invoked to denounce discrimination against African Americans in Hollywood, against women in Silicon Valley tech giants, or in an ironic twist, to claim special rights as in the case of white supremacists in the United States. In general terms a political polarization of two forces can be discerned in the variety of such movements: one giving primacy to corporate control, another trying to shake off precisely such corporate controls. In this context, many authors, among them Marc Augé, have tried to add precision to the concept of diversity.

One of the negative effects of globalization, according to Augé, is the shift from thinking about the universal towards a statistical and quantitative perception of human societies, while fitting in the paradigm of difference (Augé 2017: 88). This perspective is loosely theorized and disseminated under the banner of "respect for differences". "However, the paradigm of difference, in reifying[8] cultures, tends to ignore or deny the difference and autonomy of individuals" (ibid.). By showing the close connection of the statute of individuality and the definition of the social, Augé posits that "the question of respect for cultural differences tends to be uninfluenced by and hence to cause people to take no account of both the individual differences and the existence of the human species, the human race, over and above culture" (Augé 2017: 89).[9]

8.6 Culture, Postcolonialism and Imperial Eyes

Heightened interest in global issues also led to the proliferation of new vocabularies designed to re-situate culture in the neo-liberal framework of global development. Cultural capitalism, cultural internationalism, the political economy of culture, international cultural society and cultural democracy are among the main concepts proposed. It is beyond the scope of this book to review all these proposals, so only those that illustrate culture's connections will be mentioned.

Postcolonial studies, in particular, postulated a more historicized version of the location of culture in history and proposed taking a perspective from the global South on cosmopolitanism. Homi Bhabha argued that an interest in 'slower' art should not be seen as "an obscurantist celebration of tradition or a revival of the past", but as having the purpose of creating a "vernacular cosmopolitanism" (Bhabha 2007: 10). In his view, "A renewed sense of civic or civil belonging in a globalised age demands a language of intercultural interpretation as well as the policy-oriented prose of social integration" (ibid.: 15).

Paul Gilroy, in taking up multiculturalism and conviviality and responding to James Clifford, refers to a "vernacular sociality" and argues that "culture and nationality" are projections of a "racial nomos" which received its imprimatur in the colonization of the New World (Gilroy 2007: 126). His proposal is to generate counter-histories of the convivial social interactions practised ordinarily, in order to shift attention from the interpretive focus on cultural resources to the cosmopolitan and transcultural imagination. He takes up the question of converts to Islam and argues that the push towards political Islam, occultism and drugs has happened because of the meaninglessness of the consumer world, allied to conservative religion. In other words, nihilism, especially in young people, is attributed to loss of morals, but this loss comes from the loss of a political ideology that might overarch the teeming, foaming proliferation of non-belonging and frivolity. He concludes "Adherents and practitioners [of political Islam] may not even be aware of the

[8]Marc Augé uses the term 'substantitalisant' in French, which seems to me is best translated by 'reified' in English.

[9]"Le thème du respect des différences culturelles tend à ignorer et à faire ignorer d'une part, les différences individuelles à l'intérieur des cultures et d'autre part, l'existence de l'espèce humaine, du genre humain, au-delà des cultures."

history of the gestures they have been repeating, of revolutionary conservatism, ultra-nationalism and occultism, all this within civilization" (ibid.: 138). In fact, while defending the Enlightenment, Gilroy defines 'civilizationism' as linked to fascism and supports the dynamic of the European cultural tradition which repudiates civilizationism and must be supported for reconstruction (ibid.: 127).

In the Conference on "Anthropology in the World" organized by the Royal Anthropological Institute at the British Museum on 8–12 June 2012, which I attended, discussions were not so much about the vocabulary necessary for updating dialogues on culture in international and public debates as about providing a syntax and theory to give new meaning to an anthropology of a globalizing world. Lissant Bolton began by restating the principle that anthropology aims to explain why we are human and to describe to ourselves what it means to be human. In the ensuing discussion, Marilyn Strathern chose to highlight that the purpose of anthropology is to provide the collective capacity for understanding the bigger picture when looking at specific phenomena.

In another session, Nigel Rapport stated that anthropology embodies "A Premise and a Promise" (his paper's title) of ethics against an ancient regime of hierarchical divisions. He made a strong case for the need for anthropology to reclaim the human rather than a communitarian ethos, arguing that a Kantian anthropology is the best way to approach the human. Kant considered anthropology as a tool for cosmopolitanism, that just society encompassing all human beings. Rapport then brought up the need to look at the epistemology of human capacities. Anthropology is a chimera, a generalizing science. It slides between difference and sameness, the momentariness of experiences and the unity of humans through a constant movement between different poles. The truth of social life is always moving, always in transition. Subjective knowledge must not be reduced to abstraction but must capture the generality of complex human singularity and plurality. By maintaining the tension between singularity and plurality, we can arrive at truth by movement. The nature–nurture truth is more fluid, a move of particularity and generality of juxtaposed capacities as against substantiations. In one of his final points, Rapport stated that anthropology accounts for the self-creation of human lives.

One of the most important books on how "zones of contact" generated cultural representations is Mary Louise Pratt's analysis in her 1992 book, *Imperial Eyes: Travel Literature and Transculturation* (Pratt 2010). Her heuristic tool is not culture but the process whereby, in "zones of contact", Europeans and native peoples of colonized regions co-produced representations that gave the former the power to name, classify and give scientific validity to their observations of nature and natives, while the natives themselves created representations of themselves through "self-ethnography", based on the colonizers' own terms.

Pratt follows the 'transculturation' of the Americas in particular, as Alexander Humboldt's naturalist studies opened the door to modernism among the elites, by nurturing romanticism, not as a change in artistic expression (as in Europe) but in the development of ideas of liberation in political movements seeking independence from Spain. Soon after, the "capitalist vanguard", as Pratt describes it, left behind the explorers' and adventurers' studies of nature, replacing them with a new representation of colonized areas as landscapes full of resources which could be commercially exploited for the market. Ideologies, literature and the values of elites thus all converged along this axis.

Pratt also supplies an original viewpoint by studying gender in the literature of travellers, highlighting the writings of Flora Tristan and Mary Graham. In contrast to the texts written by the men of the capitalist vanguard, Tristan's and Graham's work took up other themes, emphasizing the social situation of women, as well as economic justice.

For individuals, and especially for women, to live in a situation of justice, they must develop capabilities that allow them to become creative and cooperative through the choices they are able to make. The next section reviews the most important proposals in this highly influential international policy instrument, the "Capabilities Approach".

8.7 Human Development and the Capabilities Approach

As explained in Chap. 3, the World Commission on Culture and Development was given the mandate to provide a framework and policy guidelines for culture understood as a dimension of development. At that time, a new paradigm was in the making: human development as an alternative to the utility theory to achieve economic growth. The basic assumption for this paradigm, in Amartya Sen's formulation, is that "development can be seen … as a process of expanding the real freedoms that people enjoy" (Sen 1999: 1). Growth of GNP, he argues, can be very important as a means of expanding the freedoms enjoyed by the members of society. "But freedoms depend also on other determinants, such as social and economic arrangements (for example, facilities for education and health care) as well as political and civil rights …" (ibid.). He includes among such arrangements political freedoms, economic facilities, social opportunities, a guarantee of transparency, and protective security as distinct types of rights and opportunities, specifying that "each of [these] rights and opportunities helps to advance the general capability of a person" (Sen 1999: 10). It was this theoretical framework for Human Development (which has now been reformulated as Human Sustainable Development) that opened up the possibility of discussing culture beyond the limited debate about whether it enhanced or obstructed development, and allowed for a multidimensional analysis of its relations with cultural pluralism, women, heritage and so on in *Our Creative Diversity*.

The added relevance for this book of Amartya Sen's view of development is the emphasis he places on connections. He sees development as "an integrated process of expansion of substantive freedoms that connect with one another and allow people to help themselves and also to influence the world" (Sen 1999: 18). Culture is also implied when he argues that people should have the opportunity to achieve the outcomes that they value and have reason to value and which will allow them to develop their functioning and capabilities. It is this functioning and these capabilities that I believe are the inherent components of cultural evolution; they must be considered as central to sustainability transition.

The transactions on capabilities have been mentioned several times throughout the book. At the Lillehammer discussion of Our Creative Diversity, following Amartya Sen's formulation, Andreas Follesdal suggested three areas for further research: the first two are measuring how one's capabilities relate to the decrease of those of others and ensuring (as feminists have argued) that people must be aware of the option of freedom to develop their capabilities. His third suggestion is important for current debates on culture since he highlights the fact that human rights must be a means through which people protect their culture but without giving them the means to abuse their fellows, so that they do not become the right simply to preserve cultures.

Creating images through digital technologies: a viewer taking a picture of an Aztec dance. *Source* Photo by Lourdes Arizpe

As the policy application of Human Sustainable Development theory, the "Capabilities Approach" focuses on the substantive freedoms that people should have to fulfil the life choice that they have made. From Martha Nussbaum's perspective, which is important for the cosmopolitical outlook of this book, capabilities can be the "objects of an overlapping consensus among people who otherwise have very different comprehensive conceptions of the good" (Nussbaum 2000: 5).

Additionally, she proposes that capabilities be included in the "constitutional principles [that] should be respected and implemented by the governments of all nations, as a bare minimum of what respect for human dignity requires" (Nussbaum 2000: 5).

In particular, as just noted, Nussbaum emphasizes capabilities as a basis for solidarity among women having different cultural and religious backgrounds. She starts out by observing that "women in much of the world lack support for fundamental functions of a human life" (Nussbaum 2000: 1). This happens especially when gender inequality is combined with poverty, leading to an acute failure of central human capabilities. Nussbaum adopts the principle that each person has their own distinct capability, giving women the possibility of living lives for their own ends.

Nussbaum's use of the term "living lives" is, I believe, very important, and will be taken up in the conclusions of this book. To describe these central human functional capabilities, Nussbaum draws up a list which is quoted here because, in a sense, it reads like the possible content of any given culture. As contentious as many of its specific points may be, it is very useful for moving discussions forward. These central human functional capabilities, in Nussbaum's writing, are as follows:

1. Life.
2. Bodily health.
3. Bodily integrity.
4. Senses, imagination, and thought.
5. Emotions.
6. Practical reason.
7. Affiliation: (a) being able to live with and toward others; (b) having the social bases of self-respect and non-humiliation.
8. Other species.
9. Play.
10. Control over one's environment: (a) political; (b) material (Nussbaum 1999: 41–42).

The theory and research work on Sustainable Human Development, as well as on the Capabilities Approach, has been one of the most important sources of discussions both for the 2000 Millennium Development Goals and for the Sustainable Development Goals adopted by the United Nations in 2015. Sakiko Fukuda Parr, who led the UNDP Human Development Reports from 1995 to 2004, contrasts the two normative documents in relation to gender, arguing that "… the SDGs address several of the key shortcomings of the MDGs and incorporate a broader and more transformative agenda that more adequately reflects the complex challenges of the 21st century, and the need for structural reforms in the global economy. The SDGs also reverse the MDG approach to global goal setting and the misplaced belief in the virtues of simplicity, concreteness, and quantification. While the SDGs promise the potential for a more *transformative agenda*, implementation will depend on continued advocacy on each of the targets to hold authorities to account" (Fukuda-Parr 2016: 43).

At the United Nations High-Level Political Forum in July 2017, Fukuda-Parr began by stating that the adoption of the 2030 agenda was a historic achievement. It is a significant advance in our thinking about development and reflects the realities of the twenty-first century. In a clear departure from the Millennium Development Goals that envision development as meeting basic needs in developing countries, the 2030 agenda reconceptualizes development as sustainable development and as a *universal challenge*.

Fukuda-Parr states that some of the important new elements of this concept include recognizing the ecological limits of the earth, achieving consensus on inclusion and equity; and, as a political priority, leaving no one behind, consistent with human rights principles and addressing the systemic causes of poverty, inequality and environmental destruction (Fukuda-Parr 2017: 25). What is new, she believes, is that these priorities are now integrated in a single agenda, UN2030. This will require "a *holistic* review with due attention to key issues of *interlinkages*", while taking into account that indicators are merely a representation of social phenomena and that not all social phenomena can be translated into a single number (ibid.). She goes on to specify that data and measurements are technical issues but they are also deeply political and need to be scrutinized and interpreted with care.

8.8 Anthropology and Cosmopolitanism: "Spaces of Cultural Improvisation"

Discussions about culture and globalization in anthropology, as mentioned in the last section. have been closely entangled with those on cosmopolitanism. One of the most important is *Anthropology and the New Cosmopolitanism: Rooted, Feminist and Vernacular Perspectives*, edited by Werbner (2008). "Against globalisation", Werbner writes, "a term implying the free movement of capital and the global (mainly Western) spread of ideas and practices, cosmpolitanism is a word used by the new cosmopolitans to emphasise empathy, toleration and respect for other cultures and values" (Werbner 2008: 2). Importantly for this book on culture, Werbner specifies that "whatever the definition, and whether we are talking of rooted, vernacular or elite interpretations of the term, cosmpolitanism has to be grasped as an ethical horizon—an aspirational outlook and mode of practice" (ibid.).

Three discernible strands in this normative cosmopolitanism are mentioned, which "have converged as empirical research reveals the complexity of the contemporary global public sphere" (Werbner 2008: 2–3). This is the point where international debates on culture and on the global public sphere often overlap when addressing subjects such as multiculturality, democracy, the West and cultural rights. Without further exploring the very insightful analyses in Werbner's book, it is worth highlighting some of the main points which have a bearing on international transactions on culture.

The first strand of the new normative cosmopolitanism relates to global governance and, as David Graeber argues, the provenance of concepts associated with "Western freedoms", such as democracy, which he demonstrates are the product of "endless entanglements within and beyond the West" (Werbner 2008: 3). He goes on to state that "As European states expanded and the Atlantic system came to encompass the world, all sorts of global influences appear to have coalesced into European capital, and to have been reabsorbed with the tradition that eventually came to be known as 'Western'" (Werbner 2008: 291), a process that is probably impossible to reconstruct. One could say that, in cultural terms, this infinitely complex system of exchange is, indeed, totally impossible to reconstruct, throughout human history but also in the last four centuries of colonialism. I further agree with Graeber that the framing of this debate in terms of the West vs. the Rest "tends to blind us to the actual historical dynamics at play" (Werbner 2008: 295). He gives as an example the Zapatista uprising (my own choice of term would be "neo-Zapatista uprising", since the two movements, that of 1910 and that of 1994, are very different). While I agree with Graeber that this movement, which became global, is a "space of democratic improvisation", I would only add, based on my own research in the Lacandon rainforest, that the agency of such communities has always been there. It has, as he says, been obscured by the politically discursive methods used to strengthen centralized control by the provincial state—very much so in the case of Chiapas, which was the only state in which the Mexican Revolution was not triumphant—and by the national state (Arizpe 1993).

Graeber also strongly criticizes the neo-liberal solution to the need for planetary decision-making structures by declaring the market "the only form of public deliberation and restricting the state almost exclusively to its coercive functions" (Werbner 2008: 299), a position I also agree with. "In this context, the Zapatista response … to refound democracy in the self-organisation of autonomous communities makes perfect sense" (ibid.). His major point, which I wish to highlight, is that "democracy, then, seems for the moment to be returning to the spaces in which it originated: cosmopolitan spaces—the spaces in between" (ibid.).

The second strand of the new normative cosmopolitanism identified by Pnina Werbner was developed by Jürgen Habermas's revision of Kant's *Perpetual Peace*. Habermas argues that international non-governmental organizations have created an alternative "global public sphere" of debate and advocacy (Habermas 1999: 176, cited in Werbner 2008: 5). Finally, the third strand, Webner continues, is related to Ulrich Beck's theorizing on what may be termed 'reflexive' globalization: his insight that "What is new is not forced mixing but awareness of it, its self-conscious political affirmation, its reflection and recognition before a global public" (Beck 2006: 2; Werbner 2008: 5). An interesting parallel with culture is the convergence of

Marshall Sahlins's affirming that culture has now become a conscious project with Beck's recognition of self-conscious political affirmation of cosmopolitanization.

An interesting paradox of cosmopolitanism, pointed out by Sally Engle Merry who is cited in Werbner's book, is how "in the aftermath of the Holocaust, states are no longer trusted by the international community to govern their own citizens without international oversight. On the other hand, the focus of much human rights activism is the state. Sometimes the state is the human rights violator, when it subjects its citizens to torture or extrajudicial killings, for example. Ironically it is also the agent for carrying out human rights reforms in many cases. Social and economic rights, such as the right to development or the right to adequate housing, require state action, as does the provision of many civil and political rights … Thus, human rights activism ends up demanding more state regulation and service" (Merry 2006: 5, cited in Werbner 2008: 7).

Maila Stevens, in her chapter, points out that "feminisms have engaged both theoretically and practically with many of the besetting difficulties with the cosmopolitanism debates …—universalism, ethnocentricity, neo-imperialism—" (Werbner 2008: 19). Against the criticism that cosmopolitans are uprooted elite travellers, several authors in the book mention "rooted cosmopolitanism" as a way in which internationalized elites retain their connections to their vernacular communities, but also, as Merry explains, with the possibility that locals adopt ideas that are then reinterpreted and translated by local activists. Inevitably, divisions between actors may be difficult to bridge, as several authors also mention. Werbner agrees with Kwame Anthony Appiah that in some 'conversations' about cosmopolitanism, protagonists can only agree to disagree (Appiah 2006: 78, cited in Werbner 2008: 21), and endemic ethnic conflict, gender inequalities and imposed religious dogma do require a committed and genuine attempt to arrive at agreed strategies for living in amity, without denying differences (ibid.). Very importantly, Appiah also warns that "not all boundary-crossing, globally oriented groups are cosmopolitan", as Pnina Werbner notes. He labels religiously exclusivist global movements such as al-Qaeda, that appear to espouse a singular truth and demonize the Other, at times violently, as "counter-cosmopolitan, despite these movements' transnational networks and global moral aspirations" (Werbner 2008: 11).

In her final comment, Werbner states that anthropologists themselves become cosmopolitan by engaging in comparative argumentation and critical debate and, significantly, by being hosted by people who are willing to share a cosmopolitan dialogue. "Cosmopolitanism, she states, is dialogical—a collective, creative endeavour, beyond the individual" (Werbner 2008: 25).

The framework of cosmopolitanism, then, which has always been present in anthropological research, becomes ever more important in recasting culture towards a future which requires cooperation in the production of knowledge and a reconceptualization of the relations of humans to non-humans and to the Earth.

8.8.1 Reconceptualizing Worlds: Animisms, 'Planetarisation' and the Identity of 'Earthlings'

Not all cultures in the world recognize the continuities and discontinuities built by naturalism in Europe after the eighteenth century, anthropologist Philippe Descola posits (Descola 2016): nature and culture, a distinction separating the body and mind in Christianity, and the conceptualization of nature as opposed to humankind. Descola calls it naturalism, as opposed to forms of thinking about this, such as animism, that occur in many cultures around the world, based on legitimizing the dominion of Man over the goods and services of the environment in terms of ownership. At present, Descola states, we need new concepts to build different 'worlds'—'mondes' in French. How to go about doing this, he cautions, is very complicated (ibid.).

Descola mentions the "rights of the environment". He is against them because he thinks there is no way in which they could be enforced. Instead, his position is to develop the rights to an interaction such as the ecosystem in which the Achuar, the people he studied in the Amazon basin, live. In such a case, humans would be the guardians of the ecosystem. He notes the opposition between human rights and the rights of nature (ibid.). Possessive individualism gave rise to the rights in the Social Contract, based on appropriation by humankind. A very different view emerges if one starts out with the idea that the Earth is the source of rights, not humans.

The world is at present in an emergency involving massive extinction of species, pollution of the sea and so on, which requires, Descola says, that the intellectual tools which were used in the three last centuries and then had positive results must now be reconceptualized in terms of the relationships of humans to non-humans. This means giving ontological value to animisms, that is, the belief that both humans and non-humans are embedded in nature. Anthropology must show that there is another way to build 'worlds'. Since Descola uses the French term 'mondes' in the plural, he is arguing that, having created different kinds of 'worlds', humans will be able to create new forms, no single one of which will be the only model for the future. However, he points out that the present situation is very different from the restricted spaces of tribes, and that consequently, we have to invent new worlds with new concepts (ibid.).

Related to this, another transaction that is being put forward on the international stage is focused specifically on what anthropologist Marc Augé has called 'planetarisation'. Even if new images, such as the "Blue Planet" photograph of the Earth taken from space, reinforce our perception of the entirety of humankind as bound within the same planetary limits, Augé states that globalization has only partially brought the world together. Highlighting the changes of scale, he argues that we need greater precision in our vocabulary. Thus, he reserves 'mondialisation', originally proposed as the French term corresponding to but contrasting with 'globalization' in English, to refer to the plurality of political regimes that throughout history have tried to govern the world they dominated, which was never coterminous with the planet (Augé 2017: 123).

Augé considers that today's globalization is driven by technology and by the economy. While noting that certain political regimes limit access to the world web of communications and that the supposedly free circulation of commodities, capital, and financial flows faces obstacles in many parts of the world, he points out that these two processes still dominate the world. In this sense, a plurality of 'worlds' ('mondes') such as those of finance, work, sports, literature and so on, as well as those of Shiites, Sunnis, Catholics and Protestants, exist within globalization. He brings up the practical point that these 'worlds' are often closed and reserved for specialists or adherents, but could become accessible to all if education could really open up their codes and languages. "In this way", Augé continues, "a real equality could be achieved; not as a uniformity of competences and behaviours but an equal capability of listening to and understanding the challenges of each 'world'; every individual would effectively be free to develop their own view of the world" (Augé 2017: 121). Such an expectation has also been prefigured for social media, yet, he warns, this will not come about if all that we do is store up knowledge in electronic databases without creating the conditions for an education that will allow everyone to be informed effectively.

In his 2017 book, *L'avenir des terriens: Fin de la préhistoire de l'humanité comme société planétaire* (*The Future of the Earthlings: the end of the prehistory of humanity as a planetary society*), Augé goes further, linking his arguments to many of the propositions that are discussed in the next chapter, on the Anthropocene. Given the constraints of the terms *globalization* and *mondialisation*, he in turn proposes the term 'planetarisation' (Augé 2017: 123). Its two main references are geographical, in that this concept refers to the concrete extent and limits of any political and technological *mondialisation*; and ecological, referring to the planet as a physical entity, to the planet as a common good of humanity, and to the collectivity of living beings, open to the risks that human behaviour is imposing on it through accelerated demographic change (Augé 2017: 123–124).

Interestingly, there are several initiatives around the world that are already carrying this challenge into government policy. Evo Morales, the Aymara President of Bolivia, began promoting a policy "Del Buen Vivir" ("Of Good Living") as a way of incorporating the beliefs of the indigenous peoples of the Andes into modern-day rights. One of the main beliefs concerns the rights of the "Pacha Mama", the name of Earth. Along these same lines, the new Constitution of Ecuador also refers to the "rights of nature".

How do we cope, Augé asks, with a future that pulls us through a whirlwind of unthought possibilities? He offers starting points: a method, an object and a principle (Augé 2017: 23–24). The first is directed at establishing the scientific method as a general principle of action for society. Not 'scientism', but an endeavour based on the careful verification of hypotheses. This must be so, Augé insists, because science is the only domain of human activity that one can confidently state leads to "cumulative progress" (ibid.).

The object of such an activity, Augé suggests, is knowledge as the ultimate aim of human existence, one which should inform the entirety of political, economic and social debates. This ideal of knowledge is situated over and above each individual life, within the framework of equality of dignity for all human beings (Augé 2017: 26). Humankind has a triple dimension, individual, cultural and 'generic'[10] (ibid.). "This triple dimension of humans provides the principle on which is founded the possibility of formulating the goal of true universality, and to identify the problematical place where it is to be implemented: societies in their diversity"[11] (translation by this author).

Augé notes that the disappearance of the political projections of the nineteenth century is not a sign of weakness but, instead, he continues with a very tantalizing thought: perhaps we can now learn to change the world before we imagine it, turning towards the future without obtrusively projecting our illusions.

Augé ends by saying that perhaps the only utopia which is possible for the coming centuries is the utopia of education for all, which will do away with the black utopia of an unequal, uncultured, illiterate world society, shackled to consumption and exclusion, including brutal and insidious forms of intellectual intimidation, and giving way "to all forms of proselytism leading to violence, ideological regression and, ultimately to the risk of planetary suicide" (Augé 2017: 30).

This book has shown that most transactions about culture have been activated within the boundaries of the concept. Now, two new areas of human experience will inevitably change our view on culture: discoveries about the way in which the mind-body works in relation to sociality and the explosive growth of the virtual world.

8.9 Resonance, Resilience and the "Social Brain"

Three terms seem to me to be key in relating culture to recent discoveries about the mind: the "social brain", "neuronal resilience" and emotional resonance. Social neuroscience, begun in the nineties by John Cacciopo and Gary Berntson, found that neurons, to develop, require the physical presence of others and a resonance of empathy with them. Discovering the neuroplasticity of the brain has led to a revolution in understanding how human relationships shape the "social brain". Children, especially orphans, who have been deprived of close sensorial, affective and securitizing relationships with others, may suffer brain atrophy. When they are taken to other families or foster homes where they do establish close connections to other people, a "zone of neurogenesis" in the limbic system may undo this brain

[10]The word in French is 'générique' which may be interpreted with reference to 'genus', as species, and in English may be ambiguously translated as 'generic'.

[11]"La triple dimension de l'être humain constitue donc le principe à partir duquel il devient possible de formuler sa finalité, l'universalité effective, et de repérer le lieu problématique de sa mise en oeuvre: les societés dans leur diversité" (Augé 2017: 28).

damage by creating neurons, according to Boris Cyrulnik, who considers this a revolution that strengthens the concept of "neuronal resilience" (Cyrulnik et al. 2012: 68).

There is a growing consensus among neuroscientists that our brain is neurosocial. In the words of Boris Cyrulnik, "we are no longer condemned to an 'exclusive determinism' where our genes define what we can be at the moment we are born; instead, we are now part of a social network in which our neurons 'catch' the emotions of other people" (Cyrulnik et al. 2012: 68).

Importantly, Cyrulnik emphasizes that in understanding the psychological development of adolescents, one must think in terms of a system and not of a cause that produces an effect. Particularly during adolescence, when the teenager must separate from her/his family, "if society does not propose structures, intermediate structures between the families and culture, the adolescent will have difficulties. And this will not be because of a failure of the family but the failures of a whole social organisation that does not offer possibilities of sociocultural resilience" (Cyrulnik et al. 2012: 60).

A person who does not 'belong' cannot become a person, Cyrulnik insists. There are two aspects to belonging, he says: familiarity and affiliation. Familiarity is rooted in the sensorial stimuli of domestic life. Affiliation is nurtured by culture. "When a child does not belong to anyone, he finds himself excluded from society, without the structures that should develop his feeling of familiarity and his desire for affiliation" (Cyrulnik et al. 2012: 85).

Daniel Goleman, who pioneered studies of "emotional intelligence", now refers —interestingly for this book—to *cultivating relational intelligence*. When two or more persons meet in highly convivial circumstances, their neurons create a resonance that gives them identity, reciprocity and security (Goleman 1995, cited in Cyrulnik 2017). A loop consisting of constant transfer/counter-transfer is then put in place in a process of cultivating "relational intelligence", as Goleman has called it, and this becomes the core process by which cultures are invented and transformed. In other words, cultures are the resonance chamber that helps individuals become fully active persons by relating to many others, thus providing them with the resilience to keep living in a way that is significant for them.

The importance of the environment where our capabilities develop is reflected in the detailed and complex coding of social roles, Cyrulnik adds. He also calls attention to the fact that even though the ubiquity and portability of smartphones, tablets and computers allow a constant stream of communication, "cyber-machines have amazing communication performances, but they prevent the rituals of emotional interaction" (Cyrulnik et al. 2012: 63). Young people may grow up without learning successful emotional interactions. In this dystopian situation, he asks, are we condemned to disappear due to the regression of the "social brain"? (ibid.). He reminds us that the crucial issue of the twenty-first century is to widen the circle of those we consider as 'us', and reduce the number of those we think of as 'them'.

"The wiring of our social brain relies on the core of our common humanity" (Cyrulnik et al. 2012: 76).

One of the key concepts that people say is part of culture is empathy. It is also, by the way, the core principle of anthropological research. Cyrulnik defines it from the perspective of an ethologist and psychotherapist, as a process of the projection of human sentiments about the material world, in which the imitation of another person by a child creates a feeling in her/him. This allows her/him to differentiate her/his own experience from that of others and she/he understands that these are two different but associated subjectivities (Cyrulnik 2006: 144). Far from being a rational cost–benefit analysis, empathy, Cyrulnik states, "is the emotional aptitude to allow oneself to be modified by the world of another person to whom one is attached" (ibid.).

Seen in this light, empathy is indeed part of the 'dreamcatcher' effect of the soft concept of culture. However, Cyrulnik also points to the many dangers for children and adolescents who have grown up without familial and cultural attachments and become psychologically and morally disorganized. He cites the insight of John Bowelby, President of the Royal Society of Great Britain, who noted in the seventies that "empathy is the counterpoint to aggression" (Cyrulnik 2006: 144).

This insight is important for understanding culture and its ambivalent effects in present-day societies. In cultures that are not organized by communities continuously as a base for engaging with the rest of the world, young people who feel detached from their families and communities may turn to their culture as a justification and a shield for attacking others from other cultures.

A new analytical tool, that of "social resilience" is now being used to describe such situations. Psychologists and psychotherapists have developed the concept of social representations, as proposed by Serge Moscovici, "as a kind of cognitive photosynthesis … that builds our understanding of the world and which we give back as comprehension but also as judgements, definitions and classifications (Moscovici 1984, cited in Serrano Oswald 2016). On the basis of this idea, Serrano Oswald proposes that social resilience is the capability of a collective subject or system to reconfigure its identity in the face of change (to be), solving the problems, contradictions and crises inherent in change (to continue), in order to restore balance and ensure continuity (Serrano Oswald 2016: 58). In the framework of social resilience, subjects and collectivities are seen as agents of change and equilibrium through interactions, exchanges and meanings (Serrano Oswald 2016: 4). This allows a constant process of re-encounters and feedback through which people update their views of themselves and of the future. Serrano Oswald then analyses how people in a specific community, Zacualpan in Mexico, engage in an annual festival, the Mojiganga, to sustain collective participation in intangible cultural heritage and to offset the disruptions of joblessness, migration and, now, the deportations of migrants from the United States (Serrano Oswald 2016).

When such a process of social resilience is uprooted, as noted in the sections of this book that describe "failing societies", and there is a high level of unemployment, inequality of opportunities, disruption of families, and poverty-stricken communities, children and adolescents who grow up in such conditions will engage with culture in a negative way, that is, they will use it as a weapon to attack others. In other words, culture must not be expected to make up for deleterious conditions in societies that do not even allow their members to grow up in a healthy, social environment.

8.10 Cyberculture and Artificial Intelligence

A new dawn for the human world is augured by the wildly exhilarating development of the new digital technologies. Except that, once the machines are able to learn, think, and eat, say, biomass, humans may have to shut down the dawn to survive in their own planet. This is, in fact, the plot of a highly imaginative video game called "Zero Dawn Horizon". Not having prepared a "kill switch" for AI (artificial intelligence) machines, the last scientists opt for a geoclimate shutdown to allow a few humans to survive. This emerging meme for a twenty-first century myth may be placed beside others that are becoming globalized through the convergence of technologies. Another example, in the film Blade Runner 2049, is the consoling phrase spoken to a replicant (a robot), "dying for the right cause is the most human thing anyone can do".

In this way, many of the concerns that were discussed at the meetings cited in this book are being taken up in the constellation of digital technologies. Myth-making, cultural values, ethnic and racial diversity, gender roles and so on are being given a global reach through the convergence of the Internet, films, smartphones, podcasts, television programmes and laptops. As digital technologies shift from being used as a tool for enhancing human capabilities to becoming machines with much deeper and broader competences, questions are shifting which at present have no answers, as Pierre Lévy argues.

Apart from the immediate questions of how digital technologies are having an impact on societies and on international relations, on elections, on consumer habits and so on, the central question is whether there is any sense in continuing to use culture as a concept to point pathways to the future, given its deep interaction with digital technologies. The literature on the impacts of digital technologies is strewn with ambiguous conclusions about whether their influence is positive or negative. Indeed, the short time span of their impacts, the lack of methods of collecting evidence about them, the sheer mixture of entangled activities and the assorted vested interests in demonstrating different kinds of outcomes simply overruns the traditional structures of scientific research and discourse. Central to understanding this is Arthur and Marilouise Kroker's statement that "technology is not simply technical, but has everything to do with the more powerful currents out of which contemporary society has been fashioned" (Kroker/Kroker 2013: 6). The old computing saying, GIGO (garbage in, garbage out) is relevant here. Thus, in this double helix of culture and digital technologies, it becomes impossible to separate the dynamics that drives each of them. How should we distinguish between the "dumbing down" of television programming and the 'shallows' that Nicholas Carr attributes to the use of the Internet and the social media (Carr 2010)? Kroker and Kroker in their book *Critical Digital Studies* conclude that "neither inevitably utopian nor dystopian, digital reality is an open future, determined in the final instance by the play of human interests, desires, and power that have always represented the elementary matter of the human condition" (Kroker/Kroker 2013: 3).

With this assumption in mind, we can go on to briefly summarize the main issues discussed in this field which pertain to culture.

8.10.1 "Questions Without Answers"

Beginning in the nineties, as enounced by participants in many of the meetings included in this book, alarm over the global spread of the Internet among countries of the world focused on the possible loss of their traditional cultures. The UNESCO conventions on intangible cultural heritage and on the protection of cultural expression were both outcomes of this concern. Indeed, Pierre Lévy, in his 1997 book *Cyberculture*, pointedly asked whether cyberculture threatened the values of European modernity. He answered that, on the contrary, "cyberspace may appear to be a kind of technical matérialisation of modern ideals" (Lévy 1997: 302). Cyberculture may be considered a continuation of the progressive project of the Enlightenment, he explained, as it enhances participation in communities of debate and argument, equality in reciprocal human relations, exchange of information and knowledge and liberty in communication across borders to all communities in the world (ibid.).

Pierre Lévy also asked several other "questions without answers" about content and signification in cyberculture (Lévy 1997: 288). Given the rise in concerns about the technology gap between the West and other regions of the world, Lévy queried whether cyberculture would lead to exclusion, given the "oppression, social injustice and poverty in the world" (ibid.). In recent times, the unprecedented rise in inequality has deepened exclusion, but the use of laptops, tablets and smartphones has penetrated even to the most poverty-stricken social groups. The relevant question now is how they are being used: to lessen exclusion, or to dig deeper into it?

In his second question: "Is the diversity of languages and cultures threatened by cyberculture?", he relates this to the "absence of totalisation" in cyberculture (Lévy 1997: 293). In the last fifteen years the number of web pages in different languages—even Esperanto!—and the variation in cultural beliefs and practices have invaded content on the Internet, although the original language of hardware and software is English. English's role as a lingua franca will continue to expand with the spread of digital technology though all other languages will also use niches of the Internet to strengthen their power of communication.

Lévy himself notes that the threat of uniformity is not as serious as it would seem. To begin with, producing and disseminating information for an immense international public became much more accessible, and the instruments to filter, to navigate and to provide orientation towards the contents of the Web have continued to multiply. Lévy also makes the important point that the presence of variability would be directly proportional to the active engagement and quality of contributions made by the representatives of these varied cultures (Lévy 1997: 296). This means that the vitality of cultural practices will be enhanced, but will not depend on the technology itself.

In the conclusion of his book, Lévy argues that cyberculture represents "a major mutation in the very essence of culture" (Lévy 1997: 305). This change consists, for him, in a transition of humanity to "a universality without totality" (Lévy 1997: 305), which means a "moment where our species … tends to form a single global community" (Lévy 1997: 307–308).

A different, unanswerable question is how digital technologies can help us understand the nature of the human propensity to create cultures. Can human cultures be mapped out by algorithms? Will such a thing as an "algorithmic culture" begin to evolve? Are we as prone to follow mathematical trails as, say, philosophical ones, inclined, as we think we are, to follow philosophical or ethical trails by virtue of our own free will?

Yet another, Big question, is whether, beyond creating cartographies of the human mind, human ingenuity, through digital technologies, will be able to create a 'mind', a topic that is taken up further on in terms of AI.

The last unanswerable question, if it indeed qualifies as a question, is whether computational machines would be able to create a culture. In Chap. 1 of this book several participants in international meetings were quoted who said that "culture has a heart", meaning, it seems to me, that culture belongs to humans and is also the jewel in people's most heartfelt emotions, love and empathy. If, as described in the previous section, children learn cultural meanings and attachments through mirroring how people around them behave, it does not seem possible to think that machines can learn through mechanical binary codes. This intriguing question will not be pursued here, since so much deeper research is needed to even begin to structure this discussion. That it is already a highly tantalizing social *problématique* is evident in the number of films, documentaries, television programmes and so on that are taking this up.

8.10.2 "New Connectivities"

As mentioned earlier in this chapter, anthropologist Michael Fischer avers that in cyberspace cultures may become tools for experimentation to create "new connectivities" (Fischer 2009: 37). He is very optimistic that "the Internet, networked data banks, visual icons, video clips, film animation, streams and repetition information flows are repositioning and enveloping older cultural media such as orality and literacy, reshaping the public sphere by changing power relations … mobilizing money and attention in electoral campaigns, and drawing attention to alternative geopolitical narratives" (ibid.).

What will happen to art as the digital technologies take over many of the remaining fields of art and what Nestor García Canclini has called the "aesthetic imminence"? (García Canclini 2010: 180). In recasting art beyond the traditional locations of museums, galleries and biennal exhibitions, García Canclini argues that the new visibility and comunicability is at present innovating artistic practices, placing them in "uncertain locations"—"localizaciones inciertas". The photographs of Sebastiao Salgado, for example, especially his collection of *Exodos*, portraying migrants from forty countries, create a new aesthetics of solidarity.

Pierre Lévy also has an optimistic view about these new connectivities and he sees in hypertext "a textual machine that could profoundly change writing and, therefore, thought" (Lévy 2011: 8). Arthur and Marilouise Kroker go even further in affirming that "digital culture literally remaps, rewires, and recodes life itself using complex algorithms" (Kroker/Kroker 2013: 3). In contrast, Nicholas Carr, in his book *The Shallows: What the Internet is Doing to Our Brains* considers digital technology as eroding or dispersing human memory at the individual level (Carr 2010). A way out of this conundrum is to say that "complexity is the essence of digital technology" (Kroker/Kroker 2013: 6).

However, in recent years a growing literature on the effects of such connectivities on personal lives and social mores points to dislocations. Novelist Olivia Sudjik in her article "Virtual Anxiety: the Disturbing New Reality of Life Online", published by CNN: "The idea that the internet's mission is still about connection, making our experience of the world seamless, persists in the names of the digital companies breaking down the divide between two words to make a new one. Social media still professes to support an enhanced empathy for others and a more porous, better-networked self. While this may have been the case for some (arguably people who would have been nice to strangers anyway), there are plenty of racist, sexist, xenophobic, transphobic homophobes coming out of the woodwork every day for whom Instagram has apparently done the opposite. Well, that's life, you might say. And yes, it is: Real life and digital life can no longer be considered separate" (Sudjik 2017: 2). This very wise perception inevitably makes us think about the long centuries that it took to build nation states and societies, philosophical systems and ethical rules that help contain warfare and criminality.

As Sudjik remarks, "in the so-called age of connectivity, my country [the UK] has voted to cut itself adrift like a hormonal adolescent marooning themselves in their room and I watch the news in horror at the increasing visibility of xenophobia and other divides" (Sudjik 2017: 4).

In a surprising proposition, Nicholas Christakis of Yale University has suggested that culture was the first sort of artificial intelligence created by human beings: "like the intelligence in a machine, we create culture, interact with it, are affected by it, and can even be destroyed by it. Culture applies its own logic, has a memory, endures after its makers are gone, can be repurposed in supple ways, and can induce action" (Christakis 2015). Apart from noting that Christakis is reifying culture, I would argue that, certainly, those similarities between 'culture' and artificial intelligence are there, but only if one takes an extremely iconoclastic view would one say that all human inventions are artificial. Yet the invention that is now taking public spaces by storm is, understandably, Artificial Intelligence.

8.10.3 Acting Ethically—Who is Human?[12]

In the film Transcendence, a scientist hooks his mind up to a powerful computer which allows 'him' still to be in communication with his wife and his best friend. The experiment ends when his best friend convinces his wife that that 'mind' is no longer her husband and they shut 'him' down. Fatal destiny? Petty human jealousy? Horror at losing our embodied cultures? The question that springs to mind is to what extent digital technologies will be able to substitute for human capabilities, since these are deeply embedded in cultural creations. The *Oxford Dictionary of English* (2010 edition) defines artificial intelligence as "The theory and development of computer systems able to perform tasks normally requiring human intelligence, such as visual perception, speech recognition, decision-making, and translation between languages". The Committee on Robotics and Artificial Intelligence of the UK House of Commons has defined it as "an evolving assemblage of technologies that enable computers to simulate elements of human behaviour such as learning, reasoning and classification" (Science and Technology Select Committee 2016: Sect. 1).

Learning, reasoning and classification are indeed capabilities of human beings, but it must be pointed out that, although a machine may build a picture of the world around it by learning from experience, this learning always starts out from the database and instructions given by engineers. Machines may incorporate new knowledge but it will always be based on the materials given to them. Or would it? If not, machines could take over and create a dystopian future, as brilliantly portrayed in the video game Zero Dawn Horizon.

This makes the views of the UK Select Committee very relevant. At present, artificial intelligence has been mainly used in weapons, banking and finances,

[12]For research work on this subsection, I am grateful to Mike Headon.

surveillance, natural language translation, global positioning systems (GPS) and, interestingly, opinion-forming, that is, the manipulation of social media. The UK Select Committee noted that each application is developed in isolation, so that there is no agreed national or international oversight of the consequences of producing such systems. The Committee agreed that controls need to be set up, among them taking steps to minimize bias being accidentally built into AI systems, ensuring that the decisions they make are transparent, and instigating methods that can verify that AI technology is operating as intended, and that unwanted or unpredictable behaviours are not produced. A case in point, where culture is thought of in terms of realpolitik, identified in relation to lethal autonomous weapons systems (LAWS), is: how does the weapons system distinguish between enemy personnel and innocent civilians?

The cognitive posibilities of an AI system must also concern us in terms of cultures. Artificial intelligence now builds a picture of the world around it by learning from experience. However, the learning that it carries out starts from a base of data supplied by the designers of the system. All designers, as they use their cognitive skills, have learned from specific sociocultural settings, and will carry cultural choices that are then built into the system as an invisible world view. If they are all male scientists and engineers, their blind spots related to sociality and polity from women's points of view will be absent from the system. If they predominantly belong to a given ethnic, territorial or religious group, that group's conscious or unconscious biases will also be absorbed into the AI system. It is all the more urgent, then, to make the imperceptible visible so that societies can act to spread the capacities to use and to influence AI to bring in diversity as far as possible.

AI is being developed in discrete applications, many of which concern military and surveillance operations, banking and finance, medical diagnoses, GPS, natural language translation and so on. The UK Select Committee also noted that one consequence of this was that each application is developed in isolation, so that there was no agreed national or international oversight of the consequences of producing such systems. A key point mentioned in the Select Committee's 2016 report is that "The implications of new technologies tend … to be examined and understood by policymakers too late in the day to engage with them in any significant way" (Science and Technology Select Committee 2016: Sect. 1). A second point to ponder is how, if applications of AI are being developed in such a discrete way, will international bodies be able to advance towards a common understanding of their possibilities through international transactions?

Equally important is the question of the ethics of AI systems. An interesting attempt to launch this discussion is work by Julia Bossmann, president of the Foresight Institute, listing the top nine ethical issues in artificial intelligence (Bossman 2016). These were:

1. Unemployment. What happens after the end of jobs?
2. Inequality. How do we distribute the wealth created by machines?
3. Humanity. How do machines affect our behaviour and interaction?

4. Artificial stupidity. How can we guard against mistakes?
5. Racist robots. How do we eliminate AI bias?
6. Security. How do we keep AI safe from adversaries?
7. Evil genies. How do we protect against unintended consequences?
8. Singularity. How do we stay in control of a complex intelligent system?
9. Robot rights. How do we define the humane treatment of AI?

It is very telling that all these questions, it seems to me, are issues in present-day society and in transactions on culture.

The uses of AI for opinion-forming and manipulating social media have come to the fore in the last two years, as the next section will show.

8.10.4 Memetics: The "Replicator Dynamics" in the "Spread of Ideas"? or the Threat of "Meme Warfare"? [13]

Is the meme concept a useful one? Robert Aunger asks this question in his book *Darwinizing Culture: the Status of Memetics as a Science*. "The explanatory target of memetics, at least as narrowly conceived, is culture", he explains, but, "unfortunately, there is perhaps an equal amount of controversy about what culture might mean as we have seen surrounding the concept of memes" (Aunger 2000: 7). Among the main critics of what some see as an intromission into anthropology on the part of biologists, Adam Kuper, along with Maurice Bloch, argue that "culture is not divisible into units because it is a complex, heterogeneous thing" (Aunger 2000: 21) and therefore, that the memetic project is "rather like a blunt arrow shooting into the dark" (Aunger 2000: 19).

With so vast and ambiguous a terrain, to try to make sense of it through the meme concept defined by Ryan Milner as "the linguistic, image, audio, and video texts created, circulated, and transformed by countless cultural participants across vast networks and collectives" (Milner 2016: 1) seems quite appealing. Milner argues that in our contemporary media environment, "mediated expression and debate consistently occur through and around an ever-growing collection of populist memetic texts" (Milner 2016: 2) and it is therefore important to analyse the logic of these memetic processes. Milner acknowledges that "memetic media simultaneously facilitate both empowerment and marginalization" (Milner 2016: 5) but seems rather optimistic about its power to collectivize social dialogues, saying

[13]For research work in this subsection I am grateful to Mariana Roa Oliva.

that it “isn’t always egalitarian and empowering, but it’s more open than the gatekeeping practices associated with traditional mass media” (Milner 2016: 220).

The enthusiasm which was felt at the beginning of the expansion of the Internet, which was portrayed as a land of freedom, is by the middle of this second decade now being described very differently. Mozilla, the founders of Firefox, published this text on 24 October 2017, on the entry page to their search engine: “Big corporations want to restrict entry. Fake news and filter bubbles are making it harder for us to find our way. Online bullies are silencing inspired voices. And our desire to explore is hampered by threats to our safety and privacy. It’s time to join Mozilla and do our part as digital citizens. Donate today to support programs that keep the internet healthy, free and open for us all”.

Memes, it is now said, have had a lot to do with such outcomes. Ruining democracy is not what one would expect from a kind of media like memes, which “are now born from the swamps of the internet in real time, distributed from the bottom up” (Haddow 2017), as Douglas Haddow describes them in his article “Meme warfare: how the power of mass replication has poisoned the US election.” “Meme warfare,” Haddow goes on to explain, “a term that was originally coined by social change activist Andrew Boyd” presented back in 2010 “an opportunity for individuals to seize control of the means of media production from corporate interests” (Haddow 2017). It was a viral and open-source medium that would allow individuals to compete for attention against “the all-consuming hydra of advertising, marketing and public relations” (Haddow 2017). But what social change activists like himself and his team at Adbusters did not predict was that the use of “shareable images and ideas in an effort to engender real political change” (Haddow 2017) could just as easily be taken up by the right.

In her article “Forget fake news—alt-right memes could do more damage to democracy”, Sophia A. McClennen warns about the danger of memes through the example of a tweet by Donald Trump where he posted a meme of himself beating up CNN. Paradoxically, a kind of media that was lauded by the left at first as a promising tool for democracy is now being perceived more and more as a threat to it: “At a moment in U.S. history where trust in the news is at a record low, we are seeing the growing power of individuals outside of traditional media circuits who have built a following by suggesting that their voices are more authentic and more accurate than those of the so-called liberal, elitist, majority-silencing news” (McClennen 2017).

Moreover, the spread of alt-Right propaganda opens up another new facet in social media. As McClennen notes, there are reports that suggest that “Facebook’s internal rules for censoring hate speech [are] biased to favour both the right and celebrity users” (McClennen 2017). Social media might look like an open democracy, but the control the corporations have over it now has to be carefully analysed to understand its deeper impact.

In the conclusion of his book, Lévy argues that cyberculture represents “a major mutation in the very essence of culture” (Lévy 1997: 305). This change consists, for him, in a transition of humanity to "a universality without totality" (Lévy 1997: 305), which means a "moment where our species [...] tends to form a single global community" (Lévy 1997: 307–308). In this sense, the most important aim of digital technology is to create and shape ideas (or 'minds') capable of developing a sustainable world. This will only become possible if technological convergence contributes to a philosophical convergence of all peoples in shaping a common human destiny.

This philosophical convergence, I posit, could be the framework for people to work together in using this extraordinary cyberspace in which the rich diversity of world cultures and lifestyles may be expressed and debated. The problem which I see arising that that such debates are never transacted because there is no scaffolding or ethical rules to allow for arbitrating and consolidating consensus. As Olivia Sudjik expresses it, "we float in the cloud as we project our fantasy selves. We can forget to do the real work,—which is messy and dysfunctional, unlike a user-centric app" (Sudjik 2017: 3). Personal relationships, community collectivities, political societies are always messy and dysfunctional: this is why diplomacy, endless transactions through speech, manoeuvring and lobbying, institutional arbitration and democratic political skills are needed to avoid wars and to enhance cooperation. That all of these efforts may be spun in such a way as to bring about the opposite results to those intended only brings vividly to the fore the fact that a philosophical and ethical thrust is needed to drive cyberspace to the positive and not to the negative.

The Big Question is: Who is to do this? The heads of the large Internet corporations? Business enterprises commoditizing the Internet? The loosely organized social media communities? Individuals voting in a seemingly impossible-to-organize fair and inclusive voting system?

At the end of her article, Olivia Sudjik cites Tim Berners-Lee who has said "it has taken all of us to build a web we have, and now it is up to all of us to build the web we want—for everyone". The question that follows from this, in my mind, is how are we all to do this when transactions, as this book has demonstrated, require endless face-to-face, body-to-body transactions and arbitrations. This embodied learning and co-production of knowledge is at the core of sociality. The challenge will then be how to enhance an "augmented form of sociality" that will continue to make human 'living' the creative, future-producing endeavour that it has been for millenia.

8.11 Scales of Cultural Interaction: A Heuristic Proposal

The materials described in previous chapters as well as the present one showing the sequence of transactions on culture demonstrate that, while recasting the theory of human nature and of cosmopolitical interactions in a globalized world is a long-term undertaking, nevertheless a first heuristic proposal is needed to distinguish the two related, yet different, scales of transactions on culture that have become apparent in international debates. The first relates to *cultural politics* as the arena in which different groups invoke culture in power struggles to pursue their economic, political or social advantages. In a sense, following my own proposal when I carried out fieldwork on indigenous peoples in Mexico that ethnicity should be dealt with differently from *ethnicism*, the same distinction needs to be made between culture and *culturalism*. Much of the confusion, cross purposes and contradiction in present-day discussions on culture, especially its policy applications, stem from this lack of conceptual and theoretical clarity about the different processes being transacted.

The weakening of nation states through globalization is bringing about heightened forms of cultural politics. While this kind of strife takes the form of perennial conflicts of ethnicism in a majority of developing countries, it is very significant that, for the first time, conflicts of culturalism have also erupted in developed countries, especially in those that previously embraced policies of multiculturalism. Brexit is perhaps the most salient example of this kind of strife, in which governments struggle with the opposing demands of right-wing movements for a defence of their 'national' culture, on the one hand, and of immigrant communities for inclusion and the retention of their cultural identities, on the other. At this meso- and micro-level, culture takes the role hitherto occupied by political ideologies; in David Steigerwald's words, class consciousness has been replaced by cultural consciousness—except that, as shown in Chap. 1, according to Google's n-grams, the word 'consciousness' has long fallen out of use and been replaced by 'culture'!

Thus, as 'culture' has been adopted by the mainstream, and alternative political philosophies have been left by the wayside, it has in fact absorbed practically all the markers of chosen ideologies. Today, as an umbrella form of ethnicism, and a euphemistic term for religion—or, rather, to be consistent, for *religionism*—culture provides a formidably legitimized, bounteous form of identification and blueprints for action for, on the one hand, groups who feel that they are losing their place and their privileges in a postcolonial, neo-liberal world and, on the other, groups who see an opportunity to throw off historical shackles and carve out their new location in a purportedly borderless global society. For this scale of micro- and meso-cultural interactions which in our contemporary world takes the form of intense strife over the comparative economic and political advantages of every group declaring that it has privileges, rights or entitlements, I suggest the term *laoculture* (taking its root from the Greek word *laos*, meaning "the people of the nation").

This umbrella form should more properly be termed "cultural politics", since it is evident that "managing cultural diversity", "defending national culture" and "integrating immigrant communities" (among other frequent expressions appearing in the media) all have less to do with culture than with political structures. Thus, it becomes inescapably evident that a new theory of the national state, if this is to be the principal form of governance (in spite of claims to the contrary by the IMF and neo-liberal thinkers), is what is needed in a contemporary world in which the media insist so incessantly on the goals of multiculturalism. In other words, no amount of "managing diversity" will solve the current turmoil of retrenchment into "manufactured parochialisms". On the contrary, it will exacerbate tensions as, inevitably, any group securing advantages will deepen the feeling of injustice of other groups.

A very different heuristic and search for new theoretical avenues and strategies is required for the *construction of a new global society*, a *cosmopolitics* which, although also based on national power relationships, involves transactions that have another layer of economic and political intentions, trade-offs, diplomatic manoeuvrings and expected and unexpected outcomes. Most of the materials and debates described in this book pertain to this scale of global interactions. It is the intensity of such interactions in reference to culture and its associated concepts that has created an international transaction space in which different actors are proposing different models of a global society. In such a densely transactional world, with multimedia and the Internet offering emergent phenomena and movements related to environmental and ecological awareness, women's changing participation, inter-religious dialogues and new eco-spiritual horizons, it is no longer a question of negotiating cultures on the ground, as the obsolete metaphor of the world as a "mosaic of cultures" had suggested. Rather, it is a question of building a new cosmopolitical vision for the world.

This goes beyond political models for the world based on power—unipolar, multipolar or multiregional—towards what anthropology originally termed 'cosmovisions'. These are cognitive structures that allow individuals to understand their presence in the physical and social world so that they can live in it and interweave with it in certain ways. As 'wayfarers', according to Tim Ingold, but also as creators and cultivators of flows of meanings that make it possible for human beings to analyse these worlds and to cooperate in navigating them. To move through the rapids in a common journey whose precise routes to the future we cannot fully comprehend as yet. A journey that, at present, is being recast, as the solid framework that the world had moved in since the eighteenth century shimmers and liquefies, opening up, as in the film *Inception*, to a new multiverse, now under construction as the Anthropocene. My suggested term for this scale of cultural transactions is *ontoculture*.

Embedded deeply in this cosmoculture is the philosophical and spiritual debate about the ontology of human nature, constructed life-worlds and environmental sustainability. It is no coincidence that a broad debate was held by anthropologists on whether culture is an ontology. How are we humans to conceptualize human nature in a way that is compatible with a biosphere under threat? Is there, in fact, such a thing as a human nature? The new discoveries of evolutionary science, genetics and epigenetics, neuroscience, the mind–body connection and asking what happens when species meet are contributing to the shimmering of our own cultural creation: our reflexive image of ourselves. This image is, of course, coloured by the animistic, totemic, naturalistic or analogic conceptions of human nature in different cultures, as Philippe Descola has demonstrated (Descola 2013). Or is it, as Lewens argues, a mere illusion (Lewens 2015)?

Just as we humans have been dethroned from the centre of the universe, the pivotal sphere of our planetary system, the centre of a flat Earth, the centre of 'Creation', the core explored geographical world, and are now the reigning lords over a passive and acquiescent Earth system, the major ongoing transaction is how to recombine two opposing forces. The first is that of humans as a geological force in the history of the Earth system; the second, the extreme weakness of cosmocultural new thinking.

The task I set myself in this book was to explore the threads of cultural transactions that will drive the discussion of laocultures and cosmoculture in the next few years. This means going through the Sustainability Transition towards making the Anthropocene a better template to interact with the Earth system. A global ethic should begin by stating simply that every single individual on Earth is part of humanity and must consciously recognize and acknowledge this fact. Many recent policy attempts have ended up bloating cultural policies when the narrow institutional definition of cultural policy cannot deal with the complex and variegated issues currently influencing sustainability, conviviability and development. Cultural politics pertaining to trade and investments, to pluralistic political systems, to transmission of media content of different cultures, and to the cultural content in cyberspace (to name only the major ones) have overridden this narrow institutional definition.

Keeping cultures alive means sustaining the difference between life and living. To make culture relevant again to the emergent Earth system we need new theories to provide the basis for progress, new narrativity and mythologies to fill common human imaginaries, and new practices in cultural rhizomes for negotiation and convergence. Through the development of a cosmoculture we can reconceptualize the human place in the world as embedded in the Earth system.

As Donna Haraway argues, "no one exists in a culture of no culture" (Haraway 2003: 246). Instead, she explains that we exist in a sea of powerful stories out of which there are "many possible structures, not to mention contents, of narration" (ibid.). And so she thinks our hope and work towards a different, more livable world, is about "changing the stories, in both material and semiotic sense" (ibid.). The loyalties, belongings, rights and other attributes of our common heritage come after the fact. This may make it easier to negotiate and arbitrate many, many facets of humanity's future. On the basis of this assumption, history will no longer belong to those who first set down in writing the stories of their regions and people. It will belong, *ipso facto*, to everyone on the planet. As such, it will slowly begin to fill up, as an archipelago that becomes a continent. All the stories of all peoples, all cultural heritage, as the international conventions spell it out, are part of our heritage and our lineage—an *Earthling identity*, if you will. Already, in some languages, terms to this effect are appearing, for example, 'terricola' or 'terrígena' in Spanish and 'térrien' in French. That identity, its dependence on the relationship between culture and nature, and the seismic shifts that have occurred in that relationship during the Anthropocene era are the focus of the next chapter.

References

Amit, V. & Rapaport, N. (2002). *The trouble with community: anthropological reflections on movement, identity and collectivity*. London: Pluto Press.
Anheier, H. & Isar, Y. R. (Eds.) (2007). *Cultures and globalization: conflicts and tensions*. Los Angeles: Sage Publications.
Anheier, H. & Isar, Y. R. (Eds.) (2008). *Cultures and globalization: the cultural economy*. Los Angeles: Sage Publications.
Anheier, H. & Isar, Y. R. (Eds.) (2010). *Cultures and globalization: cultural expression, creativity and innovation*. Los Angeles: Sage Publications.
Anheier, H. & Isar, Y. R. (Eds.) (2011). *Cultures and globalization: heritage, memory and identity*. London: Sage Publications.
Anheier, H. & Isar, Y. R. (Eds.) (2012). *Cultures and globalization: cities, cultural policy and governance*. Los Angeles: Sage Publications.
Arizpe, L. & Aranda, J. (1984). "The comparative advantages of women's disadvantages: women workers in the strawberry agroindustry in agribusiness in Mexico", *Signs* (University of Chicago). Republished 1986 in: Safa, H. and E. Leacock (eds.) *Women's Work*. Nueva York: Praeger. Republished 1988 in: *On work: historical, comparative and theoretical approaches*. Oxford: Basil Blackwell: 633–651.
Arizpe, L. (1996). Social Deforestation in the Lacandon Rainforest. Ann Arbor: University of Michigan Press.
Arizpe, L. (2004). Mexicanidad, migración y globalización. In: L. Arizpe (Ed.), *Los retos culturales de México frente a la globalización* (pp. 19–48). Mexico: M.A. Porrúa.
Arizpe, L. (2014a). The dialogue of cultures and civilizations. In: L. Arizpe, *Lourdes Arizpe: pioneer in Mexican anthropology* (pp. 21–30). Heidelberg: Springer.
Arizpe, L. (2014b). On the cultural and social sustainability of world development. In: L. Arizpe, *Lourdes Arizpe: pioneer in Mexican anthropology* (pp. 31–43). Heidelberg: Springer.
Arizpe, L. (2015a). The genealogy of cultural heritage. In: L. Arizpe, *Culture, diversity and heritage: major studies* (pp. 101–117). Heidelberg: Springer.
Arizpe, L. (2015b). How to Reconceptualize Intangible Cultural Heritage, In: L. Arizpe, *Culture, diversity and heritage: major studies* (pp. 95–99). Heidelberg: Springer.
Augé, M. (2017). L'avenir des terriens: Fin de la préhistoire de l'humanité comme société planétaire. Paris: Albin Michel.
Aunger, R. (2000). *Darwinizing culture: the status of memetics as a science*. Oxford: Oxford University Press.
Beck, U. (2006) *La sociedad del riesgo: hacia una nueva modernidad*. Barcelona: Paidós Ibérica.
Bhabha, H. K. (2007). Ethics and aesthetics of globalism: a post-colonial perspective. In: A. P. Ribeiro (Ed.), *The urgency of theory* (pp. 1–20). Manchester: Carcanet Press and the Gulbenkian Foundation.
Bøås, M. & McNeill, D. (Eds.) (2004). *Global institutions and development: framing the world?* Abingdon: Routledge.

Bossman, J. (2016) *Top 9 ethical issues in artificial intelligence*. World Economic Forum website, accessed 21 October 2016; at: https://www.weforum.org/agenda/2016/10/top-10-ethical-issues-in-artificial-intelligence/.

Buzan, B. (1993). From international system to international society: structural realism and regime theory meet the English school. *International Organization*, 47(3), 327–352.

Capra, F. (1996). *The web of life: a new scientific understanding of living systems*. New York: Anchor Books.

Carr, N. (2010). *The Shallows: What the Internet is doing to our brains*. London: W. W. Norton and Company.

Christakis, N. A. (2015). "Human culture as the first artificial intelligence". Available on www.edge.org at: https://www.edge.org/response-detail/26147.

Cohen, J., Okin, S. M. et al. (1999). *Is multiculturalism bad for women?* Princeton: Princeton University Press.

Cyrulnik, B., Bustany, P., Oughourlian, J-M., Andre, C., Janssen, T. & Van Eersel, P. (2012). *Votre cerveau n'a pas fini de vous étonner*. Paris: Editions Albin Michel.

Cyrulnik, B. (2001). L'Ensorcellement du monde. Paris: Odile Jacob.

Development Alternatives for Women in a New Era. Dawnnet.org. Accessed December 5, 2017.

Descola, P. (2013). *Beyond nature and culture*. Trans. J. Lloyd. Chicago: University of Chicago Press.

Descola, P. (2016). "Avis de Tempête sur le Climat: l'Anthropocène, Par-Delà la Nature et la Culture". Conference in Radio France Culture. (Accessed 15 September 2016.)

Eagleton, T. (2000). *The idea of culture*. Oxford: Blackwell.

Englund, H. & Leach, J. (2000). Ethnography and meta-narratives of modernity. *Current Anthropology*, 41(2), 225–248.

Eriksen, T. H. (2000). Introduction. In: T. H. Eriksen (Ed.), *Globalisation: studies in anthropology* (pp. 1–17). London: Pluto Press.

Escobar, A. (1995). *Encountering development: the making and unmaking of the Third World*. Princeton: Princeton University Press.

Ferguson, J. (1994). *The anti-politics machine: "development", depoliticization, and bureaucratic power in Lesotho*. Cambridge: Cambridge University Press.

Fischer, M. (2007). Culture and cultural analysis as experimental systems. *Cultural Anthropology*, 22(1), 1–64.

Fischer, M. M. J. (2009). Anthropological futures. Durham, NC: Duke University Press.

Fukuda-Parr, S. (2016). "From the Millennium Development Goals to the Sustainable Development Goals: shifts in purpose, concept, and politics of global goal setting for development". *Gender and Development*, 24(1), 43–52. Available at: http://www.tandfonline.com/doi/pdf/10.1080/13552074.2016.1145895, accessed 1 September 2017.

García Canclini, N. (2010). *La Sociedad sin relato: Antropología y estética de la inminencia*. Buenos Aries: Katz Editores.

Gilroy, P. (2007). Multiculture and conviviality in post-colonial Europe. In A. P. Ribeiro (Ed.), *The urgency of theory* (pp. 125–142). Manchester: Carcanet Press and the Gulbenkian Foundation.

Haddow, D. (2017) "Meme warfare: how the power of mass replication has poisoned the US election." *The Guardian*, 4 November [Online]. Available at: https://www.theguardian.com/us-news/2016/nov/04/political-memes-2016-election-hillary-clinton-donald-trump, accessed: 18 October 2017.

Hall, P. (1989). *The political power of economic ideas: Keynesianism across nations*. Princeton: Princeton University Press.

Hannerz, U. (2002). Notes on the global ecumene. In: J. X. Inda & R. Rosaldo (Eds.), *The anthropology of globalization: a reader* (pp. 37–45). Oxford: Blackwell.

Hannerz, U. (2006). Afterthoughts. In: A. Gingrich & M. Banks (Eds.), *Neo-nationalism in Europe and beyond: perspectives from social anthropology* (pp. 271–282). New York: Berghahn Books.
Haraway, D. J. (2003). *The Haraway reader*. New York: Routledge.
Hentz, J. (2004). The power of ideas: across the constructivist/realist divide. In: M. Bøås & D. McNeill (Eds.), *Global institutions and development: framing the world?* (pp. 190–199). Abingdon: Routledge.
Illouz, E. (2017). Interview "Le opulisme emotionnel menace la démocratie". *Le Monde, 26* July 2017: 24.
Inda, J. X. & Rosaldo, R. (Eds.) (2002). *The anthropology of globalization: a reader*. Oxford: Blackwell.
Iwabuchi, K. (2008). Symptomatic transformations: Japan in the media and cultural globalization. In: R. T. Segers (Ed.), *A new Japan for the twenty-first century: an inside overview of current fundamental changes and problems* (pp. 126–140). Abingdon: Routledge.
Jasanoff, S. (2004). The idiom of co-production. In: S. Jasanoff (Ed.), *States of knowledge: the co-production of science and the social order* (pp. 1–12). London: Routledge.
Kahn, J. S. (2001). *Culture, multiculture, postculture*. London: Sage Publications.
Klotz, A. (1995). Norm reconstituting interests: global racial equality and United Nations sanctions against South Africa. *International Organization*, 49, 451–478.
Krasner, S. D. (1985). *Structural conflict: the Third World against global liberalism*. Berkeley: University of California Press.
Kroker, A. & Kroker, M. (2013). Critical digital studies: a reader. Toronto: University of Toronto Press.
Kuper, A. (2003). *Culture: the Anthropologists' Account*. Cambridge, MA: Harvard University Press.
Lévy, P. (1997). *Cyberculture. Rapport au Conseil de l'Europe dans le cadre du projet "Nouvelles technologie: coopération culturelle et communication"*. Paris: Odile Jacob. [English version (2001): Cyberculture: electronic mediations. Minneapolis, MN: University of Minnesota Press.]
Lewellen, T. (1999). *The anthropology of globalization: cultural anthropology enters the 21st century*. New York: Bergin and Garvey.
Lewens, T. (2015). Human nature: the very idea. *Philosophy and Technology*, 25, 459–474.
Marchand, M. & Runyan, A. S. (2010). *Gender and global restructuring: sightings, sites and resistances*. London: Routledge.
Maturana, H. & Varela F. (1987). *The tree of knowledge*. Boston: Shambhala.
McClennen, S. A. (2017). "Forget fake news—alt-right memes could do more damage to democracy." *Salon*, 7 August [Online]. Available at: https://www.salon.com/2017/07/08/forget-fake-news-alt-right-memes-could-do-more-damage-to-democracy/, accessed: 18 October 2017.
Milner, R. M. (2016) *The world made meme: public conversations and participatory media*. Cambridge, MA: MIT Press.
Moscovici, S. (1984). The Phenomenon of Social Representations. In: R. M. Farr & S. Moscovici (Eds.), *Social Representations*. Cambridge: Cambridge University.
Nora, P. (2011). Foreword. In: H. Anheier & Y. R. Isar (Eds.), *Cultures and globalization: heritage, memory and identity* (pp. ix–xii). London: Sage Publications.
Nussbaum, M. C. (1999). *Sex and Social Justice*. Oxford and New York: Oxford University Press.
Nussbaum, M. C. (2000). *Women and human development: the capabilities approach*. Cambridge and New York: Cambridge University Press.
Oswald Spring, U. & Brauch, H. G. (2009a) A HUGE Gender Security Approach: Towards Human, Gender, and Environmental Security. In: H. G. Brauch et al. (Eds.), *Facing Global Environmental Change*. Berlin and Heidelberg: Springer (pp. 1157–1181).

Oswald Spring, Ú. (2009b). “Social Vulnerability and geopolitical conficts fue to socio-environmental Migration in Mexico”. In: Oswald Spring, Ú. & Brauch, H. G. (Eds.), *Climate Change: Global Risks, Challenges and Decisions* (Mexico: National University of Mexico).

Oswald Spring, Ú. (2017). “Reduccion de riesgos por desastres y resiliencia social: superando la discriminación, la vulnerabilidad social y los desastres en el mundo y México”. Mimeo.

Pieterse, J. N. (2004). *Globalization and culture: global mélange.* New York: Rowman and Littlefield.

Pratt, A. C. (2007). The state of the cultural economy: the rise of the cultural economy and the challenges to cultural policy making. In: A. P. Ribeiro (Ed.), *The urgency of theory* (pp. 166–190). Manchester: Carcanet Press and the Gulbenkian Foundation.

Pratt, M. L. (2010). *Imperial eyes: travel literature and transculturation* (2nd edn). London and New York: Routledge.

Putnam, R. D. (1995). Bowling alone: America’s declining social capital. *Journal of Democracy*, 6 (1), 65–78.

Robertson, R. (1992). *Globalization social theory and global culture*. London: Sage.

Sahlins, M. (1994). A brief cultural history of “culture”. Unpublished paper presented to the World Commission on Culture and Development.

Sassen, S. (1999). “Culture beyond gender” in: Cohen, J. & Okin, S. M., *Is Multiculturalism Bad for Women?* Princeton, Princeton University Press.

Science and Technology Select Committee (2016). Report of the Science and Technology Select Committee of the House of Commons on Robotics and Artificial Intelligence 2016, available at: https://publications.parliament.uk/pa/cm201617/cmselect/cmsctech/145/14502.htm.

Sen, A. (1999). *Development as freedom.* New York: Oxford University Press.

Serrano Oswald, S. E. (2016). “Social resilience and intangible cultural heritage: a mutually fertilizing potential seen in a case study in Mexico”. In: Ú. Oswald Spring, H. G. Brauch, S. E. Serrano Oswald, J. Bennett, (Eds.), *Regional ecological challenges for peace in Africa, the Middle East, Latin America and Asia Pacific*. Cham: Springer (pp. 57–90). Available online at: https://link.springer.com/content/pdf/10.1007%2F978-3-319-30560-8_4.pdf, accessed 24 October 2017.

Steigerwald, D. (2004). *Culture’s vanities: the paradox of cultural diversity in a global world.* Lanham, MD: Rowman & Littlefield.

Stengers, I. (2005). The cosmopolitical proposal. In: B. Latour & P. Weibel (Eds.), *Making things public: atmospheres of democracy* (pp. 994–1003). Cambridge, MA: MIT Press.

Strathern, M. (1991). *Partial connections*. Savage, MD: Rowman and Littlefield.

Streeten, P. (1994). Globalization. UNDP working paper.

Sudjik, Olivia. 2017. “Virtual anxiety: The disturbing new reality of life online”. In: *CNN Style* on edition.cnn.com. Available at: http://edition.cnn.com/style/article/olivia-sudjic-disturbing-new-reality-of-life-online/index.html.

UNESCO (2009a). *Towards a new cultural policy framework: global synthesis of papers commissioned by UNESCO.* Paris: UNESCO.

UNESCO (2009b). *UNESCO world report: investing in cultural diversity and intercultural dialogue*. Paris: UNESCO.

UNESCO (2016). *Roadmap: the rapprochement of cultures*. Paris: UNESCO.

Vertovec, S. (2011). The cultural politics of nation and migration. *Annual Review of Anthropology*, 40, 241–256.

Werbner, P. (2008). *Anthropology and the New Cosmopolitanism: rooted, feminist and vernacular perspectives*. Oxford: Berg.

Williams, R. (1958). *Culture and society, 1780–1950.* London: Chatto & Windus.

Culture and Global Change: Social Perceptions of Deforestation in the Lacandon Rainforest, by Lourdes Arizpe, Fernanda Paz and Margarita Velazquez, published by University of Michigan Press in English, 1966. Also published in Spanish by the National University of Mexico, 1995

Chapter 9
Culture and the Anthropocene

There is no border where evolution ends and history begins, where genes stop and environment takes up, where culture rules and nature submits, or vice versa (Donna Haraway 2003: 2).

The current debate on establishing a scientific and intellectual framework called the Anthropocene to examine the impact of humans as a geological force in transforming Earth has opened up a broader scenario in terms of both time and space. This chapter deals with the discussions taking place at present among geologists, biologists and other specialized scientists, with contributions from archaeologists and social scientists, on defining this new epoch. As a social anthropologist, my interest is the way in which a new geostory is generating ideas and intellectual exchanges to form a bridge between the "two cultures" of science and social science. I believe that an intentional narrative must be built into this geostory, in which all world actors may participate, to direct the divergent human actions towards reconstituting the Earth, and us as one of its inhabitant species. There is urgent need for interdisciplinary collaboration that may help to concentrate intellectual, political and cultural efforts on creating a sustainable future for all.

How can the polysemic concept of culture contribute to this unfolding scientific and intellectual revolution? In the previous chapter I showed how the substantive issues and heuristics relating to culture taken up in the last two decades have helped in understanding how culture has become a predominant concept in thinking about sociality, human rights, gender equity and political management in a post-postmodern world. Here, I will explore the evolutionary cultural processes that may help or hinder our understanding of current world phenomena.

9.1 From Human Dimensions of Global Change to the Anthropocene

The announcement by Paul Crutzen and Eugene Stoermer in 2000 of their formulation of the term 'Anthropocene' was the culmination of the evolution of ideas launched by the International Council of Scientific Unions (ICSU, now the World

L. Arizpe Schlosser, *Culture, International Transactions and the Anthropocene*,
The Anthropocene: Politik—Economics—Society—Science 17,
https://doi.org/10.1007/978-3-642-41602-6_9

Science Organization) through their International Geosphere–Biosphere Programme (IGBP) in 1986, and by the International Social Science Council (ISSC)'s own programme of research, Human Dimensions of Global Environmental Change (HDGEC), established in 1988. As President of the International Union of Anthropological and Ethnological Sciences and a member of the ISSC Executive Committee, I was then invited to participate in the Standing Committee on this new social science perspective of the human dimensions of global change (see Sect. 9.1.1).

Roads and cattle-ranching show the spread of deforestation in an aerial photograph of the Lacandon rainforest in Marqués de Comillas, Chiapas, Mexico, 1993. *Source* Photo by Lourdes Arizpe

Questions about the effect that humans were having on the earth had been raised since the middle of the twentieth century. In 1955 one of the important events in research related to the environment, a conference on "Man's Role in Changing the Face of the Earth", was held in Princeton University and set the research guidelines for a new field of human and environmental interactions (Thomas 1956). The conference established the groundwork for pioneering anthropological and archaeological studies that consolidated this interdisciplinary field. Before that time, research was generally framed in terms of "Man and His Habitat", given that most small-scale societies studied by anthropologists had close interactions with bio-physical and ecological processes, with their cultures reflecting this embeddedness.

From the mid-fifties, however, in parallel with studies in other social sciences, research expanded to show the range and diversity of human activities affecting the ecosystems of the Earth (Turner et al. 1993). The IGBP established the collaborative and deliberative work of natural scientists of many countries to provide comparative scientific methods and measurements on environmental global change.

One early issue was how to define the processes being studied in 'global change'. García (1986) proposed a three-tiered structure linking micro- to macro-processes. To analyse the dynamics of such processes, he stressed that the property of the system is not found in its components but in the relationships between them. Decision-making then becomes important to understand, for example, the emergence of famines. Researchers led by Gilberto Gallopin at the Bariloche Foundation gave a different perspective on global reality, proposing the concept of "global impoverishment" to encompass both ecological and economic impoverishment as the central process of global change (Gallopin et al. 1989; Leff 1986). Thus, by the end of the 1980s, two very different meanings of 'global change' existed, promoted by different constituencies in the social and natural sciences (Price 1989). One could be described as anthropocentric, emphasizing the interactions between people and their institutions, primarily at scales extending to decades. The other could be described as geocentric, emphasizing the interacting processes of the Earth's atmosphere, biosphere, geosphere and hydrosphere, typically over far longer timescales.

Social scientists were also considering the nature of social process in relation to the environmental global change. Roberta Miller of the US National Science Foundation noted the need for social science research programmes that should not only "feed into the physical and natural science activities" (i.e. the emerging IGBP) but also "be concerned with those elements of the global change research agenda that are purely social and economic but that ultimately are as powerful determinants of environmental change as ongoing physical and biological processes" (Miller 1989: 86–87).

I remember thinking, when Crutzen and Stoermer first coined the term 'Anthropocene', that this name could perhaps be the answer to enabling the natural science and social science perspectives to develop jointly. Natural scientists were so steeped in their discoveries on the geosphere and biosphere that they found it difficult to insert the concerns brought up by human perspectives of geography, anthropology and psychology. In particular, my mind went back to a joint ICSU–ISSC meeting in 2002 where hardly any mention had been made of human actions in describing global change. Near the end of the session, I asked Thomas Roswall, then President of ICSU and director of the IGBP, if he could he tell me more precisely what percentage of the phenomena that were being described in the various research projects as global change could be attributed to anthropogenic actions. He answered, very forthrightly, "Eighty per cent". Somehow, I felt that the veiled wall that had separated social scientists from biophysical scientists had suddenly became more transparent.

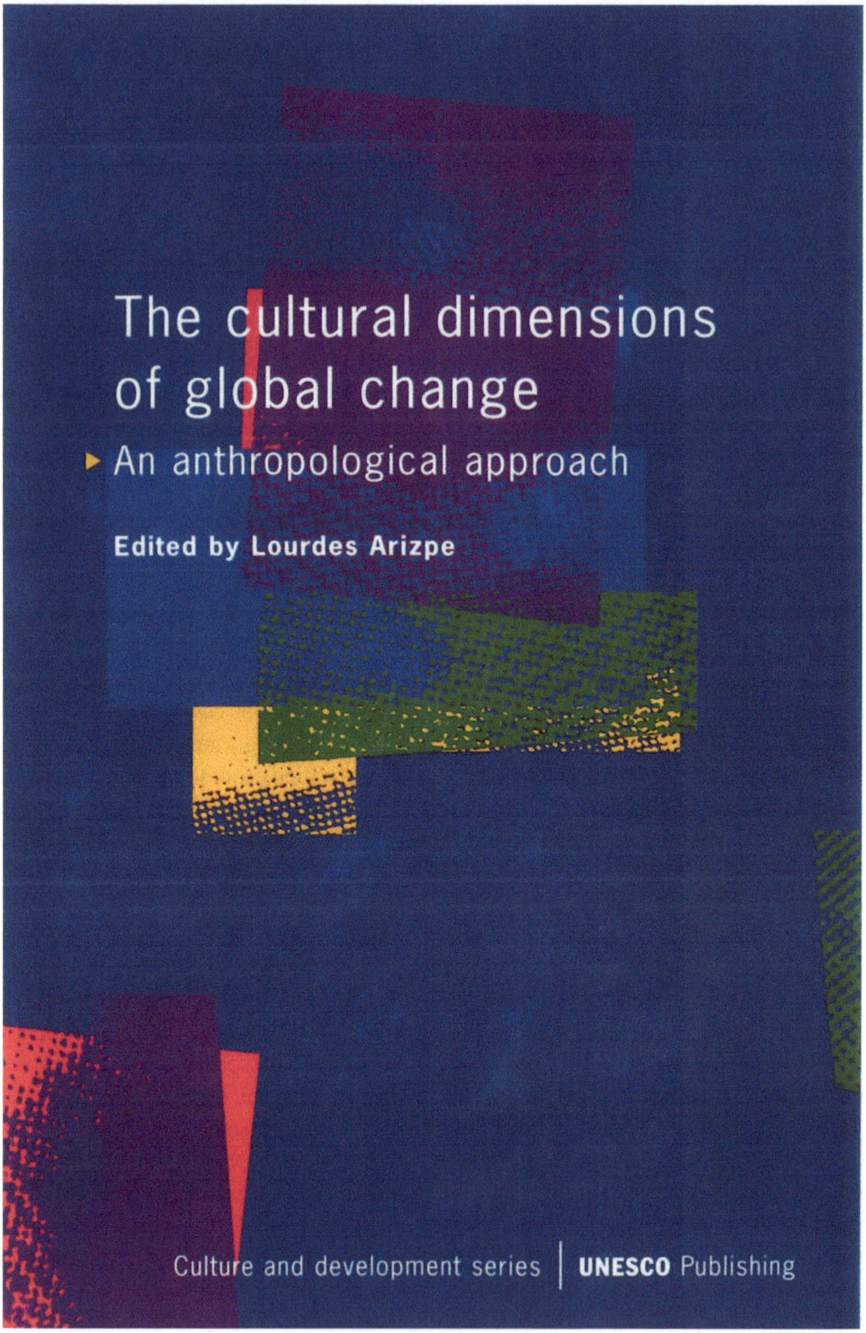

The Cultural Dimensions of Global Change: an Anthropological Approach edited by Lourdes Arizpe, UNESCO publication, 1996

9.1.1 The International Social Science Council Human Dimensions Programme

In 1987, the ISSC joined with the Institute for International Assistance and Solidarity (IFIAS) and the United Nations University (UNU) to develop a Human Response to Global Change Programme, which soon became the Human Dimensions of Global Environmental Change Programme (HDGECP). The following year, in December, the ISSC established a Standing Committee on the Human Dimensions of Global Environmental Change, in which I participated as President of the International Union of Anthropological and Ethnological Sciences. The Standing Committee produced *A Framework for Research on the Human Dimensions of Global Environmental Change* (Jacobson/Price 1990), which identified seven broad areas in which research should be carried out:

- social dimensions of resource use;
- perception and assessment of global environmental conditions and change;
- impacts of local, national, and international social, economic and political structures and institutions;
- land use;
- energy conversion and consumption;
- industrial growth; and
- environmental security and sustainable development.

The first three topics encompass the central issues involved in understanding HDGEC (Jacobson 1992). The next three themes were recognized as major proximate causes of global change and jointly responsible for the majority of greenhouse gas emissions. Understanding the dynamics of these activities was considered crucial to fully understanding the dynamics of global change. The framework explicitly linked the last topic, environmental security and sustainable development, to the report of the World Commission on Environment and Development; this was seen as embodying the normative tradition in the social science, namely, exploring the values involved in human actions.

It was stressed that these investigations should involve description, analysis and prescription and would be an essential component of social science research on global environmental change. The costs and benefits of various strategies to adapt to global change or to mitigate its effects should be explored, and norms for dealing with issues of international and intergenerational equity should also be taken into account.

Following the creation of the ISSC's Human Dimensions of Global (Environmental) Change Programme (HDP), many member associations formed committees, study groups and research networks to study global environmental change and to increase the number of exchanges with natural scientists. Numerous interdisciplinary meetings and research projects were held to develop vocabularies, theories and heuristics on global environmental change. I, with Wolfgang Lutz and Robert Constanza, argued that, in such an interdisciplinary effort, "To achieve a sustainable pattern of resource use and population we must understand and control the interactions of population and per capita resources as mediated by technology, culture and values" (Arizpe et al. 1992: 61).

In the last two decades countless international meetings have taken up these fields of research with the aim of rebuilding policies towards development and sustainability. A major achievement in this direction has been the Handbook on Sustainability Transition and Sustainable Peace in which authors and editors discuss how a process contributing to a transition toward sustainability with a global transformation of the energy system and the economy through a decarbonization process could avert the negative consequences of current risks (Brauch et al. 2016). Such risks include the probability of increasing resource conflicts and the possibility of climate-induced violent outcomes (ibid.). Given this context, the new term of Anthropocene is focusing attention on the specific processes that have given rise to the anthropogenic present.

9.2 Defining the Anthropocene

The notion of the Anthropocene, as happens with many scientific terms, evolved over many decades, as it was used, variously and simultaneously, by scientists beginning to grasp processes that were beyond the phenomena being observed. Antonio Stoppani, an Italian geologist, tried to define an "anthropozoic era" in 1873. Just over half a century later, in 1926, the Russian geochemist Vladimir Vernadsky proposed three terms to refer to the biological, ecological and philosophical domains of Earth: the geosphere, the biosphere and the noosphere. He wrote: "Chemically, the face of our planet, the biosphere, is being sharply changed by man, consciously, and even more so, unconsciously. The aerial envelope of the land as well as its natural waters are changed both physically and chemically by man" (Vernadsky 1945: 9). A few years later Vernadsky was invited by Édouard le Roy and Teilhard de Chardin to jointly publish a more developed version of this tripartite nomenclature, and the terms became part of the Western canon of science.

To return to the genealogy of the term 'Anthropocene', it was proposed by Paul Crutzen, a Nobel prizewinner in chemistry at the Cuernavaca meeting in 2000, but he then discovered that Eugene F. Stoermer had already mentioned it a few years earlier in one of his research papers. As a result, Crutzen invited Stoermer to co-author a short account of the Anthropocene for one of the IGBP Newsletters in 2000 (Crutzen/Stoermer 2000). Crutzen considered it appropriate to assign the term 'Anthropocene' to the present geological epoch, which has been dominated by humans in many ways (Crutzen 2002: 23). He proposed that it should follow the Pleistocene, the first epoch of the Quaternary, which began 2.5 million years ago, and the Holocene, which is judged to have started 11,500 years ago. Crutzen's concept was welcomed by an editorial in *Nature* in 2011 ("The Human Epoch") as a proposal that could "encourage a mindset that will be important not only to fully understand the transformation now occurring but to take action to control it" (*Nature* 2011: 254).

Today, as stated by Jan Zalasciewicz and his colleagues in the Anthropocene Working Group (AWG), anthropogenic changes to the Earth's climate, land, oceans and biosphere are now so great and so rapid that the concept of a new geological epoch defined by the action of humans—the Anthropocene—is widely and seriously debated (Zalasciewicz et al. 2011: 835). On current evidence, the Anthropocene seems to show global change consistent with the suggestion that an epoch-scale boundary has been crossed within the last two centuries (ibid.). According to the AWG, the major factors that account for the beginning of human-induced changes in the Earth system are tools, the demise of megafauna, agriculture, clearance of forests leading to elevated carbon dioxide levels in the atmosphere, and permanent settlements in villages and towns (ibid.: 836). While the concept of the Anthropocene continues to be discussed in terms of its formalization as a new epoch with the International Chronostratigraphic Chart, it is considered that a sufficient number of relevant studies have already been published for an interim assessment of this question (Zalasiewicz et al. 2015: 197).

Unlike a geological stratigraphic time boundary, which, according to the AWG needs to be as far as possible "singular and globally synchronous", evolutionary processes of ideas and concepts are taken forward in time within a mesh of concepts in which particular concepts slowly emerges as the central "Spirit of the Age" (ibid.). The AWG calls attention to the paramount importance of the fact that the geostratigraphic and geoatmospheric indicators of a new epoch may correlate with the evolution of the statistical relevance of particular ideas and concepts. My contention is that the rise of the idea of 'culture' to encompass many others that were used in both science and public discourse has had a parallel evolution with the processes that are mentioned as characteristic of the Anthropocene.

In establishing the initial boundary of the Anthropocene, the AWG has considered the means by which this widely used term may acquire a stable and consistent meaning with regard to its inception and they have therefore proposed a specific boundary level (ibid.). Taking into consideration the dual hierarchy of time units in Earth history, as formalized by the International Chronostratigraphic Chart, they speak of the history of the Anthropocene within the Anthropocene Epoch but also of its material record presented within the deposits of an Anthropocene Series, although they question the need for the second, given its brevity.

In the discussion on the Anthropocene there are many dissenting views. Walker and others have proposed that it should be designated as a unit of lesser rank than an epoch: that is, an age or a stage, a subdivision of the Holocene (Walker et al. 2015). The use of the suffix—*cene* is also questioned since, in the Geological Time Scale, it has hitherto been reserved for units of epoch status such as Eocene and Oligocene. Others cite the brevity of the Anthropocene as making it a premature concept that needs to be developed, with the added consideration that so many more anthropogenic impacts will be affecting the Earth system (Wolff 2014). Some scientists refer to the low utility of such a concept in terms of geological, geoatmospheric or biosystemic time or series. And yet, human actions and their probability of causing irreversible damage to all those times and series do indicate the importance of creating a geobiosystemic framework that human beings can actually grasp in terms of their capacities for collective interventions.

Additionally, critics of the term 'Anthropocene' argue that numerous references such as "Medieval Period" and "Little Ice Age" are not defined as formal units of geological time (Walker et al. 2015). They conclude their comments by pointing out that a fifty-year interval is unlikely to represent more than a few centimetres in discernible strata and that human-made materials are insufficient justification for recognizing a high-status geological interval based on this type of evidence (ibid.: 206). For the Anthropocene to be designated an epoch, and hence the equivalent of the Holocene, or as an age defined as subset of the Holocene, a stratotype distinguishing it from the Holocene must be identified (International Subcommission on Stratigraphic Classification 1994; NACSN 2005; Autin/Holbrook 2012: 60).

The name proposed for the Anthropocene has also come under fire from several other authors, among them Donna Haraway who argues that when humans "can no longer figure themselves as individuals and societies of individuals", "such a transformative time … must not be named the Anthropocene" (Haraway 2016: 35). Instead she proposes "a name for an elsewhere and elsewhen that was, still is, and might yet be: the Chthulucene" (ibid.). This unpronounceable name comes from an eight-legged arachnid, implying that these 'tentacles' will allow us to tangle with "string figuring". This is the sympoietic form that she suggests for "collectively producing systems that do not have self-defined spatial or temporal boundaries. Information and control are distributed among components" (Haraway 2016: 37).

How far, indeed, is this defining perspective from the distinctive, hard boundaries that the debates on cultural diversity have produced. Such systems, in Beth Dempster's view, are 'autopoietic', 'self-producing' autonomous units "with self-defined spatial or temporal boundaries that tend to be centrally controlled, homeostatic and predictable" (Dempster 1998, cited in ibid.: 37). In her very provocative manner, Haraway then throws down the gauntlet by asking why the epochal name of the Anthropos appeared "at just the time when understandings and knowledge practices about and within symbiogenesis and sympoetics are wildly and wonderfully available and generative in … arts, sciences, and politics" (Haraway 2016: 61). Could these sympoietic systems become a major reference for the culture maelstrom of the Anthropocene? It is such innovative debates as these that are one of the main reasons why the proposal for the Anthropocene is already raising the possibility of shattering the glass ceiling of our understanding of culture.

Questions about the Anthropocene have also been formulated by authors on the left who find the 'Capitalocene' a better title for our current age. Taking up Raymond Williams's historical semantics about culture in *Keywords*, Daniel Hartley highlights the importance of culture in justifying the theoretical and political superiority of the term Capitalocene. "Strictly speaking," he states, "the Anthropocene is cultureless: it is the result of 'man' and technology, or 'man' insofar as he develops and wields technology" (Hartley 2016: 164).

In a world operating according to an "endless accumulation of capital", Jason Moore takes a different high ground, embedding the Anthropocene in the framework of 'world-ecology'. In his aptly titled article "The rise of cheap nature", he clarifies that this term does not refer to the "ecology of the world" because it "is not the ecology of Nature—with upper-case N—but the ecology of the *oikeios*: that creative, generative, and multilayered relation of life-making, of species and of environments" (Moore 2016: 79). The problem today, he argues, is not the march of the Anthropocene but the end of the Capitalocene, "not one of humanity 'overwhelming the great forces of nature' … but rather the exhaustion of its Cheap Nature strategy" (ibid.: 113). In the Capitalocene, according to Elmar Altvater, "nature has been transformed into a capital asset" (Altvater 2016: 145).

Bonneuil and Fressoz, in their book *The Shock of the Anthropocene*, provide evidence that the last three centuries have been characterized by an extraordinary accumulation of capital, which grew by a factor of 134 between 1700 and 2008 (Bonneuil/Fressoz 2016: 222). They propose that, instead of looking at the great universals of 'capital' or "human species", we should investigate the historical metabolisms of the capitalist "world systems" of the last quarter-millennium and their effects on the Earth system in a way that illuminates both the history of capitalism and the genesis of the Anthropocene (ibid.: 223).

A key consideration in the debate is the idea that the Anthropocene should be seen as an event rather than a thing (Bonneuil/Fressoz 2016: xiii). Bonneuil and Fressoz insist that it is not enough to measure and chronicle the natural history of interactions between the human species and the Earth system. Instead, they advocate "deconstructing the official account in its managerial and non-conflictual variants and forging new narratives for the Anthropocene and thus new imaginaries. Rethinking the past to open up the future" (ibid.: xiii).

On the basis of the evidence given above, the AWG considers that "the Anthropocene already has a robust geological basis that is in widespread use and indeed is becoming a central, integrating concept in the consideration of global change" (Zalasiewicz et al. 2015: 198–9). They consider—rightly, it seems to me—that an interim definition of duration for the Anthropocene as a unit, regardless of formal status, would be useful for clarity and communication. The definition that they have now proposed to the International Geological Congress is one based on a new interval in geological history, as measured in strata studied by geologists, and marked by the introduction of radioactive traces caused by the nuclear tests of the 1950s.

A second definition of the Anthropocene comes from Earth system science, which draws expertise from many different disciplines: climatology, global ecology, geochemistry, atmospheric chemistry, oceanography, geology and more. As noted by Jan Zalasiewicz, chair of the AWG, "the Anthropocene is not about being able to detect human influence in stratigraphy, but reflects a change in the Earth system" (cited in Hamilton et al. 2015: 2). Hamilton and Grinevald put it similarly: "We are no longer talking about the spread of human influence across 'the face of the Earth' (the geographical and ecological approaches that predominated in the twentieth century) but of a shift in the total system" (cited in ibid.: 3).

A third definition of the Anthropocene, which is the frame of reference for the analysis in this book, describes what James Syvitski calls "the cumulative impact of civilisation" (cited in ibid.). It encompasses changes in land use and cultivation, industrial metabolism, urbanization, production of greenhouse gases and so on, all driven by human beings. It is worth noting that the description of the main elements in this definition are almost exactly the same as were identified by the HDGEC programme in our 1992 publication. An important difference is that Syvitski places these multiple processes within a framework focused on the cumulative impact of human actions across long-term time spans.

The historical and social sciences have always tried to establish boundaries and thresholds to contain the process under scrutiny. Importantly, such time and space classifications are vital for political history. Many, in spite of scholarly diffidence, become the accepted norm: for example, 'Prehistory' as contrasted with 'History'; the "Age of Discovery", whose reverse side is the "Age of Colonialism"; and so on. Processes that changed the course of events are also classified according to their intrinsic characteristics such as capitalism, modernity, postmodernity and post-colonialism. Markers in our present age are proliferating. In recent times we have been told that we live in the era of the Microelectronics Revolution, the Age of Technology, the Post-Capitalist Era, the Post-Postmodern Era, the Global Age and the Age of Extinction, to name only a few. Most of these ages are defined by one particular trend or process, which in their popular rendering are sometimes referred to as 'turns'. After the "cultural turn" of Postmodernity, in this second decade of the millennium, proposed terms are the "post-human turn", the "human turn" and, more recently, the "non-human turn".

My belief is that one of the most important effects of creating the concept of the Anthropocene—quite independent of how it is finally defined in chronostratigraphic geological time—is its strong impact in fostering interdisciplinary exchanges. This supports the AWG's remark that "such a boundary selection, the Anthropocene Working Group proposes, may open possibilities of historical fields other than Earth history (geology) to more easily engage in the emerging interdisciplinary science base of the Anthropocene" (Zalasiewicz et al. 2015: 201).

9.3 When Did the Anthropocene Begin?

In 2015, the AWG described three major proposals for the duration of the Anthropocene. The first is laid out in terms of the early human impacts on landscape, habitat modification, terrestrial biotic change and altered carbon dioxide levels related to land use changes and agriculture between 3700 and 2000 BP (Zalasiewicz et al. 2015: 198). Extinctions of large mammals and the wider consequences of the global carbon cycle, although spanning some fifty millennia of the Late Peistocene and Holocene, are also mentioned (ibid.).

The second proposal was put forward by Crutzen and Stoermer in 2000, and suggested that the Anthropocene began with the Industrial Revolution, around the beginning of the nineteenth century (Crutzen/Stoermer 2000). The third proposal focuses on more recent environmental history as a phase of enhanced population, industrial metabolism and global economic growth, termed by Steffen and his co-authors the “Great Acceleration” (Steffen et al. 2007). The threshold is set in 1945, a date which is particularly important for the discussion in this book, since the concept of culture became an emergent trend in the 1950s.

While scientific debates continue as the International Commission on Stratigraphy and other international commissions consider proposals to designate the Anthropocene as a new geological epoch, the field has become interdisciplinary, as other scientists, social scientists and humanities scholars have begun to participate in the discussions. The archaeologist Todd Braje notes that academics quickly adopted Crutzen and Stoermer’s Anthropocene in the geologic fields, as well as in palaeontological, palaeoclimatic and archaeological analysis of its chronostrata, golden spikes and boundaries (Braje 2015: 374). He reports 12,900 results on the Anthropocene in Google Scholar when accessed on 3 October 2014. Braje believes that this measure shows that the Anthropocene has become an intrinsic part of academic discourse, despite the fact that there is no agreement on when it began or how to define it (ibid.). It is remarkable to note that just two years later, at the end of September 2016, the number of citations of ‘Anthropocene’ had more than doubled, to 28,400.

As Braje notes, the complicated discussion of the Anthropocene, which will probably continue for many years, has centred on “what markers should take precedence (biotic, stratigraphic, atmospheric, etc.) and when the age of humans began” (ibid.: 376). In contrast to scientists who focus on biotic, stratigraphic or atmospheric markers, those scientists who view the human domination of the Earth systems as a process, “one that has played out over thousands of years … believe the Anthropocene boundary should mark the human that set these processes in motion” (ibid.).

Following this view, the archaeologists Bruce Smith and Melinda Zeder argue that the Anthropocene should be defined by the causes of global change rather than the effects (Smith/Zeder 2013: 11, cited in Braje 2015: 378). The Anthropocene marker would be the domestication of plants and animals between 11000 and 9000 BP. Similarly, taking into account the declines in native plant and animal biodiversity, the archaeologist Kent Lightfoot and his colleagues argue that European colonialism from about 1400 CE to the early 1800s marked a global transformation in which foreign species were taken around the world, core native species were exterminated and native habitats were disrupted (Lightfoot et al. 2013: 112, cited in Braje 2015: 378).

Braje notes that Lightfoot and his colleagues do not explicitly set a Columbian Exchange boundary date for the Anthropocene. Nevertheless, this biotic boundary may be important for the cultural analysis in the present book, since there is also a boundary to be noted in the flow of ideas, as new concepts on human nature began to be rehearsed, alongside interests and responsibilities in dealing with non-European peoples and cultures, as European intellectual and political mindsets adapted in the light of colonial expansion. The vocabularies of those mindsets, words that flattened cultural textures and placed together peoples of very different intellectual and cultural trajectories—for example, ‘Indian’ for natives, or ‘autochthonous’—are still in use in contemporary lexicons. This is one of the reasons why 'indigenous' groups the world over are changing their names to ones that are derived from their own languages and that separate them out from the general descriptive terms imposed by the colonists.

Despite the widespread interdisciplinary and ongoing debate, Braje believes that there is a growing consensus, at least among many members of the AWG, to forgo establishing a golden spike and to adopt a beginning date for the Anthropocene in the last 250 to 50 years (Braje 2015: 379). He cites researchers who, taking their lead from Crutzen and Stoermer, have identified rapid and wide-ranging impacts of humans on Earth systems since 1750 CE, continuing with the "Great Acceleration" after the Second World War. For example, Stephen Foley and his co-authors propose 1780 as the starting point of the 'Palaeoanthropocene', which they intend, not as a geological epoch, but as a descriptive term that refers to human impacts that are local or regional rather than global in scale (Foley et al. 2013). These include global increases in greenhouse gases, human populations, fertilizer use, land surface alternations, global transportation networks, freshwater consumption, urbanized populations, damming of rivers and much more (Ellis et al. 2013, cited in Hamilton et al. 2015: 204).

In an audio interview between Geoff Marsh and the ecologist Simon Lewis of University College London, on the *Nature* website, Lewis spoke of two dates for the beginning of the Anthropocene—1610 and 1964; he has deliberated at great length over which of these is preferable, but has ultimately chosen the former (Marsh 2015). Lewis argues that macro-scale changes occurred in 1610 through exchanges of species and genes between the Old and New Worlds which led to new biological and ecological evolutionary trajectories. At the same time, Europeans brought diseases that killed about fifty million native Amerindians, most of whom had been farmers; their lands were left fallow and the dip in carbon dioxide can be traced as a geoatmospheric marker. The second date, 1964, is the year in which the fallout of atomic tests reached its peak. Following a series of international agreements, nuclear bomb testing ceased in 1963, so that a clear boundary is evident in the amount of fallout in 1964, before and after the ban.

Marsh asked Lewis what would happen with the Holocene if the Anthropocene is formalized as an epoch. Lewis's suggestion is to call the early part of the Holocene, which only lasted 11,500 years, the last stage of the Pleistocene, and thus go directly from the Pleistocene to the Anthropocene. Decisions must be taken on the macro-scale chronology, because the date that is chosen changes the story. If 1610 is selected, then the story is about species transformations, trade and colonialism. If 1964 is chosen instead, the story will be about technology, its enhancement and harnessing. If, on the contrary, the Anthropocene is defined as an event, the question of when it began loses importance, since, as many scientists have already suggested, several dates may hold as a continuity of thresholds, among the most important being 14000 BP and 1510, 1650, 1750, 1800, 1870 and 1945 CE.

In what may be understood (to paraphrase Sheila Jasanoff) as the co-production of geological and human processes, Bonneuil and Fressoz make a fascinating case for rethinking the Anthropocene both in geobiospheric and sociospheric terms. In a framework which they call the "history of the whole system", they compare statistical socio-economic trends and Earth system trends to show how, rather uncannily, all of them rose steeply at the beginning of the eighteenth century and continued along a very steep rise after 1950 (Bonneuil/Fressoz 2016: 10–11). Under Earth system trends they compare statistical measurements of the rise in carbon dioxide, nitrous oxide, methane, stratospheric ozone, surface temperature, ocean acidification, marine fish capture, shrimp aquaculture, nitrogen in the coastal zones, tropical forest loss, domestication of land and terrestrial biosphere degradation. In socio-economic trends, they consider population, real GDP, direct foreign investment, urban population, primary energy use, fertilizer consumption, large dams, water use, paper production, transportation, telecommunications and international tourism. Although the choice of certain items, and the way in which they and others were measured, may be debatable, there is no denying the steep rise in most cases, many of them notably since the 1950s, which have surpassed all previous historical precedents. To paraphrase the prophetic words of Paul Crutzen and Will Steffen, the Earth is currently operating in a state without previous analogy, driven by anthropogenic processes (Crutzen/Steffen 2003: 253).

9.4 The Anthropocene as Political Event

At the same time as being a geological event, the Anthropocene is also a political event, as Bonneuil and Fressoz highlight (Bonneuil/Fressoz 2016: 24). With a foreseeable increase in average temperature of 3.7 °C by 2100, the Earth will be warmer than it has been for fifteen million years, with a concomitant extinction of biodiversity unmatched for sixty-five million years (ibid.: 24; Kolbert 2014). They contend that the genus *Homo* is now facing up to a world it has never experienced, and for which, therefore it is neither biologically adapted nor culturally prepared. The Anthropocene thus brings a new situation for humanity, a new human condition (Bonneuil/Fressoz 2016: 24).

Jedediah Purdy states that "The Anthropocene begins amid a threefold crisis of ecology, economics and politics" (Purdy 2015: 17). I would add to these a cultural crisis of ethnic and religious violence and the emergence of alternative spiritual currents. Purdy's conviction is that the only way to build a shared living place deliberately is through politics. He believes—and I endorse his view—that "collective, binding decisions are how people can give the world a shape that we intend" (ibid.: 17).

A first, fundamental assumption that must be made, then, is that the Anthropocene is a political issue as well as a category in Earth science. As Bonneuil and Fressoz put it,

> We are therefore not in the peaceful and infra-political problematic of a reconciliation of humans with nature: the Anthropocene is political inasmuch as it requires arbitrating between various conflicting human forcings on the planet, between the footprints of different human groups (classes, nations), between different technological and industrial options, or between different ways of life and consumption (Bonneuil and Fressoz 2015: 26).

This assumption is central to my argument, since, if we take a perspective in which culture is as much an experimental method as an arbitrating mechanism between inherited and newly created knowledge, then the main task for the anthropologist is to analyse this interface. That is, it is the anthropologist's job to record experiences in order to drive forward the choice and arbitration of multiple and diverse actions to support the transition to sustainability.

Although few take this view, Jens Kersten believes that we "must move beyond our understanding of ourselves as self-victimizing creatures, a victimization for which we seek compensation, especially compensation from global nature" (Kersten 2013: 51). He argues, following Peter Sloterdijk, that we need to stop considering ourselves as a poor species and adopt a "de-escalation approach", developing an Anthropocene culture in which we enjoy complexity in our individual lives as well as in our "passive lives in collective infrastructures" (ibid.: 51–2). This accords with Bonneuil and Fressoz's contention that "the Anthropocene challenges certain distinctions that were formerly deemed fundamental to the modern West: human exceptionalism and the ontological break between the human being as subject of entitlement and object of nature" (Bonneuil/Fressoz 2016: 40).

To return to the question of how this affects our political choices, Purdy couches this in terms of the kind of democracy that is needed to address the Anthropocene challenge. His answer is a democracy capable of self-restraint, which would establish limits that give the world some definite shape, some stability in its climate and seasons, some diversity in species and habitats and landscapes. In his view, "the image of social life as too complex to be managed from any one perspective resembles the picture of the natural world that ecology introduced, where everything is connected to everything else in subtle and hidden ways" (Purdy 2015: 262).

However, he cautions that the way forward is not to attempt a utopian blueprint of global Anthropocene democracy, but to name some attitudes that would bend human society toward it. The first attitude that he proposes is scepticism toward all those political approaches to the Anthropocene that patently flout a democratic future, including the neo-liberal approach (ibid.: 269). The second attitude is more active and constructive, and includes questions that would be genuinely democratic, promoting self-awareness and collective engagement. The third is an attitude of intense interest in every effort that brings some people, or many people, closer to a shared identity or compatible and overlapping identities, to a sense of solidarity and to political and institutional resources needed for directing that sense toward shaping future world (ibid.: 271).

Existing normative and organizational frameworks for making public policy choices are now seen as inadequate for solving the kinds of problems that humanity faces. Part of the problem is keeping the social sciences away from speculative and propositional analyses of contemporary phenomena, and, specifically, away from more imaginative theories and explorations beyond the phenomena that frame the dominant political and scientific paradigms. As an example of this new type of structure, Clark A. Miller describes how the Intergovernmental Panel on Climate Change (IPCC) avoided framing climate change as changes in the weather in specific locales. Instead, it focused on climate change as a risk to the global environment, thus globalizing the atmosphere. and, finally, the IPCC created a discourse that globalized the atmosphere. In order to reinforce its authority, the IPCC then articulated a new model of science and politics in which experts and expert knowledge were given the opportunity to define problems that affected global policies (Miller 2004: 47).

The IPCC appears in this account as both an agent and a product of co-production. Miller states that this is not a contradiction: "A co-productionist idiom attunes the analyst to ways in which micro- and macro-categories, actors and dynamics connect up, directing careful attention to how, in their day-to-day routines and practices, institutions like the IPCC simultaneously reconfigure their ideas, their institutional forms, and the cognitive and social landscapes they inhabit" (ibid.: 48). Such an approach requires a rethinking of the perspectives of anthropology and the social sciences, as we shall now see.

9.4.1 The Great Acceleration

In Chap. 2, culture was shown to have risen to prominence in public discourse and in research since the 1940s for three reasons. The first was the collapse of the ideology based on 'Kultur', in its German rendering, as a higher-order racist covenant as held by the National Socialists during the Second World War. The second was the use of the term 'culture' in anthropology, a use which became generalized during this period as, among other things, a substitute for 'race', a shift also accelerated as a countermeasure to the racist claims which led to the Holocaust. The third was the politicization of the culture as one emblematic discourse of colonized peoples and nations to claims a presence, voice and entitlement with respect to the colonial powers. As decolonization proceeded after the end of the war, and as economic development theories began to be applied in newly independent countries, evidence from many sources, including n-grams, shows that, by that time, 'culture' had replaced 'civilization', 'mankind', 'humanity', 'tribes', 'Indians' and many other alternative terms in the coalescing world discourse of the United Nations.

At the same time, 1945 has been posited as one of the starting points for the Anthropocene. After the end of the Second World War, the world entered a period of historically exceptional growth, marking what has been called the "Great Acceleration". Between 1945 and 1973 an annual rise of 4.48% in the use of fossil fuels, not to mention increasing use of uranium, corresponded to an economic growth of 4.18% a year (Bonneuil/Fressoz 2016: 10–11). Between 1950 and 1970, the world population grew by 46%, world GDP by 260%, the consumption of minerals and mining products for industry by 308%, and that of construction material by 294% (ibid.).

The Great Acceleration thus corresponds to a capture by the Western industrial countries of the ecological surpluses of the developing world. An ecological gap between Western economies and the global South generated great wealth without subjecting the West to excessive ecological impacts, while the rest of the world economies were burdened by a heavy footprint on their territory. In other words, rich countries could not have succeeded in industrializing nor have attained their affluent post-war society without the unequal ecological exchange with the rest of the world. This has led Bonneuil and Fressoz to contend that “a rematerialized and ecologized history of capitalism appears as the indispensable partner of the Earth system sciences in order to understand out new epoch” (ibid.: 252).

For this reason, some authors believe that is better to speak of a ‘Capitalocene’ rather than an Anthropocene, as is discussed above. Focusing on capitalism means analysing the unequal ecological exchange which, in combination with economic and political factors, has created the inequalities of income and consumption that are causing such political upheavals in the Anthropocene.

9.4.2 The Anthropocene and Narrativity

At the 2014 Conference of the American Anthropological Association in Washington DC, Bruno Latour took a positive view of the term ‘Anthropocene’ when he stated that “the great philosophical contribution of the Anthropocene is that ‘narrativity’, which I call geostory, is not a layer added to the brutal ‘physical reality’ but what the world itself is made of” (Latour 2014: 10). He contended that, when James Lovelock and Lynn Margulis said that the planet is alive (the Gaia hypothesis), it does not mean that there is one big organism that is to be called Earth, but that its many ingredients are all building their own world. Latour argued that ‘connected’ does not mean ‘holistic’, any more than ‘animated’ means “having a soul” (ibid.).

In Latour’s view, this may be a gift or a poisoned chalice, for several reasons. The name ‘Anthropocene’ fulfils the historical intent of anthropology to place human agency at the centre of world processes; it also brings together physical and cultural anthropology; and it reopens “the key question of what is common and what is specific in the various ways of inhabiting the earth” (ibid.: 7). All anthropologists become, in this sense, “anthropologues engagés”. Yet the moment that the phrase “Anthropic origin of climate transformation” is uttered, “no amount of naturalization will clean this little statement from being read as an attribution of responsibility that requires action and probably a fight” (ibid.: 4). Meanwhile, “as soon as historians, philosophers, anthropologists and activists stumbled on the name ‘Anthropocene’, they immediately realized that there was no sense whatsoever in lumping into one undifferentiated ‘anthropos’ all the human agents responsible for shaping the planet. Amazonian Indians, Alaskan seal hunters, Shanghai tycoons, Enron executives, and slum dwellers of Valparaiso could not be ascribed the same responsibility in this newly defined ‘geological force’” (ibid.: 5). This is what Latour finds fascinating in this discussion and in his grand scheme, “the paradox of the Anthropocene”. He concludes that “‘politics’ is now the common situation for all collectives. There is no common world, and yet it has to be composed, nonetheless” (ibid.: 12).

This has certainly been the case in the last decade, during which there has been a plethora of publications in philosophy and the social sciences that have explored the possibilities of formulating a world ethos or global ethic to encompass the rich openings that the term 'Anthropocene' has created. Claude Levi-Strauss once wrote, in contrast to Bronislaw Malinowski's explanation that totems (beliefs that certain animals closely influence human lives) are widely disseminated in human societies because they are "good to eat", that totems are "good to think". In the same way it could be said that the Anthropocene is good to think. Many scholars see the Anthropocene as an opportunity to think about how the increasing awareness of our present global civilization acts as a driving force for the planet in the twenty-first century. Refusing the doomsday scenario and apocalyptic predictions of the end of the human species, Christian Schwägerl insists that one of the most exciting issues on the debate surrounding the geological epoch is that humans have the capacity to become the new global force that replaces blind chance in the way that life forms change and new life forms develop, with conscious human actions being used to find solutions to our present dilemmas and challenges (Schwägerl 2014: 151).

9.4.3 *Cosmopolitics*

How humans develop a new awareness of their capacity to become a global force—not blind but driven by deliberation and purpose—will strongly depend on the new cosmopolitical structures. At this point, it is useful to take a step back to consider the historical progression of such structures. Michael Fischer mentions that "at least three relationships (macro, micro, and meso) are said to have changed in cosmopolitics since Kant's nine theses in search of a global patchwork of confederated republican-style governance" (presented in his "Idea for Universal History with a Cosmopolitan Purpose" of 1784). The macro level covers the fact that states are no longer necessarily the places where questions of national economy, health policy, environmental policy or foreign policy are resolved. At the other end of the scale, "what once were called new social movements have morphed into highly mediated networks, able to cope distance and diversity, and facilitating constant organizational tinkering and experimentation with alliances and bottom-up decentralized decision making" (Fischer 2009: 253–4). Between these two, Fischer proposes, works of art, music, literature and film continue to disrupt mainstream ideas. Sometimes these developments are positive, as Kant had believed that they would be within the realm of republican democracies. Others, such as religious fundamentalism, take retrogressive steps not towards a past state but rather towards a changed state. Positive adaptation of our behaviour comes, in Europe, from the "overlapping weave" of the EU, NATO, the Organisation for Economic Co-operation and Development, the UN, the World Trade Organization and the World Court, alongside networks such as the Association of Southeast Asian Nations, the Organization of Islamic Countries, the Organization of American

States and others. Of at least equal importance are the roles placed by NGOs, the media, capital, states and international organizations (ibid.: 255–6).

Fischer summarizes three temporal and spatial relations of cosmopolitanism that "need to interact as differential critiques of one another, juxtaposing trifocal sensibilities: the humanistic textures of literature, film, music, and philosophy; the social anthropologies of living with diversity and alterity across differing historical and cultural experiences; and technoscientific knowledges and infrastructures that facilitate and constrain flexibility, creativity, and freedom" (ibid.: 270).

9.5 The Anthropocene as a Cultural Event

9.5.1 Exploring the Anthropocene: The Human, the Post-human and the Non-human Turn

A contrasting debate is at present being held between those who believe that, in philosophical terms, the age of humanism has been left behind and so we are in a post-human condition; those who think that, on the contrary, we have entered a "human turn"; and, even more radically, those who propound that we need to redefine our world from a 'non-human' point of view.

The Danish philosopher Sverre Raffnsøe traces the history of how humans have considered their place within their environment, beginning in Antiquity, when human beings conceived of humanity as minor and inferior, existing with and exposed to a local limited and limiting world that extended around their dwelling places of humans. He argues that the Western tradition has, however, adhered to Protagoras' statement that the human being (*anthropos*) was "the measure of all things"—a scandalous provocation in his time. This had developed by Kant's era to the notion that the individual subject was obliged to start from the human being itself and its proportions, "for here he could find a 'basic measure' when concerned with establishing order in, and conducting measurement of, the world" (Raffnsøe 2016: 23). In 1887, in *The Genealogy of Morals*, Nietzsche sought to show how humankind had "stepped forward in an interactive relationship with, and had exerted its influence upon, its own dynamic habitat" (ibid.: 21).

Thus, humans were placed at the centre of the universe; all turned towards the human being. Now, in the Anthropocene era, the human species takes on a "colossal size and status and sets new agendas, not merely in and for a multitude of contexts around the globe but also for itself, as a species that is constantly getting ahead of itself" (ibid.: 21). In other words, Raffnsøe argues that humans now inhabit the world in a different way, and at the same time are responsible for how their habitat is transformed.

An alternative view is that of the non-human turn, bringing together a diversity of theoretical, critical and philosophical approaches. According to Richard Grusin, these are "engaged in decentering the human in favour of a turn toward and concern for the nonhuman, understood variously in terms of animals, affectivity, bodies, organic and geophysical systems, materiality or technologies" (Grusin 2015: vii). For myself, I wonder how such disparate concepts and phenomena can be dealt with in co-producing new knowledge. Grusin does note that

> Even the new paradigm of the Anthropocene, which names the human as the dominant influence on climate since industrialism, participates in the nonhuman turn in its recognition that humans must now be understood as climatological or geological forces on the planet that operate just as nonhumans would, independent of human will, belief or desires (ibid.).

The last part of his statement is, indeed, very daring and will no doubt be challenged. As has been noted in previous chapters, the place of humans in this new, experimental framework of the Anthropocene is still very much under construction.

9.5.2 Exploring the Anthropocene: Cultural Evolution

The perennial debate on the primacy of culture or nature still surfaces in discussions on how to redefine the term culture. Terry Eagleton, in saying that 'culture' is "one of the two or three most complex words in the English language", goes on to remark that, in fact, "the term which is sometimes considered to be its opposite—nature—is commonly awarded the accolade of being the most complex of all" (Eagleton 2000: 1). He contends that both words can be both descriptive and evaluative terms, "meaning what has actually evolved, as well as what ought to" and that this is "relevant to a refusal of both naturalism and idealism" (ibid.: 5). However, we should be equally wary of both determinism and voluntarism in our use of these concepts, since "human beings are not mere products of their environs, but neither are those environs sheer clay for their arbitrary self-fashioning" (ibid.: 5).

Building the Anthropocene rather than designing a blueprint for it is also strongly supported by authors taking an evolutionary perspective. As Matt Ridley explains it, "the way in which … streams of human culture flow is gradual, incremental, undirected, emergent and driven by natural selection among competing ideas" (Ridley 2015: 2). He argues that human history is taught with too much emphasis on design, direction and planning and far too little on evolution, saying that this thinking is also transferred to nature, which is analysed looking for purposeful design rather than emergent evolution (ibid.: 5).

Ridley gives as an example the striking parallel between the evolution of DNA sequences and the evolution of written and spoken language. Both consist of lineal digital codes. Both evolve by selective survival of sequences generated by at least partly random variation; both are combinatorial systems capable of generating effectively infinite diversity from a small number of discrete elements. "Languages mutate, diversify, evolve by descent with modification and merge in a ballet of unplanned beauty" (ibid.: 79). Yet the result is structure, and rules of grammar and syntax as rigid and formal as you could want.

As an anthropologist, and as a participant in many national and international movements and policy applications, I agree with Ridley that, when there are no initiatives clearly explaining events and leading towards solving problems, individuals will move and will bring along others in creating new, important movements. Yet, in my experience, this is a natural process that is constantly re-enacted in small-scale societies and local communities. However, the moment that this process moves into more complex societies or, at another scale, into international affairs, if there is no leadership and no arbitration to arrive at consensus, political initiatives and movements tends to fall back upon themselves, winding up as conflicting versions and finally creating clashes that lead to stalemate. In a sense, this is what I see happening at present in the development of the concept of the Anthropocene: a lack of political and ethical ideas that self-organizing groups could latch on to.

Archaeological and anthropological studies show that there was a gradual, cumulative evolution in many different human groups migrating to different continents. It was not, as was insistently highlighted until recently, cultural, since primates and other animals also exhibit basic traits of culture. Nor was it based in language, since it has been discovered that whales, birds and other kinds of animals have communicative codes. Instead, the key factor that gave humans a fundamental advantage was the ability to accumulate meanings, knowledge and social and technical abilities, and to share, copy and adopt these from other groups. Compared to genetic descent, this is a second system of inheritance, which, according to Mark Pagel, is something akin to a software "operating system", defining who we are and acting as our internal voice (Pagel 2012: 6).

Ridley calls the emergence of human traits a "cultural evolution", kicked off by the habit of exchange and specialization feeding back upon itself and breeding innovation (Ridley 2015: 8). Whether this was made possible by a mutation in a gene or some other biological process is still under discussion. Yet no genetic mutation or biophysical trait in humans would expand at the cumulative rate in which multiple and layered cultural changes appeared in human groups all over the world. Ridley also argues that "the genetic mutations associated with facilitating our skill with language … were more likely the genetic responses to the fact that we were speaking". He therefore concludes that "In people, genes are probably the slaves, not the masters, of culture" (ibid.: 84).

9.5.3 Exploring the Anthropocene: Nature and Culture Transformed

The shifting definition of nature and culture is the third and perhaps most decisive change being brought about by exploring the Anthropocene. A world-ecology perspective has now been formulated in which, according to Jason Moore, human activity is environment-making and turns nature from noun ("the environment") to verb ("environment-making"). He makes the broad philosophical argument that "this is the double internality of historical change—humanity inside nature, nature inside humanity" (Moore 2016: 79). Yet within the bounds of necessity is "'Nature as wily coyote', Donna Haraway's lovely metaphor, revitalizing an old Native American insight, characterization, and mytheme" (cited in Fischer 2009: xiv).

Christian Schwägerl believes that another Charles Darwin or Ernst Mayr is needed to develop the intellectual and emotional understanding for the new nature of the Anthropocene: "The next Darwin will have to make sense of a world in which nature is transformed into culture and people test out the role of creator. He will be an explorer of the Anthropocene" (Schwägerl 2014: 167). This view is opposed by thinkers such as Luc Ferry, for whom the idea of man is as a "being of anti-nature", basking in liberty as a "glorification of uprootedness or innovation" (Ferry 1995). In his view, the idea that humans were made "in the image of God" can only be sustained by keeping humanity as a separate entity, following the dictates of this God creator.

In the new ecopolitical perspective now being put forward, human beings have in fact been creating and destroying species and eco-ecosystems for centuries. They have combined cultivation, domination, enhancement and even creation. Or, in the more transformative terms of contemporary performance, they have been recombining, looping and remixing the elements they find in the physical and social world, conjuring them up through culture. In today's techno-biological revolution, humans can alter genetic codes to create new animal species and clones of human organs and can reconstruct bodies as cyborgs, among many other possibilities. Attempts are still being made, however, to keep human life within the bounds of genetic and biological determinism, limiting human behaviour to being stewards or custodians of nature.

The most open interpretation is that of Marshall Sahlins, in his statement that "there is no selective pressure for violence or any other genetic disposition, nor could genes track the behavioural values varying rapidly and independently of them. *It follows that what is strongly selected for in human beings is the ability to realize innate biological dispositions in a variety of meaningful ways, by a great number of cultural means*" (Sahlins 2000: X01).

Thus, inevitably, as Schwägerl insists, "DNA reductionism—the idea that life is controlled just by the sequence of base pairs—is on its way out and a much richer picture emerges", one in which "entire complexes of protein react to psychological and physical stimuli from the brain and environment, thereby acting as intermediaries between the outside world and genetic code" (Schwägerl 2014: 170–171). This is the margin of action that humans have been using to influence the course of evolutionary processes.

One last statement by Schwägerl opens up a wide range of possibilities for thinking about culture in the Anthropocene. He highlights that our appreciation of life will be challenged in the Anthropocene and asks: "Will we subjugate it to purely utilitarian thinking or enhance it in other ways, thus developing it for a world beyond ours in which nature has become culture?" (ibid.: 172). Is this one of the underlying reasons why culture has become a central concept in discussion of globalization and cosmopolitics? Since the 1970s, while explicit debates on the supposed opposition between nature and culture have focused mainly on explaining human behaviour—for example, the social construction of gender, the kind of responsibility humans have in taking care of nature—other disciplines, such as biological anthropology, have been creating bridges between these two worlds.

Sustainability no doubt generated such debates, while it drove incremental changes in people's behaviour towards nature and the environment. In many practices linked to sustainability, a cultural mindset became the guide: for example, in reconstructing a landscape, in devising new ways to conserve forests and the people who live within them, and even in creating attractive scenery for all kinds of activities, from educational centres to commercial venues. No doubt, in people's minds, the idea that culture should now be applied to nature in ways that had not been envisaged in previous historical epochs made culture an especially relevant concept to apply to this new vision of the environment. Importantly, it has also been more malleable than others such as nation and religion (popular in the 1960s, though the latter may be on the rise again), soul and spirit (commonly used in the 1970s) and society (the favoured term of the 1980s). In the 1990s, we had no vocabulary to discuss the emerging trends in humanity's relationship to the environment. Yet people were already changing their behaviours in many different ways and so developing a new mindset in which culture became a handier "tool for thinking".

Donna Haraway epithomizes the new openness of imagination when she says "There is no border where evolution ends and history begins, where genes stop and environment takes up, where culture rules and nature submits, or vice versa." (Haraway 2003: 2). She then uses a new word, 'naturculture' to refer to this new viewpoint. In a radical flourish Haraway states that her *Kinship of Feminist Figurations* are "made up of the florid machinic, organic, and textual entities with which we share the earth and our flesh" (Haraway 2003: 1).

9.6 Summary

In the current debate on the Anthropocene, what guidelines can be extracted from an anthropological perspective to give substance to thinking about culture in this new age? Rachel Carson left us with the first value that must be made explicit in our thinking about the human future: the profound recognition that we share the planet with other life forms upon which our own biophysical make-up depends—even microbes. As Philippe Descola has demonstrated, many historical laocultures have conceived this connection as part of their own human nature. A second value, given by Elizabeth Kolbert in her book on the sixth extinction, is to incorporate into all thinking and planning towards the future the realization that the decisions taken today will determine the course of life and living long after people who have written about it have disappeared and other species have inherited the earth.

A third value would incorporate my arguments about the need for conviviability into a broader framework. As explained in previous chapters, my purpose in developing this concept is to state clearly that there can be no sustainability without conviviability: that is, the more that conflicts and wars among countries and peoples undermine the capacity to cooperate, the less agreements can be reached on climate change, biodiversity and specific environmental concerns. The time has come to incorporate this value in the framework of "cultivating *response-ability*" as proposed by Donna Haraway (Haraway 2016: 38). If and when this broader objective is to be achieved through political cooperation, it will be driven by the urgency of achieving priority results in the Sustainability Transition.

If anthropogenic global change has been driven by actions taken by humans and these are necessarily informed by cultural configurations, be they only reactive or intentional, then those configurations must be analysed as variant forms of an ontological global cosmovision, to be built through a wide diversity of explorations around the globe. Such a cosmovision will never be homogeneous, as the experience of the 1990s showed, when, instead of producing homogeneity, globalization began to foster even greater cultural diversity. Nevertheless, if not homogeneous, or identical in terms, a cosmopolitical framework may provide a single-minded unity of purpose, based, at the very least, on one single recognition: that every human alive is a native of the planet, and that every living person is capable of reaching out to a global collective identity.

References

Altvater, E. (2016). The Capitalocene, or, geoengineering against capitalism's planetary boundaries. In: J.W. Moore (Ed.), *Anthropocene or Capitalocene? Nature, history, and the crisis of capitalism* (pp. 138–150). Oakland, CA: Kairos PM.

Arizpe, L. (1989). On the cultural and social sustainability of world development. In: L. Emmerij (Ed.), *One world or several?* (pp. 45–61). Paris: OECD Development Centre. Republished in *Development*, 1997, 43–57. Republished in L. Arizpe, *Lourdes Arizpe: a Mexican pioneer in anthropology* (pp. 31–42). Heidelberg: Springer, 2014.

Arizpe, L. (1991). "The global cube" in global environmental change. *International Social Science Journal* 43(4), 599–608. Republished in L. Arizpe, *Lourdes Arizpe: a Mexican pioneer in anthropology* (pp. 43–55). Heidelberg: Springer, 2014.

Arizpe, L., Lutz, W. & Constanza, R. (1992). Population and natural resource use. In: J. C. I. Dooge et al. (Eds.), *An agenda of science for environment and development into the 21st century* (pp. 61–68). Cambridge: Cambridge University Press.

Arizpe, L., Price, M. F. & Worcester, R. (2016). The first decade of initiatives for research on the human dimensions of global (environmental) change (1986–1995). In: H. G. Brauch, Ú. Oswald Spring, J. Grin & J. Scheffran (Eds.), *Handbook on sustainability transition and sustainable peace* (pp. 349–358). Heidelberg: Springer.

Autin, W. J. & Holbrook, J. M. (2012). Is the Anthropocene an issue of stratigraphy or pop culture? *GSA Today*, 22, 60–61.

Bonneuil, C. & Fressoz, J.-B. (2016). *The shock of the Anthropocene: the Earth, history and us* (trans: Fernbach, D.). London: Verso.

Braje, T. J. (2015). Earth systems, human agency and the Anthropocene: planet Earth in the human age. *Journal of Archaeological Research*, 23, 369–396.
Brauch, H. G., Oswald Spring, Ú., Grin, J. et al. (Eds.) (2009). *Facing global environmental change: environmental human energy food health and water security concepts*. Heidelberg: Springer.
Brauch, H. G, Oswald Spring, Ú. et al. (Eds.) (2016). *Handbook on Sustainability Transition and Sustainable Peace*. Heidelberg: Springer.
Clark, W. C. (1988). The human dimensions of global environmental change. In: Committee on Global Change, *Toward an understanding of global change* (pp. 134–200). Washington, DC: National Academy Press.
Crutzen, P. J. (2002). Geology of mankind. *Nature*, 415, 23.
Crutzen, P. J. & Steffen, W. (2003). How long have we been in the Anthropocene Era? *Climatic Change*, 61, 251–7.
Crutzen, P. J. & Stoermer, E. F. (2000). The "Anthropocene". *Global Change Newsletter*, 41, 17–18.
Dempster, B. M. (1998). A self-organizing systems perspective on planning for sustainability. MA thesis, Department of Environmental Studies, University of Waterloo.
Eagleton, T. (2000). *The idea of culture*. Oxford: Blackwell.
Escobar, A. (1996). *Encountering Development: the Making and Unmaking of the Third World*. Princeton: Princeton Univerity Press.
Ellis, E. C., Fuller, D. Q., Kaplan, J. O. & Lutters, W. G. (2013). Dating the Anthropocene: towards an empirical global history of human transformation of the terrestrial biosphere. *Elementa: Science of the Anthropocene*. https://doi.org/10.12952/journal.elementa.000018.
Ferry, L. (1995). *The new ecological order*. Chicago: Chicago University Press.
Fischer, M. (2009). *Anthropological futures*. Durham, NC: Duke University Press.
Foley, S. F., Gronenborn, D., Andreae, M. O. et al. (2013). The Palaeoanthropocene: the beginning of anthropogenic environmental change. *Anthropocene*, 3, 83–88.
Gallopin, G. C., Gutman, P. & Maletta, H. (1989). Global impoverishment, sustainable development and the environment: a conceptual approach. *International Social Science Journal*, 121, 375–397.
García, R. (1986). Conceptos básicos para el estudio de sistemas complejos. In: E. Leff (Ed.), *Los problemas del conocimiento y la perspectiva ambiental del desarrollo* (pp. 381–409). Mexico: Siglo XXI.
Gardner, K. (1995). Anthropology, Development and the Post-Modern Challenge. London: Pluto Press.
Grusin, R. (Ed.) (2015). *The nonhuman turn*. Minneapolis: University of Minnesota Press.
Hamilton, C., Bonneuil, C. & Gemenne, F. (2015). Thinking the Anthropocene. In: C. Hamilton, C. Bonneuil & F. Gemenne (Eds.) (2015). *The Anthropocene and the global environmental crisis: rethinking modernity in a new epoch* (pp. 1–13). London and New York: Routledge.
Haraway, D. J. (2003). *The Haraway Reader*. New York: Routledge. Available at: https://www.docdroid.net/XoY2wMN/129046282-donna-haraway-the-haraway-reader.pdf.
Haraway, D. J. (2016). Staying with the trouble: Anthropocene, Capitalocene, Chthulucene. In: J. W. Moore (Ed.), *Anthropocene or Capitalocene? Nature, history, and the crisis of capitalism* (pp. 34–77). Oakland, CA: Kairos PM.
Hartley, D. (2016). Anthropocene, Capitalocene, and the problem of culture. In: J.W. Moore (Ed.), *Anthropocene or capitalocene? Nature, history, and the crisis of capitalism* (pp. 154–166). Oakland, CA: Kairos PM.
International Social Science Council (1990). *Human Dimensions of Global Change Work Program, 1991–1992*. Paris: UNESCO/ISSC.
International Subcommission on Stratigraphic Classification (1994). *International stratigraphic guide* (2nd edn). Trondheim: International Union of Geological Sciences.
Jacobson, H. (1992). Human Dimensions of Global Environmental Change Programme. *Environment*, 34(5), 44.

Jacobson, H. K. & Price, M. F. (1990). *A framework for research on the human dimensions of global environmental change*. Paris: ISSC/UNESCO.
Kersten, J. (2013). The enjoyment of complexity: a new political anthropology for the Anthropocene? In: H. Trischler (Ed.), Anthropocene: exploring the future of the age of humans, special issue of *RCC Perspectives*, 3, 39–52.
Kolbert, E. (2014). *The sixth extinction: an unnatural history*. New York: Picador, Henry Hold and Company.
Latour, B. (2014). Anthropology at the time of the Anthropocene: a personal view of what is to be studied. Lecture at the American Association of Anthropologists meeting in Washington DC, December 2014 (draft for comments).
Leff, E. (Ed.) (1986). Los problemas del conocimiento y la perspectiva ambiental del desarrollo. Mexico: Siglo XXI.
Lightfoot, K. G., Panich, L. M., Schneider, T. D. & Gonzalez, S. L. (2013). European colonialism and the Anthropocene: a view from the Pacific Coast of North America. *Anthropocene*, 4, 101–15.
Marsh, G. (2015). Audio interview with Simon Lewis. http://www.nature.com/nature/journal/v519/n7542/pdf/nature14258.pdf, accessed 3 October 2016.
Miller, C. A. (2004). Climate science and the making of a global political order. In S. Jasanoff (Ed.), *States of knowledge: the co-production of science and the social order* (pp. 46–66). London: Routledge.
Miller, R. B. (1989). Human dimensions of global environmental change. In: R. S. DeFries and T. F. Malone (Eds.), *Global change and our common future* (pp. 84–89). Washington, DC: National Academy Press.
Miller, R. B. (1994). Interactions and collaborations in global change across the social and natural sciences. *Ambio*, 23(1): 19–24.
Moore, J. W. (2016). The rise of cheap nature. In: J. W. Moore (Ed.), *Anthropocene or Capitalocene? Nature, history, and the crisis of capitalism* (pp. 78–115). Oakland, CA: Kairos PM.
NACSN (North American Commission on Stratigraphic Nomenclature) (2005). North American stratigraphic code. *AAPG Bulletin*, 89: 1547–1591.
Nature (2011). The human epoch. *Nature*, 473, 19 May 2011, 254.
Pagel, M. (2012). *Wired for culture: origins of the human social mind*. London: W. W. Norton and Company.
Price, M. F. (1989). Global change: defining the ill-defined. *Environment*, 31(8): 18–20, 42–44.
Price, M. F. (1990). Humankind in the biosphere: the evolution of international interdisciplinary research. *Global Environmental Change: Human and Policy Dimensions*, 1, 3–13.
Price, M. F. (1994). *Options for EC-level activities on the human dimensions of global change*. Brussels: European Commission Directorate-General for Science, Research and Development.
Purdy, J. (2015). *After nature: a politics for the Anthropocene*. Cambridge, MA: Harvard University Press.
Ridley, M. (2015). *The evolution of everything: how new ideas emerge*. London: HarperCollins.
Sahlins, M. (2000). Jungle Fever. *Washington Post, Book World*, 10 December: X01.
Schwägerl, C. (2014). *The Anthropocene: the human era and how it shapes our planet* (trans: Renner, L.). Santa Fe and London: Synergetic Press.
Smith, B. D. & Zeder, M. A. (2013). The onset of the Anthropocene. *Anthropocene*, 4, 8–13.
Steffen, W., Crutzen, P. J. & McNeill, J. R. (2007). The Anthropocene: are humans now overwhelming the great forces of nature? *Ambio*, 36, 614–621.
Thomas, W. (1956). *Man's role in changing the face of the Earth*. Chicago: University of Chicago Press.
Turner, B. L., Moss, R. H. & Skole, D. L. (Eds.) (1993). *Relating land use and global land-cover change: a proposal for an IGBP–HDP core project*. Report from the IGBP–HDP Working Group on Land-Use/Land-Cover Change. Joint publication of the International Geosphere-Biosphere Programme (Report No. 24) and the Human Dimensions of Global

Environmental Change Programme (Report No. 5). Stockholm: Royal Swedish Academy of Sciences.

Vernadsky, V. (1945). The biosphere and the noösphere. *American Scientist*, 33(1), 1–12.

Walker, M., Gibbard, P. & Lowe, J. (2015). Comment on "When did the Anthropocene begin? A mid-twentieth century boundary is stratigraphically optimal" by Jan Zalasiewicz et al. (2015), *Quaternary International*, 383, 196–203. *Quaternary International*, 383, 204–207.

Worcester, R. M. & Barnes, S. H. (1991). *Dynamics of societal learning about global environmental change*. Paris: ISSC/UNESCO.

Wolff, E. (2014). Ice sheets and the Anthropocene. In: C. N. Waters, J. Zalasiewicz, M. Williams, M. A. Ellis & A. Snelling (Eds.), *A stratigraphical basis for the Anthropocene* (pp. 255–263). London: Geological Society of London.

Zalasiewicz, J., Williams, W., Haywood, I. & Ellis, M. (2011). The Anthropocene: a new epoch of geological time? *Philosophical Transactions of the Royal Society A*, 369, 835–841.

Zalasiewicz, J., Waters, C. N., Williams, M. et al. (2015). When did the Anthropocene begin? A mid-twentieth century boundary level is stratigraphically optimal. *Quaternary International*, 383, 196–203.

Chapter 10
Conclusions

Before, culture was just lived. Now it becomes a self-conscious project (Marshall Sahlins 1994).

10.1 Culture: A Template for Transactions

Since the creation of the United Nations allowed the development of a space for dialogue and negotiations on international development in the 1950s, the peoples of the world have been engaged in ever-expanding transactions about who they are and who they will be in a global age. Culture has emerged as the concept of choice in resetting local, national and regional identities, as have globalized cultural spaces in contexts of power. Is it a coincidence that the Anthropocene Working Group has also recently proposed that the 1950s be the starting point for this new geological era so decisively marked by human thinking and the actions derived from it that affect the geophysical and bioecological environment?

Why culture? Because every human being on the planet has the capacity to create language and ideas. "Living culture" still continues to be the guiding debate, the comforting space of belonging and the field of scientific and political experiments of humanity. However, when culture is turned into a political ideology, as in recent years, it may become a weapon to impose group interests, religious doctrines or exclusive rights and privileges against other groups. A new analysis, as many speakers, participants of meetings and authors put forth, is needed to understand the paradoxes and contradictions evident in recent international policy events.

The richness of statements made about cultural practices in the many meetings and comments described in this book attests to the great interest now centred on an expanded and vibrant use of this term. It is impossible to summarize those usages, or even to do justice to the major transactional advances made in this global collective experiment, though I will, in these Conclusions, attempt to draw attention to vocabularies, scenarios and foresights that give some order to this vast discussion. As has become evident from the data and many statements made in the

L. Arizpe Schlosser, *Culture, International Transactions and the Anthropocene*,
The Anthropocene: Politik—Economics—Society—Science 17,
https://doi.org/10.1007/978-3-642-41602-6_10

meetings and academic debates described in this book, the major actors in constructing cosmopolitical space—governments, United Nations institutions, scientists and social science researchers, cultural practitioners and stakeholders, intellectuals, artists and religious leaders—have fully participated in opening channels for a vast transactional process as the world becomes global. The question in all of our minds, given the proliferation of large and small conflicts associated with cultures, religions and racially defined attitudes, is whether such conflicts are manifestations of what will become successful renegotiations of global relationships, or whether they are driving a regression in world affairs back to old phenomena of "ethnic cleansing", destruction of cultural heritage and rape as a weapon in ethnic wars. According to the news media, 'xenophobia', "new tribalisms", "gender crimes" and "hate crimes" are on the rise, with religiously inspired terrorism in the background, all echoes of former historical times. The political philosophies of the Enlightenment and American-inspired democracy helped establish the rule of law of nation states and the international treaties, conventions and policies that have contained such conflicts. What has happened, then, in the last twenty years so that the 'dreamcatcher' attribute of culture that was so prominent until the beginning of this century has now exacerbated such conflicts? None of these latter trends will ensure a sustainability transition in the Anthropocene that will allow humankind to build a future of peace. It is therefore a matter of urgency that we develop an understanding of how different actors have debated and transacted the perception of culture in global affairs, and have proposed new avenues for reorganizing culture and national societies.

This book is an intellectual history of international transactions in the field of culture from the 1990s to 2014.[1] The empirical data collected are based on a multi-situated ethnography bringing together data from my direct anthropological observation of and participation (including decision-making) in institutional meetings in United Nations settings, government cultural policy conferences, scientific organizations, global economic forums and a few national and local meetings where questions of culture were discussed. I have aimed to do three things. My first aim has been to create an archive of the ways in which a vocabulary and grammar of culture grew as a force in development and in international relations as part of a process of global change, for the first time incorporating all nations and cultures in an open, international dialogue. This is important for setting precedents and marking ascendant steps in a world in which people feel increasingly insecure and believe that retrenchment into past identities will bring back security. As a result, however, instead of allowing for readjustment in new political and social settings, they have created intractable conflicts.

My third aim has been to explain why the concept of culture is now such a popular term in a surprising variety of areas of political and social life, as well as in international political and normative assignments. As the Introduction showed,

[1]A first analysis of this process is summarized in Chap. 2, from an article published in Arizpe (2004).

Google n-grams demonstrate that, while the preferred terms for referring to the human species were formerly 'mankind' and 'humanity', by 1900 both had been overtaken by 'culture'. Others terms whose usage has declined with respect to 'culture' are 'tribes', 'conscience', 'Indians', 'progress', 'nation', 'religion' and 'civilization'. While there are methodological issues with the use of n-grams, the emergence of their use in mainstream intellectual and policy work is evident in the rise in the number of intellectual and academic texts referring to the concept.

One conclusion I wish to highlight at this initial stage is how awesome it is that so many people having dissimilar historical lineages, nationalities, cultural and spiritual affiliations, whose statements and transactions are reflected in this book, have all converged in wanting the same thing: a collective human understanding for a world which must become sustainable.

10.2 International Cultural Politics

The material described in this book has illustrated a very interesting trajectory in the use of the word 'culture' in international politics since the 1990s and the end of the Cold War. Culture was recognized as an important factor in fostering the internal cohesion of groups and nations, in resisting cultural encroachment through market and political globalization, and in searching for new social and institutional innovations that could be used to adapt to rapidly accelerating world changes. Culture has always been recognized as the path to the highest attainment in art and in intellectual fulfilment, as well as the joy of collective celebrations and festive occasions. The time was then ripe for a new international normativity, called for by decolonized and developing countries, to link economic development and culture in the setting of globalization and the 'Capitalocene'.

From the 1950s culture began to be used in research and policy arenas along three different axes. Firstly, as explained in the first chapters of this book, and lucidly discussed by Kuper (2003), culture became the concept of choice among anthropologists for theoretical reasons, in order to eradicate the possibility of genocides being committed on the basis of 'race', as had happened during the Second World War in the hands of Nazis and fascism, and for methodological convenience in carrying out fieldwork.

Secondly, culture was taken up in the following decades in theories of economic development in the United Nations as a way of affirming universalism and human rights around the world. Culture, in particular, became the world stage where peoples and individuals could be made to feel part of humanity. The 1972 Convention Concerning the Protection of the World Cultural and Natural Heritage, in particular, helped establish this "pride of all", as a cultural activist expressed it, that fosters an international feeling of belonging. Culture became a broad, open concept of choice to set normative guidelines for the future.

Thirdly, as development brought countries into world markets, many countries in Africa, Latin America and Asia reacted by trying to incorporate their historical, ethnic and cultural traditions into development. The Group of 77, with the support of the Nordic and many other countries, created the programme of the Decade on Culture and Development (1987–97) at the United Nations. After the first few years of this programme, a new resolution was adopted that established the World Commission on Culture and Development, with UNESCO as lead agency, with the specific mandate of proposing a new normativity for policies concerning culture and development.

The Report of that Commission, *Our Creative Diversity*, defined culture as "ways of living together". The most important legacy of the Commission's work was in demonstrating that many new claims and questions about culture and development in all regions of the world had moved into a global age—claims and questions which could now be represented, debated and transacted on a world stage. Almost one hundred and fifty cultural practitioners, government cultural functionaries, researchers and local and indigenous peoples participated in the nine consultations of the Commission.

Our Creative Diversity recognized culture as a vital aspect of development and argued forcefully that cultural pluralism must be embedded in a global ethic. Importantly, the five ethical principles that the Commission submitted in its first chapter of *Our Creative Diversity* were: human rights, equality, accountability, sustainability, and equity between generations and between women and men. This is, *that cultures can be respected only if they in turn respect other cultures*. Ignoring this principle has led to a state of affairs that has allowed some countries to support religious terrorism, or to flout democracy and some cultural leaders to use their cultural traditions to deny the same cultural rights to other groups.

After the report *Our Creative Diversity* was presented and discussed in over seventy countries, there began the intense politics of international culture, which pulled in major political forces and mobilized national, local, and indigenous peoples' movements. Culture, already defined by anthropologists as a "site of contention", had moved into the international global arena.

The protection of intangible cultural heritage was the next step, as the rapid acceleration of telecommunications, markets and migrations uprooted people and cultures around the world. A vigorous campaign by countries concerned by the loss of their cultural traditions and the survival of indigenous groups led to the adoption in 2003 of the UNESCO Convention for the Safeguarding of the Intangible Cultural Heritage. This Convention has been the most rapidly ratified international convention, generating great enthusiasm as well as intricate transactions on cultural representations and claims, and intellectual property, especially that of local peoples. It is worth noting, however, that the least developed domain of this Convention has been the one concerning practices related to the natural world and its resources. Intellectual property, however, was not included in the 2003 Convention as trade-related aspects of intellectual property rights are dealt with by the TRIPS agreement between the World Intellectual Property Organization (an agency of the UN) and the World Trade Organization. The recent trend in commercializing ethnic and aboriginal art and performance and in developing digital industries in this field now forms part of policy initiatives to shift development towards a creative economy.

The 2005 UNESCO Convention on the Protection and Promotion of the Diversity of Cultural Expressions has also had a wide-ranging impact in cultural as well as economic and technological sectors. This has been reinforced by UNESCO's 2015 report *Reshaping Cultural Policies* and its recent publications *Investing in Cultural Diversity and Intercultural Dialogue* and the *Creative Economy* reports.

10.3 The Complexities of Culture

The polysemy of the term culture, that is, its open definition that includes everything human beings have created, gives it a very appealing 'dreamcatcher' effect. References to culture in most international conferences and in policy documents overwhelmingly highlight the desirable qualities of culture. Yet the failure to recognize that using the term 'culture' at certain scales and settings often leads to paradoxical or contradictory outcomes that create divisive political and social trends.

To recapitulate what many researchers and participants in international transactions have said, culture is not a precise term—it is an abstraction based on countless generalizations. It requires so many assumptions and qualifications that it runs into the same dilemmas as the "species problem" in biology. No definitive boundaries can be established, neither for species nor for cultures, except through pragmatic stipulations that set one group apart from another. This being so, just like 'species', 'culture' doesn't really mean anything in and by itself. Differences exist in cultural practices but considering them as 'differences' rather than 'variants' of a human attribute is a theoretical choice whose assumptions are frequently not well understood.

At the end of last century, the theoretical choice to interpret such variations as 'differences' has been used to define a "politics of difference". Is it a coincidence that, twenty years later, the impact of the dissemination of this trend has led, as Klaus Schwab, founder of the Global Economic Forum of Davos, stated in February 2018, to "a divided world"? Many other world leaders acknowledge this rising phenomenon, a state of affairs that I remember UNESCO ambassadors from pluricultural countries such as India, Nigeria and Indonesia, since the nineties kept warning about. Their countries has lived through historical periods of communalism, racial strife which are still present in their countries. As described in previous chapters, similar misgivings were also expressed at the presentations of Our Creative Diversity, especially in relation to cultural rights. How has it happened then that divisiveness based on cultural markings has risen to become a major world concern? This is discussed in the following section.

In terms of this section on the complexities of culture, I will contend that when culture as a descriptive term is turned into an ideological concept then its application creates paradoxes and contradictions. The first problem is the failure to recognize that the positive assumption that "culture unites" has, paradoxically, a dividing effect when moved out of the discourse of universal liberalism, to policy contexts where defending the unity of one group *ipso facto* discriminates against other groups.

The positive meaning of culture, as a force that unites, grew out of the belief that humanity is endowed with unique attributes that lead to goodwill and cooperation. This envelops every person in the world in the comforting knowledge that they belong to humanity and they can empathize with all other people. Such a belief is foundational in many cultures as well as religions, except, among the latter, since for those that separate believers from infidels with a mandate to proselytize or kill those who do not believe.

As explained in this book, identity is relational. Relations between people can only be outlined by contrasting one's cultural practices against those of others, with finely tuned differences being highlighted according to the historical, political or social nearness of the other groups. In this context, being human is not an identity, it is the acknowledgement of human commonality through living. The assumption, then, that "culture unites" is only true if it is applied at the ontocultural level, that is, to the whole of humanity.

A significant problem about the entitlements of humans is now emerging from scientific discoveries and ecological movements around the world. The boundary between humanity and other species of animals and plants is not present in cultures having animistic, totemic and similar kinds of cultures, as Philippe Descola, among many other anthropologists, has shown. In the Western world, instead, this boundary was established by naturalism and by the doctrines of some religions, notably Christian churches. All other sentient and existent beings on the planet were left out of the commonality of humans. This boundary is now beginning to shift. Not only because of what was cited in the chapter on the Anthropocene, whereby culture, the jewel of the human creations, is partly shared by primates and other intelligent animals, but because our own bodies are ecosystems and any separation is no longer possible.

It follows, then, that the "human turn" and the "non-human turn", ideas that the authors cited in this book have been developing, only confirms that a philosophical recasting of the place of humans in the planet is overdue.

For all the constant mention of the richness of cultural content and expressions in the practices of people around the world, cultural programmes and policies tend not to take into account the internal textures of cultures. This means that dissidence, alternative practices and distinct aspirations of practitioners are not part of the debates on how to safeguard practices and how to ensure that the impacts of such practices improve the life of its practitioners. Importantly, this sometimes means that only the voices of the gatekeepers are heard as representatives of their cultures. What happens when, in safeguarding their way of life, practitioners end up retrenching back into values and habits that had marginalized them from world progress? In particular, those cultures that repress personal freedom, especially for women, reject scientific explanations, that refuse adaptive behaviour to their circumstances are contradicting the very basis that gives intellectual and political support to international and national cultural policies. In other words, cultural practitioners must be accountable to the international and the nation state communities in terms of democracy, equality and sustainability, as established in the global ethic advocated by the World Commission on Culture and Development. This is especially pertinent to ensure that cultural, ethnic and religious groups that demand entitlements and rights are able to implement the practices or changes needed for their participation in the economic and political systems of the nation state.

In terms of development practices, the following question must be raised. If countries are assessed and ranked according to their fulfilment of international development principles and goals, why aren't other organized groups, including ethnic, cultural, religious and indigenous entities, assessed under the same terms? This partly explains why, while the goals of sustainability are pulling the world towards concerted collective action, contradictions may arise at the laocultural level because some practices of cultural and religious groups may oppose or even reject those international principles and actions. A case in point is that of honour killings of young women at the hands of their Muslim fathers or brothers when they find friends or spouses across the boundary of their religious groups.

Such dilemmas bring up difficult questions: if diversity is to be respected how far must divergences of groups be accepted in international or national political structures? This highlights the real question: where are the juridical and anthropological theories and proposals that will allow a rigorous analysis in such cases, one that will bring compatibility to the rule of law and juridical practices of nation states and consuetudinary laws and customs?

This is why I advocate that new ranking systems be devised so that the positive contribution of cultures be given the acknowledgement they must have in the transformative engagement towards a sustainable horizon.

The "flat cultures" view not only disregards the contributions of different cultures to the Human Project internal inconsistencies and dissensions—which we anthropologists recognize all too well as an ethnographic fact—, deep historical resentments and the ambiguous impacts of affirmative action are rarely taken into account. Yet many, if not all, projects of applied cultural policy run into the problems created by such dissensions and ambiguities. To these must be added, in recent years, the distortions that arise when greed, which may be necessary to bring down barriers and create new productive or organizational trends, becomes abuse in unregulated fields. Culture may even be invoked to justify this abuse when hierarchical controls or corruption turn it into a "traditional way to do business" or "do governance" or ignore legitimate cultural demands.

Finding new ways to recast the concept of humanity, to anticipate the paradoxical outcomes of applying a flat concept of culture to political transactions, and to create new theories and terms for the many unprecedented cultural trends accelerating through markets, governance and social organization must become a priority in development and in achieving the United Nations Global Goals.

In this context another term which is also facing challenges is 'diversity'; this is taken up in the next section.

10.4 Ungrounded Diversity

It is creativity, this unique faculty of humans acquired through living, that brings about a cascade of new ways of speaking, relating, loving, making, trading and organizing things and people. In other words, human living constantly spawns variations in cultural thinking and practice. This is why I believe that the loss of human diversity is impossible: it is underwritten by genetic codes and epigenetic influences. Yet this creativity will only be present if and when individuals have access to emotional resonance with others, cumulative learning and to freedom to experiment, and therefore, to create the new. Any homogenous society will inevitably becomes more heterogeneous because of this faculty.

This being the case, diversity becomes another word for human freedom. A problem arises when the concept of diversity is applied at the laocultural level as a political ideology in nation states that are pluricultural.

Biological variance in life forms is already a formidable process to encompass in human perception. Even more incommensurable is the level of variation that neuroscience has now confirmed, of humans reflecting on what they perceive—consciousness—and deciding to reformulate this perception or changing their behaviour to respond in many different ways to those stimuli. At the ontological level, as Mark Pagel has insisted, humans are "wired for culture" (Pagel 2012). Thus culture is, by definition, an instrument of freedom from the constraints of geo-biological systems, which seem to be overcome to an astonishing degree as we become cyborgs, no longer only cultivators but creators too. This is why I believe that the freedom to create is by far the deepest core attribute of humans.

DNA studies showing that every human being shares an identical genetic makeup confirm the extraordinary unity that ratifies the common foundation from which all peoples create their cultures. In our contemporary world the possibilities now exist of using the new CRISPR-Cas9 technology to edit the genome by removing, adding or altering the DNA sequence, of creating cyborgs by attaching machines to human bodies, and of enhancing the capabilities of the mind with new learning techniques and drug applications. This means that the capacity of human beings to *produce diversity* is increasing exponentially in historically unprecedented ways. In this context, I argue that the capabilities of human beings will increase, as cultivators and as creators of species but more crucially, as *transformers* of the Earth system. Thus, a new paradigm for knowledge and action is needed in which diversity must be reconceptualized, crossing the traditional human and non-human boundary.

A first step might be to recognize that given the common DNA foundation of all humans, it follows that all cultural groups are divergent rather than 'different'. An assumption may then be made that 'diversity' refers uniquely to possibly epigenetic influences or to the incessant exchanges of norms, goods and ideas among peoples of the world over millennia. If we make the assumption that practically all cultural groups in the world have exchanged knowledge as well as societal norms and are continuing to do so—in excess, some would say—through new digital technologies today, it follows that, as some authors have argued, all cultures are hybrids or palimpsests of previous cultural practices. Understanding their claims, then, has more to do with detecting how and *why certain elements of their cultural archive and repertoire are activated in a given space and at a given time*. They become activated for a variety of reasons but very often because the group who holds such cultural elements in their archive need to reaffirm their belonging, their entitlements and claims, or to strike a contrast towards another group. Such reactions become exacerbated if they feel that their way of life is being menaced and thus they need to set up fences and shields against encroaching external conditions, invasions or coercions. Of course, the same reasons apply in cases where internal dissident groups feel threatened.

Recognizing diversity as a basic atribute of human groups, however, is very different from reifying diversity to turn it into a political ideology. Multiculturalism has brandished cultural rights whose boundaries have never been explored, much less explicitly explained as to their origins, co-evolution, political legitimacy and clearly defined membership. The lack of a theory of such processes has led to infinite strife between rival gatekeepers, cultural practitioners, stakeholders, and dissident and progressive members in such undefined groups. Ethnicity, a natural attribute of any human associative group, has been turned into ethnicism, that is, the intentional use of ethnicity for explicit political purposes. This also applies to religiosity when it is turned into religionism as well as when representations of real or imagined attributes of a group are turned into racism.

As mentioned in Sect. 10.3. on the paradoxes of culture, every boundary affirming one's identity towards the inside *ipso facto* creates a barrier towards other groups. It is important to realise that this perception only exists when it is expressed and acted upon in sociopolitical terms. In every day life most cultural groups overlap and have constant exchanges in their members. In this strife for recognition and rights at a laocultural level, people must then decide what primacy they give to one identity, normally defined with regard to another identity, or within the framework of other, larger-scale identities.

Since the nineties, unsettling trends leading to conflicts attributed to culture have been taken up by politicians and social scientists. Many speakers referred to this in the meetings and books described in this volume. In this strife for recognition and rights, laocultures, as I have defined them, are important for people in deciding what primacy they accord to one identity, normally defined with regard to another identity, or within the framework of other, larger-scale identities. Serge Gruzinski warned that "the rapid disappearance of traditional models [was] encouraging a reversion to attitudes which belong to the past: those attitudes exploit the ideological exaltation of old models which are turned into the foundation of an aggressive and destructive policy" (Gruzinski 1993: 21). Amartya Sen, when asked "Does the championing of cultural diversity demand support for cultural conservatism, asking that people should stick to their own cultural background and not try to move to other lifestyles?" replied that such a position would "immediately deliver us to an anti-freedom position … diversity will then be achieved at the cost of cultural liberty" (Sen 1998: 6.).

In the 2000 UNESCO Universal Declaration on Cultural Diversity, two of its main authors, Katerina Steniou and Arjun Appadurai, were concerned (and rightly so) about "the necessity for government policies to be put in place to support a wide diversity of cultures in the face of market forces".[2] The reasons for protecting cultural diversity evoked in the Declaration highlighted the value of culture both in terms of creativity and aesthetic satisfaction as well as of its role in the knowledge economy, in fostering competitivity and the creation of wealth. While the authors of the Declaration were careful to set out the conditions for the application of the term "cultural diversity", the way in which it was applied in subsequent years denied its status as a general principle, and the dynamics of the exchanges, imitations and overlaps between cultures in the context of neo-liberal economic policies ceased to be explored. UNESCO, through several meetings described in Chap. 8, tried to counter this tendency, for example through the 2008 meeting that discussed difference and complementarity between the International Conventions of 2003 and 2005. Another meeting held at UNESCO in 2009 debated the conceptual and operational guidelines for integrating cultural diversity with the principles of intercultural dialogue. As mentioned in Chap. 7, in the preliminary programme of that meeting the following topics were noted for discussion: "no one may invoke cultural diversity to infringe upon human rights guaranteed by international law, nor to limit their scope" (from the UNESCO Universal Declaration on Cultural Diversity); pluralism is the political response to cultural diversity; the recognition of multiple belongings and identities; culturalization of social issues and the risk of segregation in the name of particularism. Global ethics was also briefly mentioned, as well as cultural nationalism. A clear statement was made, referring to "*cultural diversity as a connecting or divisive force within and between societies*".

A second problem arises when cultural belonging is turned into a political ideology in the context of the rule of law and citizenship rights in nation states, as an emblem against other groups living in the same setting. Advocates of multiculturalism have deployed an aggressive campaign against the term 'humanity' and against any other that takes a unified view of humankind; they have also been militant in affording ethnicity and religiosity greater legitimacy as emblems of political organization than the traditional affiliations of nationality, citizenship, historical lineages and so on.

Two effects multiply the contradictions in such claims at the laocultural level. The first has already been mentioned; if the positive meaning of culture as unity is applied in pluricultural settings, a clash may appear between citizens' rights, human rights and the rights that specific groups of cultural practitioners demand.

[2]The full text of the Declaration is available at: http://www.unesco.org/education/imld_2002/unversal_decla.shtml.

This becomes particularly critical in the case of recent migrant and diasporic communities who demand the right to re-establish their cultural and institutional ways of life in the host country. The relationships of such groups to the receiving countries varies enormously and, therefore, cannot be counted on to exhibit a single kind of cultural boundary with the host society. Moreover, as we wrote in the 2001 UNESCO World Report, if the opportunities and benefits of economic growth and political freedom are denied to such groups, the recognition of cultural boundaries that was supposed to help indigenous or migrant peoples will fall along the fault lines of joblessness, poverty and exclusion. The positive meaning of culture will then only provide a veil of unreality and of a comforting sentiment of belonging in conditions where the harsh reality of lack of economic growth will double the sense of exclusion and poverty.

Applying the concept of culture in contemporary situations becomes more complex with the rise of new cultural, spiritual and lifestyle movements that cross over traditional cultural patterns. For example those developing among neo-indigenous people in Mexico, Brazil and other Latin American countries as well as in countries in Africa and other regions. Other examples are the new spiritual movements that merge a wide range of beliefs and practices based on ecology and belief systems such as Buddhism and Yoga and that may be sowing the seeds for future belief systems. Lifestyle and sexual orientation are also creating new identities, including, for example, queer identity in the LBGTQ movements. Life style and consumer preferences, moving very rapidly through markets and the Internet are also spawning new kinds of cultural groups in all ranges of art and audiovisual creations. The emergence of such a wide-ranging diversity of new groups shows that doors keep opening for experiments some of which will become the new cultural variants of the future.

I believe the only solution may be found along the lines of the proposal of the UNESCO World Culture Report of 2000, arguing that diversity—including not only diversity related to culture but also diversity linked to gender, race and sexual preference—need not threaten stability, *as long as citizens are able to adhere sufficiently to national values and cultural practices to secure general compliance and support for the institutions of governance*. There have been very few subsequent studies, however, analysing policy applications and measures to ensure that consuetudinary law or customs are compatible with national legislations. Cultural rights in many countries have simply slipped into a grey area of vague juridical definitions and competing political procedures.

In this context, a change in recruitment policies for posts in United Nations agencies implemented by some governments since 2000 has had a significant effect. In UNESCO, as well as in some other United Nations agencies, the meetings between philosophers and social scientists that were traditionally held to discuss programmes, especially those related to education and culture, have been quietly set aside. Instead, attention has been increasingly focused on the juridical, legislative and operational policy aspects of cultural diversity—and the same is true of other topics in all sectors. At the same time, UNESCO then focused on new partnerships involving jurists, lawyers, legislators and policy 'experts' appointed by governments. As a result, following decisions taken by a few powerful member states of the United Nations system, the work carried out in its agencies, notably in UNESCO, became more technical, and almost mechanical, in the development of proposals placed on the table by governments' appointed experts. In my view, this has greatly impoverished in-depth philosophical and political exploration of new ideas in United Nations institutions, with the result that the interests of powerful countries and very notably, the new lobbyists for the business corporations, conservative religious institutions and organizations and other interest groups, are now practically carried unopposed at the institutional level.

Part of the problem in generating new theories about culture and diversity has to do with the extraordinary possibilities of communicating data and ideas through the Internet yet that are difficult to bring together in coherent form to influence decision-making at the international level. This is the new challenge that digital technologies bring to culture and the management of diversity, as briefly examined in the next section.

10.5 Minds, Memes and Machine Intelligence

Will the concept of culture continue to be a reference term now that cultural production and reproduction of cultural elements is moving across digital technologies? Cognitive, social and now political thresholds are emerging as people communicate, learn and adapt to new knowledge through the Internet. It is not clear as yet which cultural activities will be most enhanced or, on the contrary, wiped out by the new information and communication technologies. As stated by Kroker and Kroker, "neither inevitably utopian nor dystopian, digital reality is an open future, determined in the final instance by the play of human interests, desires, and power that have always represented the elementary matter of the human condition" (Kroker and Kroker 2013: 3). Interestingly, interactions on the Internet are still driven by the desires, interests and ambitions generated in embodied society and in the power play of political forces. Has this not been the case for all major human inventions? Yet it has taken centuries of collective efforts over centuries to distil the best effects of such inventions while organizing new patterns of behaviour and interaction, for example, how will digital communications affect basic human rights, the rule of law, and international agreements on cultural production?

There is no doubt that the new information and communications technologies are increasing the speed of cognitive exchanges, the range and depth of knowledge made available and the complexity of interactions among individuals and peoples of different cultures. Yet all these have to be processed by users who have the same capacity to carry out cognitive, emotional and moral activities, except that these are activated in a virtual sphere where the traditional rules of engagement do not hold. The rush for unfettered innovation is, of course, the appeal and the magic of the Internet. Yet the structure and clarity that cultural practices give to the jumble of data and feelings that traverse our daily lives is not there. Steps to act against fake news, intentional manipulation of data, trolling, especially against women's participation in digital projects, and communications between members of terrorist cells are only beginning to be discussed.

Since the nineties, as propounded by participants in many of the meetings described in this book, concern over the global spread of the Internet among countries of the world has focused on the possible loss of their traditional cultures. The UNESCO conventions on intangible cultural heritage and on the protection of cultural expression were both outcomes of this concern. In his 1997 book *Cyberculture*, Pierre Lévy pointedly asked whether cyberculture threatened the values of European modernity. He answered that, on the contrary, "cyberspace may appear to be a kind of technical materialization of modern ideals" (Lévy 1997: 302). Cyberculture may be considered a continuation of the progressive project of the Enlightenment, he explained, as it enhances participation in communities of debate and argumentation, equality in reciprocal human relations, exchange of information and knowledge and liberty in communication across borders to all communities in the world (ibid.).

According to Michael Fisher, such unprecedented conditions will turn cultures into tools for experimentation, keeping formalization but inserting new forms of cultural understanding. In this sense, freedom to create is the path forwards in the development of what he calls "new connectivities" (Fischer 2009: 37). Fischer is optimistically referring to the way "the Internet, networked data banks, visual icons, video clips, film animation, streams and repetition information flows are repositioning and enveloping older cultural media such as orality and literacy, reshaping the public sphere by changing power relations … mobilizing money and attention in electoral campaigns, and drawing attention to alternative geopolitical narratives" (ibid.). However, I strongly agree with Fischer when he notes that people's deepest sentiments, as with all other inventions, are always present behind each electronic message, with all their inherited feelings about power, wealth, social position and identity leading to dramas and dilemmas. The next step is to rapidly create a new vocabulary to try to keep up with what is happening on the Internet.

Is the concept of 'meme', taken from anthropology becoming a major descriptive term to classify and analyse cybertexts? Anthropologists, as explained in previous chapters, are understandably wary of using this concept of memes in this way, since it is an abbreviation of 'culturemes' and in the Internet it is being used to refer to an exceptional phrase or expression that then becomes viral. Ryan Milner argues that in our contemporary media environment, "mediated expression and debate consistently occur through and around an ever-growing collection of populist memetic texts" (Milner 2016: 2) and it is therefore important to analyse the logic of these memetic processes.

The effects of digital connectivity on psychological and affective life are only beginning to be debated. It took long centuries to build the rule of law nation states, political philosophies, ethical norms and the juridical instances to contain hate crimes, racism, sexism, xenophobia, criminality and so on. How will such rules and devices be developed inside the Internet, by whose laws and through which democratic procedures and consultative processes?

Moreover, memetics may be ruining democracy, as argued by Douglas Harrow in his article "Meme warfare: how the power of mass replication has poisoned the US election." (Haddow 2017). "Forget Fake News—Alt-Right Memes could do more Damage to Democracy," reads the title of an article by Sophia A. Mcclennen as she points to reports that suggest that "Facebook's internal rules for censoring hate speech is biased to favour both the Right and celebrity users" (Mcclennen 2017). The scandal over the appropriation of personal data of Facebook users and their friends by Cambridge Analytical to target news and videos during the 2016 elections in the United States brings up new key aspects about cultural production and manipulation. "Are people so open to influence by information techniques that they lose the capacity to think for themselves? Is there a new understanding of algorithmic culture that anthropology should be taking up?"

The cognitive possibilities of artificial intelligence must also concern us, which I will only refer to in relation to the human-to-human learning process. Machines are now able to build a picture of the world around them by learning from experience. However, the learning systems they follow start from a base of data supplied by the designers of the system. In their cognitive make-up, all designers use skills learned from specific sociocultural settings, so that their cultural choices that are then built into the system as an imperceptible world view. It is all the more urgent, then, to make the imperceptible visible so that people may become conscious of such choices and bring in diversity at all levels of digital technologies.

In the conclusion of his book, Lévy argues that cyberculture represents "a major mutation in the very essence of culture" (Lévy 1997: 305). This change consists, for him, in a transition of humanity to "a universality without totality" (Lévy 1997: 305), which would mean a "moment where our species … tends to form a single global community" (Lévy 1997: 307–308). In this sense, one of the most important aims of digital technology must be to create and shape ideas (or 'minds') capable of developing a sustainable world.

This will only become possible if technological convergence contributes to what Michael Carrithers has called the "Human Project". Such a framework would give a green light for people to work together in using this extraordinary cyberspace to ensure that the richness of world cultures and lifestyles is included and debated. However, debates on the Internet generally are short-lived, and so they rarely arrive at a resolution because there is no structure for arbitrating and consolidating consensus. Diplomacy, endless transactions through fact-finding, speech, and lobbying, as well as the more highly developed institutional arbitration techniques, are indispensable to avert hard conflicts and hot wars through enhanced agreements and partnerships. A philosophical and ethical roadmap is needed in cyberspace to give form to the freedom of thought and expression that may create an improved reality in the physical world.

10.6 Ethnography of Transactions: Follow the Rethinking

Ethnography is now being reconsidered as a uniquely powerful research means of describing and understanding specific processes in the contemporary world, yielding highly validated, precise data and analyses. However, the ethnographic method, by its very nature, tends to freeze reality at a given moment and place of observation and participation. Multi-situated ethnography, in all its varieties, has helped to diversify the territories, modalities and scales of observation, including new types of self-ethnography. In spite of this, no matter how many multi-situated sites are brought into the picture, no matter how many ethnographies of different spaces and timescales are layered into the analyses, and no matter how many differing interpretations by informants about their cultural practices are disentangled and then re-entwined by the ethnographer to create a model to explain them, there is something valuable that is lost in keeping to an ethnographic 'present' and structuring it according to theoretical coordinates.

What is lost, in my view, is the argumentative, reciprocal engagement designed to convince or impose ideas that change actors' decisions or realign them with emerging new relationships, thus creating patterns of co-evolution. This is what I have called transactions. As I began to write this book, I realized how important the recording of emerging patterns of cultural ideas and policies was, in what became an unprecedented open world dialogue. The core of this process of debating and reformulating positions involves transactions, that is, at the end of every meeting there are subtle shifts in meaning and exchanges. Such exchanges cannot be reduced to reordering in negotiations and arbitration. By opening up international spaces to discuss how people place themselves in the field of culture, new global patterns of understanding, confronting and working out new political arrangements became possible.

In this exercise in ethnography, first-hand experience is as important as evidence in scientific research, especially in the social sciences. Digitalizing human experience, as vast, exciting and forward-looking as it is today, is no substitute for real-life experience. Not allowing people to experience things at first hand, as Hannah Arendt once argued, is the first step in closing down the minds of followers so that they become cogs in the banalization of evil. This begins with the banalization of culture, not in the sense that it becomes commonplace, following Arendt, but that it may go so far as to normalize murder and cruelty.

Anthropologists and ethnographers are uniquely trained to observe, participate, make decisions and record people's real-life experiences. In the looming world of lies and fake social media, they become even more important in the process of ascertaining what is happening on the ground and placing it in the laocultural or the ontocultural theories of our eco-cosmopolitical futures.

10.7 Culture as a Template for Global Transactions in the Anthropocene

In recent years, the most influential turn in debates on culture has been the deep new involvement of social scientists with natural scientists in research on the Earth system, which strengthens my conviction that a new cosmopolitics cannot be conceptualized without rethinking the terms 'culture' and 'nature'. As explained in Chap. 9, since the eighties, social scientists have worked closely with natural scientists in trying to find common theoretical and methodological grounds to set up research on the social aspects of environmental change (see Arizpe et al. 2012). At present the Earth System Science Partnership has been consolidated and, among many other proposals, the suggestion that the Anthropocene be recognized as a new geological age may provide the framework for a new conceptualization of an eco-cosmopolis. Recent findings and proposals on how culture may contribute and be integrated into the debates on the Anthropocene are summarized in Chap. 9.[3] However, while natural scientists have been hotly debating the name, duration and time of beginning of the Anthropocene over the last decade, social scientists are asking whether the Anthropocene is a cultural idea.

For all its scientific trappings, Jedediah Purdy does believe that the Anthropocene is mainly a cultural idea, and that its potential is political and ethical. I couldn't agree more. He calls for taking responsibility for what we make, as well as for what we destroy. "It is the starting place for a new politics of nature, a politics more encompassing and imaginative than what we have come to know as environmentalism" (Purdy 2015: 17). Similarly, Dipesh Chakrabarty, who has been very influential in opening the debate on "provincializing Europe", argues that the Anthropocene entails a "new universal history of humans" (Chakrabarty 2009: 221).

In his book on ecocriticism, Timothy Clark states bluntly that "the Anthropocene names a newly recognized context that entails a chastening recognition of the limits of cultural representation as a force of change in human affairs, as compared to the numerous economic, meteorological, geographical and microbiological factors and population dynamics, as well as scale effects, such as the law of large numbers, that arise from trying to think on a planetary scale" (Clark 2015: 21). He goes so far as to ask whether the Anthropocene is "a threshold at which art and literature touch limits to the human psyche and imagination themselves" as he discusses Barbara Kingsolver's attempt to look at the problem of scale in her 2012 novel *Flight Behaviour* (ibid.: 176). He questions whether "the human imagination is really so depressingly enclosed, able to be captivated only by immediate images of itself", and whether the psychology of narrative—of what makes a credible or compelling story—is itself a problem for representations of the Anthropocene (ibid.: 178). These questions construct a Gordian knot which ties culture to the capacity of the

[3]For a complete history and review by numerous authors in this scientific field, see Brauch et al. (2016).

human mind and which opens up the startling issue of how to expand that capacity extremely rapidly.

The Anthropocene as a threshold concept opens up a "politics of cognition" (to use Tom Cohen's term), where two different readings are confronted. One reading uses and adapts the inherited conception of the human—the social, cultural and so on; according to the other, the environmental crisis questions the seeming self-evidence or coherence of such basic conceptions (Cohen, cited in Clark 2015: 20). Debating the Anthropocene from a social science perspective opens up a host of interlinkages that have to be researched. This is all the more pressing because, since 2000, increasing securitization of the world's resources has been driven by climate change, water scarcity, degradation and stress, as well as by soil erosion and desertification, turning these pressures into major security dangers and concerns in the new era of Earth history (see Brauch 2009: 1311).

Finally, on a very different level, the debate on the Anthropocene will generate new questions of intentionality and responsibility that may make people unequivocally perceive that they are ultimately accountable for the future of the thin layer of environmental conditions that allow us to be creative living organisms. Or perhaps it is more accurate to state that a majority of people around the world, many of them already suffering from the effects of climate change, already recognize this responsibility, but their actions to address it are prevented by the powerful minority interests of the Capitalocene. In any case, involving ourselves in developing a new world is at the same time the reason and the outcome of our living.

10.8 Pathways to the Future: An Agenda for Living

How does one, in the term used by Tim Ingold, "become human" in today's world? Neurocognitive discoveries, genetic engineering and planetary geo-ecological challenges require changes in concepts, ideas and policies which are under way yet in very unconnected ways. How do we establish "nature/culture connections" as proposed by Marilyn Strathern or rethink the place of humans in a realm that is, at the same time, culture and nature, as proponents of the Anthropocene are suggesting? This is clearly a major challenge for the social sciences, especially for anthropology with its global and holistic perspective. They cannot just continue to report what is happening at ground level. Social scientists must engage in new theoretical enquiries which, in many ways, must be highly hypothetical, imaginative and resourceful.

The multilayered transactions on culture described in this book, and the recent finding reviewed about culture, diversity, gender, cosmopolitanism as well as cognition and cyberculture indicate that new proposals are needed in all these fields. In what follows, five pathways may be proposed, linking research and policy actions in these areas.

Firstly, the starting point for building a new framework for an eco-cosmopolis is recognizing that we share life with all existing and sentient beings but what we do not share with other beings is *creative living*. Creative living is unique to humans and is based on four capabilities that I contend must be given a central role in reconceptualizing the process of living as a human: *the will to know, the desire to love, the capacity to transform* and *the drive to exchange*. The first leads to science and philosophy. The second creates societies as sexual drives give way to the desire to live together in affective relationships, to take care of offspring and to enhance the psychosocial transformation of the self. The third pushes people towards altering our entanglement with ecosystems and the planet. In other words, scientific enquiry, cognitive and emotional resonance and cumulative learning sculpt human beings in every generation. Art is the primary representation of this cluster of actions. The fourth kind of capability, that of the will to exchange, fosters the constant swapping of ideas and experiences, the building of rules for trade, and the transactions through costs and benefits with economic partners.

Secondly, historical evidence indicates that peoples on this planet have survived and advanced through co-production of new processes with natural forces and the co-evolution of knowledge, organization and exchange among different culture producers. Even when cumulative learning is recognized as the central mechanism in human evolution, the major driver of this achievement is rarely made explicit. That is, that *cumulative learning only occurs through sustained dialogues with others*. In other words, it is an outcome of *human sociality*. And no other species in nature has this capability. Not even the other eusocial species have it.

There is the production of things, and there is the production of capable human beings and both must be given equal importance and recognition. Because the latter is deeply embedded in the traditional notions of the division of labour, it is called reproduction and is conceptually assigned to women.

It is necessary that greater recognition be given, on the one hand, to the human capabilities of love and reciprocity that produce healthy persons and social interactions, which could be condensed in some term, perhaps that of *Homo amator*. It is significant that, in describing the principal attributes that made humans evolve, the only two originally proposed were those implicit in the labels *Homo sapiens sapiens*, Man the Thinker, and *Homo faber*, Man the Maker. Is it a coincidence that these are the principal desirable actions of men, while those of women were left out of the historical record? Producing able, intelligent, emotionally capable human beings requires the fine tuning of collective care of a child by both women and men, as is so amply recognized in our contemporary world.

Proposing a new perspective on the primordial capabilities that structured the human world is not some arcane archaeological debate. It has a direct bearing on the imbalances in the world today. Neo-liberalism's exclusive insistence in fostering competition towards making things and services for profit has been completely neglecting the production of healthy, capable human beings. This "growth of economies and collapse of societies" was noted by researchers and policy-makers in several chapters of this book. The effect of such policies has brought about a cascade of negative social effects: lowered standards of health and education, an increase in the rates of drug-taking, criminality and suicides, including "despair deaths", the breakdown of families and kin groups, and soaring rates of psychological disorders. An attempt to reconstruct families and increase the birth of children of specific demographic groups is now being incessantly reiterated in the Hollywood cinema of love, weddings, births, and the insertion of children and adolescents into almost all stories. The sudden appearance of the "caring economy", where thousands of women from developing countries are imported to developed ones, are only one of the major indicators of this imbalances.

Thirdly, ontoculture must focus on *human 'living'* as the source of the creativity needed to overcome the challenge of survival. This means going beyond the passive defence of 'life' towards *a proactive agenda of 'living' and 'transformation'*. The capabilities for transformative action, as it has been highlighted by the United Nations in recent years, takes first place. In terms of the vision of human evolution, the historical vision of *Homo faber*, that is, of Man the Maker, must be deepened and broadened into what could be referred to as *Homo transfigurator*, referring to the human capacity to organize familial, kin and social networks in which children can "become human" and as adults can engage in the co-production of knowledge, the activation of exchange systems with other groups and, importantly, the self-transformation of every individual and every group.

Fourthly, for planetary survival conserving existing life forms and cultural legacies is a first step but it may only lead to *sustaining* what already exists. This introduces an inherent conservatism that may intensify divisive trends. Sustaining must be clearly integrated with the idea of *developing* a new worlding. Young people, it seems to me, cannot possibly be fired in their imaginations and led to action if they only hear that they merely have to sustain what already exists. And they must fully understand that their own *creative living* is the path to *developing* a new biosphere/sociosphere on Earth.

Accordingly, the Sustainability Transition is, intrinsically, a *Sustainability Transformation*. In this transformation, new cosmopolitical arrangements must rebalance power relations both at the international and the national levels, people must develop new ways of living together in renewed forms of sociality and new cultural symbols and practices must be revitalized or invented. This goes far beyond development as the expansion and deepening of the capitalist model of political economy through neo-liberal economic policy. Ultimately, the goal of sustainable human development must be to *protect people's capabilities to find direct solutions to their living conditions through ensuring freedom to create.*

To sum up this review of pathways towards the future, I see culture as the expression of active living. A living unique to humans, which encompasses all the imagining, striving and achieving of minds able to engage in resonance with others and resilient enough to withstand even the greatest of challenges to our planetary survival. Yet these capabilities, it must be emphasized, can only develop through sociality, co-production of knowledge, and collective action.

10.9 Living and the Probabilities of Actions

"Living is that which is most important for us, but also, that which is most resistant to our thought" *("Vivre est ce qui nous importe le plus, mais aussi, résiste le plus à notre pensée")* (Julien 2016: 9). This is the first sentence in François Julien in his 2016 book entitled *Vivre en existant* (Living to Exist) in which he posits that neither classical philosophy nor religion, with its other world, can solve today's challenges about our lives. I agree with him, as I believe most of the scholars and speakers cited in this book would also agree, on both counts, in the quest for a new paradigm about people living on Earth. Julien contends that to ask "why I live?" is to fall within the limits, the conditions of our life. We must ask, instead, "why do I exist?", a question which takes one "to be outside". If he means this in the sense of being outside one's physical body, this could relate to the transactions on the options of soul, consciousness or culture as discussed in this book.

Existing, however, takes us only to co-existence which I contend falls short of Julien´s intention if it only implies being-one-next-to-another as physically present organisms. This is not the same thing as "living together", the key concept of the World Commission on Culture and Development which I argue represents 'con-viviability'. Instead, as Marilyn Strathern suggested, it is the *connections* that are important because, as I have reiterated in many chapters, connections have led to human societies that made all the rest possible: affective resonance, rules of behaviour, knowledge, cumulative learning, exchange, production and cooperation.

It must be understood that a person becomes human by creating a world-culture in their minds through cognitive and affective resonance with those that surround her or him. Out of this world-culture will come the safeguarding, reconfiguring and reinventing experiments that will bring about the innovations needed for the sustainability transformation. Thus, the wealth of ideas and practices of groups of cultural practitioners must allow space for the creative contributions at three levels: culture as the commonality binding humanity together; culture as a symbolic identity of nation states or internal variant cultures; and culture as an individual's irreplaceable choice of a way of life.

If we are examining issues of pre-eminence, predominance and hierarchical power in competing groups of people, then we need to apply a different analytical understanding at the laocultural level, one that incorporates political science. There is a need for a wide array of new institutions and agreements to enable such groups to agree to the rules of law and engagement in order to move on from conflicts to resolutions and to put a stop to intimidation and violence. The critical values of competition must also be extended towards productive endeavours to transform the multidirectional relationships of people with people, whatever the variant of culture they hold, and of people with the natural environment. The positive contribution that different groups make towards the achievement of the Global Sustainable Development Goals could be ranked, for example. Concerted, worldwide efforts to reconstruct the human capacity for action at this level should be the highest priority of international relations and policies.

This new understanding of humans as creators and cultivators of geophysicality and sociality needs to be central to our construction of the idea of the Anthropocene. This is all the more urgent as the possibility arises of existing gene transmission being altered through genetic and epigenetic management and of new species being created through enhanced genetic selection. If anthropogenic change continues to be driven ever more strongly by cultural impulses, then we must develop a more complex, systemic, contextualized knowledge to engage in it. This level of discussion—the ontocultural level—is immanent in every member of the human species. Each of us has to transform ourselves into a new kind of living person, capable of developing new practices of reflexivity, discernment and creativity.

Humanity is unique in its ability to gather inspiration and meaning from our own creations and to use them to provide meaning for the present and the future. As Steven Pinker has recently stated, the Enlightenment has provided the project of reason, science and humanism in which this human ability has flourished (Pinker 2018). We must follow by securing 'living', that is, the freedom to constantly create new meanings that allow us to cooperate over and above divisive ideologies. It is this ability of people to be conscious of their own capacity for thought and action and to insist on a global ethic and conviviability that can ensure a sustainable transformation towards the future.

References

Arizpe, L. (2004). The intellectual history of culture and development institutions. In V. Rao and M. Walton (Eds.), *Culture and political action* (pp. 163–184). Palo Alto: Stanford University Press.

Chakrabarty, D. (2000). Provincializing Europe: postcolonial thought and historical difference. Princeton: Princeton University Press.

Fischer, M. (2009). *Anthropological futures*. Durham, NC: Duke University Press.

Grusin, R. (Ed.) (2015). *The nonhuman turn*. Minneapolis: University of Minnesota Press.

Pagel, M. (2012). *Wired for culture: origins of the human social mind*. London: W. W. Norton and Company.

Pinker, S. (2018). Enlightment Now: the Case for Reason, Science, Humanism and Progress. UK: Allen Lane, Penguin Random House UK.

Sahlins, M. (1994). A brief cultural history of "culture". Unpublished paper presented to the World Commission on Culture and Development.

Sen, A. (1999). *Reason before identity*. Oxford: Oxford University Press.

UNESCO (2000). World culture report 2000: cultural diversity, conflict and pluralism. Paris: UNESCO.

Lourdes Arizpe at the Second International Congress to Safeguard Intangible Cultural Heritage Congress held in Zacatecas, Mexico, in 2014. *Source* From the author's personal photo collection

Universidad Nacional Autonoma de Mexico (UNAM)

The National Autonomous University of Mexico (UNAM) was founded on 21 September 1551 under the name 'Royal and Pontifical University of Mexico'. It is the biggest and most important university in Mexico and in Ibero-America.

As part of the Centennial Celebrations of Mexican Independence, the National University was officially created on 22 September 1919. With the intention of widening educational opportunities in the country, the effort to launch the National University, though often hampered by adversity, was initially spearheaded by Congressman Justo Sierra in 1881. His vision finally materialized in 1910 with the inauguration of the National University of Mexico at a ceremony held in the National Preparatory School Amphitheatre, where as Secretary of Public Instruction, Sierra told the audience that the thrust of the National University's educational project was to concentrate, systematize and disseminate knowledge in order to prepare the Mexican people for the future. One hundred years after the creation of the University, Justo Sierra's inaugural address still rings true: *we are telling the university community today that truth is unfolding: go seek it ... you have been charged with the realization of a political and social ideal which can be summed up thus: democracy and liberty.*

The fundamental aim of UNAM is to serve both the country and humanity, to train professionals, to organize and carry out research, mainly on national problems and conditions, and to offer cultural benefits in the broadest sense possible.

Website: www.unam.mx.

L. Arizpe Schlosser, *Culture, International Transactions and the Anthropocene*,
The Anthropocene: Politik—Economics—Society—Science 17,
https://doi.org/10.1007/978-3-642-41602-6

Lourdes Arizpe with her most recent publications in Spanish at the Guadalajara International Book Fair, December 2015. *Source* Photo by Miguel Angel Porrúa

Centro Regional de Investigaciones Multidisciplinarias (CRIM)

The Regional Center for Multidisciplinary Research (CRIM) is an academic institution ascribed to the Coordination of Humanities at the National Autonomous University of Mexico (UNAM). It is located in the City of Cuernavaca on the Morelos Campus of UNAM. Its objectives are:

1. Focus on multidisciplinary research in social sciences, humanities and other disciplines, mostly aimed at tackling specific problems at the local, regional, national and international levels, and their implications within globalization processes.
2. Contribute to the creation of knowledge in relevant and innovative arenas addressing social problems that require the convergence of different disciplines for their study.
3. Contribute to the development of a multidisciplinary approach to humanities, and focus on the development of innovative theoretical and methodological perspectives.
4. Participate in educational programmes so as to contribute to the academic training of professionals in social sciences, humanities and other disciplines.
5. Disseminate by all possible means the results of CRIM's research projects.

Website: http://www.crim.unam.mx.

L. Arizpe Schlosser, *Culture, International Transactions and the Anthropocene*,
The Anthropocene: Politik—Economics—Society—Science 17,
https://doi.org/10.1007/978-3-642-41602-6

A playful reproduction of Andy Warhol's painting was presented to Lourdes Arizpe Schlosser as a parting gift by her collaborators in Unesco in 1998. *Source* Photo: Rogelio Cuellar

About the Author

Lourdes Arizpe is a professor at the Regional Center for Multidisciplinary Research of the National University of Mexico. She received an MA from the National School of History and Anthropology in Mexico in 1970 and a Ph.D. in Anthropology from the London School of Economics and Political Science, UK, in 1975. She has pioneered anthropological studies on migration, gender, rural development, and global change and culture in Mexico, in Latin America, and in international research groups, both academic and policy-oriented. Professor Arizpe taught at Rutgers University through a Fulbright grant in 1979 and carried out research in India and Senegal with a John F. Guggenheim grant in 1981. She was director of the National Museum of Popular Cultures in Mexico 1985–1988. She was elected President of the National Association of Ethnologists of Mexico in 1986 and Secretary to the Mexican Science Academy in 1992. Professor Arizpe was Director of the Institute of Anthropological Studies at the National University of Mexico, was elected President of the International Union of Anthropological and Ethnological Sciences in 1988, and successfully organized its World Congress in Mexico in 1993.

Lourdes Arizpe became a member of the United Nations Commission on Culture and Development, and soon afterwards was designated Assistant Director-General for Culture at UNESCO 1994–98. She was elected President of the International Social Science Council for 2004–2008, and participated as a member of the Academic Faculty of the Global Economic Forum at Davos, Switzerland 2000–2004. At the United Nations Institute for Research on Social Development, she was Chair of the Board 2005–2011 and a member of the Committee for Development Policy of the Economic and Social Council. She is a member of the Board of Trustees of the Library of Alexandria in Egypt.

L. Arizpe Schlosser, *Culture, International Transactions and the Anthropocene*,
The Anthropocene: Politik—Economics—Society—Science 17,
https://doi.org/10.1007/978-3-642-41602-6

Lourdes Arizpe became an Honorary Member of the Royal Anthropological Institute of the UK in 1995, and has received the Order of "Palmes Académiques" from France in 2007, the Award for Academic Merit of the Universidad Veracruzana in Mexico, and an Honorary Doctorate from the University of Florida at Gainesville in 2010.

See the website on this book at:
http://afes-press-books.de/html/APESS_17.htm.

Index on Persons

L. Arizpe Schlosser, *Culture, International Transactions and the Anthropocene*, The Anthropocene: Politik—Economics—Society—Science 17,
https://doi.org/10.1007/978-3-642-41602-6

Subject Index

L. Arizpe Schlosser, *Culture, International Transactions and the Anthropocene*, The Anthropocene: Politik—Economics—Society—Science 17,
https://doi.org/10.1007/978-3-642-41602-6

MIX
Papier aus verantwortungsvollen Quellen
Paper from responsible sources
FSC® C105338

If you have any concerns about our products,
you can contact us on
ProductSafety@springernature.com

In case Publisher is established outside the EU,
the EU authorized representative is:
Springer Nature Customer Service Center GmbH
Europaplatz 3, 69115 Heidelberg, Germany

Printed by Libri Plureos GmbH
in Hamburg, Germany